NEUROMETHODS™

Series Editor
Wolfgang Walz
University of Saskatchewan
Saskatoon, SK, Canada

For other titles published in this series, go to
www.springer.com/series/7657

Photosensitive Molecules for Controlling Biological Function

Edited by

James J. Chambers

Department of Chemistry, University of Massachusetts, Amherst, MA, USA

Richard H. Kramer

Department of Molecular and Cell Biology, University of California, Berkeley, CA, USA

Editors
James J. Chambers, Ph.D.
Department of Chemistry
University of Massachusetts
Amherst, MA
USA
chambers@chem.umass.edu

Richard H. Kramer, Ph.D.
Department of Molecular and Cell Biology
University of California
Berkeley, CA
USA
rhkramer@berkeley.edu

ISSN 0893-2336 e-ISSN 1940-6045
ISBN 978-1-61779-030-0 e-ISBN 978-1-61779-031-7
DOI 10.1007/978-1-61779-031-7
Springer New York Dordrecht Heidelberg London

Library of Congress Control Number: 2011921264

Printed on acid-free paper

Humana Press is part of Springer Science+Business Media (www.springer.com)

Preface to the Series

Under the guidance of its founders Alan Boulton and Glen Baker, the Neuromethods series by Humana Press has been very successful since the first volume appeared in 1985. In about 17 years, 37 volumes have been published. In 2006, Springer Science + Business Media made a renewed commitment to this series. The new program will focus on methods that are either unique to the nervous system and excitable cells or which need special consideration to be applied to the neurosciences. The program will strike a balance between recent and exciting developments like those concerning new animal models of disease, imaging, in vivo methods, and more established techniques. These include immunocytochemistry and electrophysiological technologies. New trainees in neurosciences still need a sound footing in these older methods in order to apply a critical approach to their results. The careful application of methods is probably the most important step in the process of scientific inquiry. In the past, new methodologies led the way in developing new disciplines in the biological and medical sciences. For example, Physiology emerged out of Anatomy in the nineteenth century by harnessing new methods based on the newly discovered phenomenon of electricity. Nowadays, the relationships between disciplines and methods are more complex. Methods are now widely shared between disciplines and research areas. New developments in electronic publishing also make it possible for scientists to download chapters or protocols selectively within a very short time of encountering them. This new approach has been taken into account in the design of individual volumes and chapters in this series.

Wolfgang Walz

Preface

We have entered into a new and exciting era in the field of neurobiology. Myriad optical methods are changing the way neurobiological research is performed. Tried and true electrophysiological techniques are being challenged for their place on the stage of measuring and manipulating neuronal activity. This change is occurring rapidly and is in large part due to the development of new photochemical tools, some synthesized by chemists and some provided by nature. This book is focused on the three main classes of photochemical tools for the control of biological function. First, natural photoresponsive proteins, including channelrhodopsin-2 and halorhodopsin, can be exogenously expressed in cells and enable rapid photocontrol of action potential firing or silencing. Second, small molecule photosensitive protecting groups (cages) of neurotransmitters, including caged glutamate, are synthetic molecules that enable highly localized activation of neurotransmitter receptors in response to light. Third, synthetic small molecule photoswitches can also afford light sensitivity on native or exogenously expressed proteins, including K^+ channels and glutamate receptors, allowing photocontrol of action potential firing and synaptic events. These tools have developed at a rapid pace and are continuously being improved upon and new tools being introduced thanks to the powers of molecular biology and synthetic chemistry. The three families of photochemical tools have different capabilities and uses, but they all share in enabling precise and noninvasive exploration of neural function with light.

Beginnings

In the early days, neurophysiologists invented electrodes to learn about native electrical excitability and the functioning of neural circuits. However, it soon became apparent that the nervous system is much too complex to rely entirely on recordings from one, two, or even several neurons at a time. Even within an individual neuron, membrane potential and ion concentrations are certainly not homogeneous, limiting the usefulness of electrode-based methods that record from a single point in a cell. At least in theory, optical-based recording methods could provide a much more detailed view of the activities, either within the complex architecture of an individual neuron or across populations of neurons. The hunt for optically based neurophysiological methods was on.

The first breakthrough came from the development of optical methods for *monitoring* activity. Investigators developed a wealth of fluorescent dyes that report back on voltage, synaptic vesicle release, Ca^{2+} fluctuations, and other ions. These indicators opened new windows for observing different aspects of neuronal signaling within individual neurons and in neural circuits. Small molecule indicators, most notably for Ca^{2+}, have revolutionized our understanding of normal synaptic transmission. More recently, genetically expressed GFP-based indicators have been introduced. These reporter proteins have provided insights into many aspects of signal transduction. The search for new indicators continues at a fast pace, but there is still much room for improvement. Perhaps the most pressing need is for a genetically expressed voltage indicator that can resolve single action potentials in individual neurons that are part of a native circuit. At the same time, new developments in microscopy are allowing investigators to peer into neural tissue deeper, faster, and with better spatial resolution than ever before, allowing us to see various aspects of neural activity in real time, and, more importantly, in vivo.

Until recently, optical methods for *manipulating* neural activity lagged behind methods for measuring activity. Recently, there has been a torrent of photochemical tools that can be used for controlling neurons, and these tools are the subject of this book. Most of the tools developed to date can be placed in one of three categories: *natural photosensitive proteins*, *caged neurotransmitters*, and *small molecule photoswitches* that bestow light sensitivity on ion channels and receptors. Each family of tools has its own unique advantages and limitations. When asking a particular neurobiological question, it is important to "choose the right tool for the right job." This book offers unprecedented access to the state-of-the-art for each tool, but it is important to note that this is a rapidly developing field, and we are cataloging the available toolkit at a moment in time, knowing full well that new tools with improved properties and different functionalities are right around the corner. Available at http://www.photobio.org

The Right Tool for the Right Job

It has been suggested that neurobiologists need a "Consumers Guide" to provide an unbiased comparison of the various photochemical tools currently available for controlling neuronal activity. The reality is that all of the tools covered in this book have merits. However, choosing the right tool depends entirely on the specific question and experimental system that is being explored.

A Common Challenge for All Photochemical Tools: Delivering Light to the Nervous System

All of the photochemical tools described in this book require the effective delivery of light to the part of the nervous system being targeted for control. Projecting light onto neurons in culture or in brain slices is straightforward, but delivering light onto neurons in vivo presents a major challenge. The brain is encased in an opaque cranium that presents a formidable barrier, physically and optically. Even after removal of cranial bone and the overlying dura, brain tissue tends to scatter light, and this limits spatial precision and makes it more difficult to affect structures far from the illuminated surface.

The retina is the one part of the nervous system that is normally exposed to light, making it a useful platform for testing photochemical tools. Of course, the retina is an interesting and important part of the central nervous system in its own right, and there is great clinical interest in developing tools that can impart light sensitivity on retinal neurons that are not normally photosensitive. Retinitis pigmentosa and macular degeneration are degenerative blinding diseases in which the normal rod and cone photoreceptors are destroyed, leaving the retina with no effective way to signal the visual cortex about light. Expression of ChR2 in either retinal ganglion cells or bipolar cells can restore visual sensitivity to retinas of animals with mutations that cause rods and cone degeneration.

Expression of melanopsin or halorhodopsin is also effective. Photoregulation of all of these tools require high intensity light, and azobenzene-based photoswitches require short wavelength illumination, which can be damaging over a prolonged time. For these reasons, there is a need for red-shifted photochemical tools that also have enhanced light sensitivity. Nevertheless, these studies provide hope that some neurological disorders might be treatable in a relatively noninvasive manner, using light to regulate activity in the parts of a neuron circuit that lie downstream from sites of damage or degeneration.

Despite the obvious difficulties, bioengineers have succeeded in delivering light into the brain with implanted fiber optics. Fiber-coupled systems have been used for optical measurement or manipulation of neural activity. Recent studies raise the possibility of substituting light for electrodes in "deep brain stimulation," a procedure that is being used increasingly for treatment of Parkinson's disease and other neuropsychiatric disorders.

Finally, the delivery of light for neural control involves an important but rarely discussed trade-off between effectiveness and precision. On one hand, a highly localized optical stimulus that illuminates part of a single neuron could ensure exclusive stimulation of only that cell. On the other hand, the light-regulated proteins are usually distributed over much of the cell surface, and more widespread illumination will activate more of these proteins resulting in a faster and more powerful effect. There has been considerable interest in developing photosensitive molecules that are highly sensitive to 2-photon illumination, because this would permit deeper and more precise photocontrol in neural tissue. However, the benefits of pinpoint accuracy will be offset by the asynchronous recruitment of photoactivated proteins as the 2-photon laser scans through a given focal plane within the tissue. New optical methods involving holographic illumination may help solve this problem by allowing simultaneous activation of distributed photosensitive molecules, with spatial and temporal precision that rivals 2-photon liberation of caged glutamate.

James J. Chambers
Richard H. Kramer

Contents

Contributors

ANDREW A. BEHARRY • *Department of Chemistry, University of Toronto, Toronto, ON, Canada*
JACOB G. BERNSTEIN • *MIT Media Lab, Department of Biological Engineering, McGovern Institute, and Department of Brain and Cognitive Sciences, Massachusetts Institute of Technology, Cambridge, MA, USA*
EDWARD S. BOYDEN • *MIT Media Lab, Department of Biological Engineering, McGovern Institute, and Department of Brain and Cognitive Sciences, Massachusetts Institute of Technology, Cambridge, MA, USA*
JAMES J. CHAMBERS • *Department of Chemistry, University of Massachusetts, Amherst, MA, USA*
BRIAN Y. CHOW • *MIT Media Lab, Department of Biological Engineering, McGovern Institute, and Department of Brain and Cognitive Sciences, Massachusetts Institute of Technology, Cambridge, MA, USA*
EUGENE F. CIVILLICO • *Department of Molecular Biology & Princeton Neuroscience Institute, Princeton University, Princeton, NJ, USA*
TIMOTHY M. DORE • *Department of Chemistry, University of Georgia, Athens, GA, USA*
GRAHAM C.R. ELLIS-DAVIES • *Department of Neuroscience, Mt Sinai School of Medicine, NY, USA*
DORIS L. FORTIN • *Department of Molecular and Cell Biology, University of California Berkeley, Berkeley, CA, USA*
PAU GOROSTIZA • *Institut de Bioenginyeria de Catalunya (IBEC), Institució Catalana de Recerca i Estudis Avançats (ICREA) and CIBER-BBN, Barcelona, Spain*
DAVINA V. GUTIERREZ • *Department of Neurosciences, School of Medicine, Case Western Reserve University, Cleveland, OH, USA*
XUE HAN • *MIT Media Lab, Department of Biological Engineering, McGovern Institute, and Department of Brain and Cognitive Sciences, Massachusetts Institute of Technology, Cambridge, MA, USA*
STEFAN HERLITZE • *Department of Neurosciences, School of Medicine, Case Western Reserve University, Cleveland, OH, USA*
EHUD Y. ISACOFF • *Department of Molecular and Cell Biology, University of California Berkeley, Berkeley, CA, USA; Divisions of Material and Physical Bioscience, Lawrence Berkeley National Laboratory, Berkeley, CA, USA*
HARALD JANOVJAK • *Department of Molecular and Cell Biology, University of California Berkeley, Berkeley, CA, USA*
RICHARD H. KRAMER • *Department of Molecular and Cell Biology, University of California, Berkeley, CA, USA*

BIN LIN • *Harvard Medical School, Massachusetts General Hospital, Boston, MA, USA*
GERARD MARRIOTT • *Department of Bioengineering, University of California Berkeley, Berkeley, CA, USA*
RICHARD H. MASLAND • *Harvard Medical School, Massachusetts General Hospital, Boston, MA, USA*
PATRICK E. MONAHAN • *MIT Media Lab, Department of Biological Engineering, McGovern Institute, and Department of Brain and Cognitive Sciences, Massachusetts Institute of Technology, Cambridge, MA, USA*
EUGENE OH • *Department of Neurosciences, School of Medicine, Case Western Reserve University, Cleveland, OH, USA*
J. PETER RICKGAUER • *Department of Molecular Biology & Princeton Neuroscience Institute, Princeton University, Princeton, NJ, USA*
SAMUEL S.-H. WANG • *Department of Molecular Biology & Princeton Neuroscience Institute, Princeton University, Princeton, NJ, USA*
HUNTER C. WILSON • *Department of Chemistry, University of Georgia, Athens, GA, USA*
G. ANDREW WOOLLEY • *Department of Chemistry, University of Toronto, Toronto, ON, Canada*
YULING YAN • *Department of Electrical Engineering, Santa Clara University, Santa Clara, CA, USA*

Part I

Photoreactive Small Molecules for Affecting Biological Function

Chapter 1

Introduction to Part I: Caged Neurotransmitters

James J. Chambers and Richard H. Kramer

Abstract

The field of organic chemistry has provided neurobiologists with the ability to release biologically active neurotransmitters at precise locations and times of their choosing. These molecules are silent before the active molecule is released by photolysis, thus allowing for very accurate measurements of biological responses when temporally accurate data is required. Some of the newest caging groups have provided the added benefit of two-photon sensitivity, thus allowing for not only time, *x*-, and *y*-dimensional precision but now *z*-direction as well.

Key words: Caged neurotransmitters, Photorelease, Photolabile protecting groups, Two-photon excitation

Caged molecules contain a photolabile protecting group that is removed by exposure to light, liberating a biologically active compound. The most widely used caged molecules in the field of neurobiology have been caged agonists for neurotransmitter receptors, although studies have also utilized caged calcium buffers, caged nucleotides, and even caged peptides that can be used to influence intracellular signal transduction pathways. The first caged neurotransmitter agonists were ortho-nitrobenzyl derivatives of carbamoylcholine, an activator of acetylcholine receptors that was released in response to ultraviolet light exposure. These molecules enabled a rapid increase in agonist concentration in response to the externally supplied light, leading to a better understanding of the kinetics of acetylcholine receptor activation. But, it was the development of caged glutamate that truly revolutionized the use of caged neurotransmitters and, to this day, continues to have major impacts on neurobiology. Dalva and Katz were the first to use a laser to locally release glutamate in an intact brain slice. Laser-induced photorelease of glutamate at presynaptic neurons revealed that the pattern of connections to

James J. Chambers and Richard H. Kramer (eds.), *Photosensitive Molecules for Controlling Biological Function*, Neuromethods, vol. 55, DOI 10.1007/978-1-61779-031-7_1, © Springer Science+Business Media, LLC 2011

visual cortical neurons changes during development, a finding that would have been difficult, if not impossible, to obtain without local glutamate photorelease.

Unfortunately, however, light scattering inherently limits the spatial precision of laser photorelease. This problem motivated the development of caged molecules that could be readily photolysed by two-photon excitation, a method that can pinpoint in three-dimensional space the photorelease of neurotransmitter to individual neurons and *even individual dendritic spines.* MNI-caged glutamate (4-methoxy-7-nitroindolinyl-caged L-glutamate) has a fairly favorable two-photon cross-section, and because of this, it is now the most popular form of caged glutamate in neurobiology. Adding to its usefulness, MNI-caged glutamate has a very low rate of spontaneous glutamate liberation in the dark and the free "cage" that is formed as a photolytic reaction byproduct has no apparent effect on neuronal function. Two-photon release of MNI-glutamate has been used to trigger electrical responses that simulate the kinetics and magnitude of individual synaptic events on single dendritic spines. Fortunately, abundant and highly active glutamate transporters rapidly remove the liberated glutamate, minimizing spillover onto neighboring spines. Photo release at single spines allows for direct comparison of spine geometry and postsynaptic responsiveness, allows precise measurement of spatial summation across neighboring spines, and removes any ambiguity in attributing plastic changes in synaptic function to the presynaptic vs. the postsynaptic cell.

Highly localized and rapid photorelease of glutamate requires very bright light and a high concentration of caged compound (millimolar range). These requirements do present potential problems of phototoxicity and off-target effects on other receptors. The development of new flavors of caged glutamate with even more favorable two-photon cross-section may help alleviate these problems. At the same time, investigators are developing forms of caged glutamate that can be readily and rapidly released by exposure to visible light. These molecules are beneficial because the optical instrumentation required for their use is significantly simpler, less expensive, and widely available. However, two-photon sensitive caged molecules are still the best-suited reagents for ensuring spatial and temporal precision.

Caged versions of many other neurotransmitters have also been synthesized including new visible light-sensitive and two-photon-sensitive forms of caged GABA, glycine, and anandamides for local activation or inhibition of endocannabinoid receptors. Novel molecular tools for studying intracellular signaling include various types of caged Ca^{2+}, a caged IP_3 that is two-photon sensitive, and caged peptides that interfere with synaptic vesicle exocytosis.

A different type of photosensitive compound can irreversibly disrupt the function of certain types of glutamate receptors in response to light. ANQX is an azide-containing analog of commonly used AMPA receptor antagonists (e.g. CNQX and DNQX). Exposure of bound ANQX to UV light results in a high-energy species and then covalently attaches to the AMPA receptor, permanently preventing the binding of glutamate or other agonists. ANQX has been useful for probing the turnover of AMPA receptors in synapses between hippocampal neurons, a process that is thought to play a crucial role in long-term synaptic plasticity and learning and memory. So far, studies utilizing ANQX have been limited to neurons in culture, but compounds with different properties, including perhaps a more favorable two-photon cross-section and solubility profile, could enable ANQX to reveal receptor trafficking in more intact preparations including brain slices. Photocrosslinker-containing derivatives of antagonists of other neurotransmitter receptors might be used in a similar manner to explore receptor turnover and its possible activity dependence.

Chapter 2

Targeting and Excitation of Photoactivatable Molecules: Design Considerations for Neurophysiology Experiments

Eugene F. Civillico, J. Peter Rickgauer, and Samuel S.-H. Wang

Abstract

Each chapter in this volume describes in detail the application of one or a group of photosensitive molecules to biological research. In this chapter, we take up general prefatory questions: how to determine which molecules are appropriate to use, and what type of compound delivery and light-targeting apparatus for photoactivation is likely to give satisfactory spatial and temporal performance. We enumerate the advantages and disadvantages of currently available "caged" and genetically encoded photosensitive molecules. We also compare current mature and emerging technologies for patterned light delivery, referring as much as possible to broadly applicable general principles. Our goal is to provide a comprehensive overview with signposts to more detailed treatments.

Key words: Caged compound, Channelrhodopsin, Scanning, AOD, Galvanometric, Holographic, Beamsteering

1. Families of Photoactivatable Molecules

Photoactivatable molecules are available that influence a wide range of extracellular and intracellular neurophysiological functions. The choice and availability of photosensitive molecule depend on the research question and will influence subsequent choices in the design of experimental apparatus. The first major choice is whether to use photolysis-activated "caged" diffusible molecules (Sect. 2.1) or light-sensitive membrane proteins (Sect. 2.2).

For activation of native receptors with a time course matching endogenously occurring binding, unbinding, and biochemical kinetics, caged compounds are preferred. In general, newer optogenetic approaches are attractive when "on–off" control of neuronal

James J. Chambers and Richard H. Kramer (eds.), *Photosensitive Molecules for Controlling Biological Function*, Neuromethods, vol. 55, DOI 10.1007/978-1-61779-031-7_2, © Springer Science+Business Media, LLC 2011

membrane potential or intracellular cascades is desired as a means of determining the downstream effects on other cells. More importantly, optogenetic probes are proteins that can be expressed specifically in genetically identifiable cell types. Because by definition these approaches involve introducing and manipulating foreign molecular machinery, the effects on the manipulated cell or cells themselves have the potential to go outside the normal range of function. Topics in single-cell physiology such as dendritic integration are still best explored with caged compounds.

1.1. Caged Compounds

1.1.1. Caged Ionotropic Receptor Agonists and Antagonists

Ionotropic neurotransmitter mechanisms consist of initiation of one or more transmembrane currents. The first caged neurotransmitters were acetylcholine receptor agonists (1, 2, 45), followed by the first caged glutamate (3). Innovation in caged glutamates has resulted in improved usability for both conventional one-photon UV illumination and IR-based two-photon activation (see Sect. 4), leading to the development of the 6-bromo-7-hydroxycoumarin-4-ylmethyl (BHC) and 4-methoxy-7-nitroindolinyl (MNI) protecting groups (4–6). The most widely used caged glutamate is MNI-glutamate, which combines a high absorption coefficient and high quantum yield (4,300 $M^{-1}cm^{-1}$ and 0.085 at 350 nm, respectively; (7)), with relatively low toxicity and interference with signaling pathways.

Glutamate receptor subtype-specific ligands have also been caged, including NMDA (8), kainate (9), and D-aspartate (10, 11). Other promising improvements are 4-carboxymethoxy-5,7-dinitroindolinyl (CDNI) with increased absorption and quantum yield and reduced nonspecific effects (12) and RuBi-glutamate, which is based on novel ruthenium-based photochemistry and is photolyzed by visible light (13).

Other useful caged transmitters include caged GABA (14), which has seen recent innovations in the form of new caging groups to reduce pharmacological side effects and improve optical properties (15–17), and coumarin-caged glycine, which can be activated by visible light (18).

Although the transmembrane currents mediated by these receptors may potentially be mimicked using light-activated engineered channels, reproducing the kinetics of native receptors is a nontrivial task. When receptor populations at single synapses are heterogeneous, the native ligand may activate multiple receptor subtypes at once. For example, channelrhodopsin-mediated currents cannot currently emulate the coincidence detection properties of a synapse that includes NMDA-type, AMPA-type, and metabotropic glutamate receptors.

1.1.2. Caged Neuromodulators

In addition to opening ion channels, neurotransmitters also trigger second messenger cascades by acting upon metabotropic receptors, thus influencing more than membrane potential. Since the signaling pathways triggered by metabotropic receptors are not necessarily all identified, a logical approach is to employ caged receptor agonists.

For purposes of classification, we use the term "neuromodulator" to denote transmitters that have been traditionally thought to act more slowly than the millisecond-scale action of fast neurotransmitters such as glutamate. Physiological studies of neuromodulator action have employed agonists added to the bath or, at best, delivered in a pulsatile fashion to a targeted region of tissue through a micropipette or capillary tube. However, these modes of presentation may be far slower than the time scale of "modulator" action in vivo. The neurotransmitter/neuromodulator distinction has nearly disappeared with the appreciation of metabotropic actions of "fast" transmitters such as glutamate, GABA, and acetylcholine; conversely, neurons of the ventral tegmental area are now known to modulate dopamine levels in the nucleus accumbens on the time scale of 0.1 s (19, 20).

Many neuromodulators have been caged including agonists at 5-HT (21), adrenergic (22), and dopamine (23) receptors; for an exhaustive list, see (24). There are even caged peptide antagonists (25). Given the possibility of fast neuromodulator action in the CNS, emulation of neuromodulator activity by caged compounds is an attractive direction for future research.

1.1.3. Caged Second Messengers

A variety of intracellular signaling pathways can be controlled by light. Caged ATP was the first caged intracellular messenger for biological research (26, 27, 45) and has been used to study muscle contraction (28) and other cellular processes (29, 30). Another target for uncaging experiments is the calcium ion, which can be effectively caged by introducing calcium chelators that when photolyzed lead to a drop in affinity, thereby releasing calcium (31–35), or an increase in affinity, thereby buffering calcium (diazo-2; (36)). Calcium signaling has also been probed using caged IP_3 (37–39), caged cyclic ADP-ribose (40), a caged SERCA pump inhibitor (162), and caged caffeine (41).

Other caged messengers include caged nitric oxide (34, 42), BHC-caged cyclic nucleotide monophosphates (43), and caged nucleotides (38). Many older compounds are reviewed in (44). A number of new and old compounds are available from Molecular Probes (now Invitrogen), Calbiochem, or Tocris (Table 1 of (12)). Many caged molecules not commercially available can be synthesized relatively inexpensively. Compounds that have only been used in a few studies or made as a proof of principle may be difficult to obtain.

1.1.4. A Quantitative Index for Photoactivatability

To be useful in biological experiments, caged compounds must meet a number of basic criteria: good solubility, lack of biological activity such as interference with receptors or toxic effects, and rapid release of ligand upon illumination (for a review, see (45)).

In addition to these properties, one key parameter is sensitivity to photoactivation. In this regard, a useful quantity is the *uncaging index* U, defined for conventional (one-photon) absorption as $U = \varepsilon\varphi$, where ε is the extinction coefficient (typically in units of $M^{-1}\,cm^{-1}$) and φ is the quantum yield, or the probability that a group will be photolyzed after absorbing a photon. The extinction coefficient ε varies as a function of illumination wavelength, but takes on very similar values for a given caged group, irrespective of the agonist that is caged. Quantum yield φ does not change with respect to wavelength as long as no other significant absorption bands are present, but does vary as a function of the identity of the caged molecule and the caging position. The higher the value of the uncaging index, the less light is needed to achieve uncaging. At a minimum, the light levels used to photolyze caged compounds should not damage or interact in unwanted ways with the biological system. This issue is especially important when using UV light, which is more likely to cause damage to the sample than visible or infrared (IR) light.

The relative merits of different cage groups are well-characterized quantitatively for caged glutamates but less so for other compounds. U values for some commonly used compounds can be found in (12, 31, 46). Also see Chap. 3 of this volume (Ellis-Davies).

1.2. Light-Sensitive Membrane Proteins

A major advance in recent years is the development and use of light-sensitive ion channels, pumps, and other signaling molecules. These proteins escape many problems associated with introducing caged compounds into neural tissue. They also carry the significant advantage of allowing targeted expression in subsets of cells using molecular methods. Light-activated channels and pumps have the potential to supplant caged neurotransmitters when the desired outcome is control of spiking in cells of a particular type.

1.2.1. Modification of Endogenous Channels

Endogenous ion channels can be made light-sensitive with chemical cofactor (48–54). A recent example is the SPARK channel, a Shaker potassium channel with a covalently attached ligand that causes the channel to open when illuminated. This channel can be exogenously expressed in mammalian neurons. A similar approach has been successful in creating a light-gated glutamate receptor, termed LiGluR (49, 50). Both SPARK and LiGluR require expression of the channel in the cell of interest followed by covalent modification. A technology that does not require exogenous gene expression is the photoswitchable affinity label (PAL) (51), which can be introduced into cells where it covalently attaches to an endogenous channel and renders it light sensitive (52, 53). This approach, reviewed in (54), is channel-specific and generally does not require continuous illumination to keep the modified channels open. In addition, because it targets potassium channels, it may allow subtle modulations of neuronal firing (55).

1.2.2. Introduction of Exogenous Channels

A recent exciting advance in probing biological tissue with light is the development of "optogenetics," the genetic expression of light-gated ion channels and transporters. The most widely used optogenetic molecule is channelrhodopsin-2 (ChR2), originally isolated from the green alga *Chlamydomonas reinhardtii*. ChR2 is a nonspecific cation channel and generates an inward (depolarizing) current (56) that can excite neurons on a time scale of milliseconds (57–59). ChR2 consists of the protein channelopsin-2 (Chop2) with a covalently linked all-*trans*-retinal molecule that acts as the phototransducing moiety. In many types of vertebrate neural tissue, this channel functions in conjunction with the endogenous all-*trans* retinal that is naturally synthesized; in other preparations, all-*trans* retinal must be added to the system (58).

Upon blue-light excitation, the ChR2 channel opens to generate current within 1 ms. While the current-conducting photocycle of ChR2 is still not completely understood, it has generally been observed that sustained illumination of a ChR2 population beyond the initial peak amplitude leads to a smaller sustained current that reflects inactivation, light-dependent recovery to the initial light-excitable state, and reopening (i.e., multiple photocycles). Photocurrents decay to zero within ~10–100 ms of light offset depending on several experimental conditions, including pH (59–62). Between light-stimulation periods, recovery to the initial light-excitable state occurs more slowly than it does under illumination, imposing a delay between trials (~10 s) to repopulate the initial excitable state.

Engineered modifications to the photocycles of ChR2 and similar molecules have produced channels with different kinetics under sustained illumination, possibly favoring longer-lasting, larger-amplitude photocurrents from a photocycling population (60, 63). Other types of ChR2 molecules have been engineered to remain in a current-conducting state for much longer (up to ~100 s), allowing switchable step-like currents to be activated by blue-light excitation (64). Spectrally shifted absorption of this long-lasting current-conducting intermediate allows the termination of these currents upon illumination with green/yellow light. Another naturally occurring light-gated ion channel, comparable in several ways to ChR2 but with a red-shifted peak excitation wavelength, has been found by searching a genomic database by predicted functional homology (65).

Under full-field illumination, illumination power as low as 0.3 mW is sufficient to drive spiking in ChR2-expressing neurons (66), although 5–10 mW is often used (57, 67). In ChR2-based studies, where the desired physiological effect is often stimulation of an action potential, one straightforward approach is to use a wide-field light source to illuminate all membrane-bound ChR2 molecules in a cell at once. Many microscopes configured for wide-field fluorescence imaging with an arc lamp can be modified at moderate cost for photostimulation experiments. Standard filter sets for imaging GFP fluorescence transmit excitation light to the sample that is near the peak wavelength for ChR2 excitation. To control the timing of illumination, fast shutters (e.g., Vincent Uniblitz VS35 shutter + VCM-D1 driver) can be introduced into the optical path, although the time required to open or close a mechanical shutter fully can, in some cases, impose a "fingerprint" on photocurrents.

An alternate approach that costs less and offers more precise temporal control over illumination is to use light from a high-power light-emitting diode (LED; e.g. LXHL-LB5C from Philips Lumileds). LEDs are available in a variety of wavelengths with narrow spectral bandwidths, and can be combined with inexpensive transistor-based gating circuits to switch light on or off very rapidly. LED illumination must be relayed to the sample with an appropriate optical configuration; for an example describing a substage configuration supplying LED-based Köhler illumination, see (68). For an excellent characterization of LEDs in general, see (69).

Inhibition is also possible using a chloride pump from *Natronomonas pharanois*, halorhodopsin (NpHR) (70), which generates a hyperpolarizing current when excited by yellow light and can be used to inhibit neural activity in dissociated culture or intact tissue (71–73). As in the case of ChR2, variants and homologs have been developed for increased current amplitude, tolerability at high expression levels, and localization to the plasma membrane (eNpHR; (74)), or increased current amplitude and accelerated trial-to-trial recovery (75). eNpHR has recently been used in mammals to inhibit firing of targeted cells in vivo (76, 77).

It is worth noting that the sensitivity of channelrhodopsins and halorhodopsins to blue and yellow light, respectively, opens the possibility of bidirectional control of neuronal excitability using multiple-wavelength illumination by combining them with one another or with caged compounds.

For additional information, we refer readers to (73), Chap. 6 of this volume (Han and Boyden), and another excellent discussion of light-sensitive channels and other photosensitive molecules (78).

1.2.3. Engineered Light-Sensitive G-Protein-Coupled Receptors

Recently, it was demonstrated that the intracellular portions of rhodopsin may be altered to couple light absorption events to intracellular second messenger activity, thus placing G-protein receptor-coupled signaling under optical control (79). Two such engineered receptors, termed optoXRs, were constructed:

opto-α_1AR, which mimics the intracellular actions of the α_{1a} adrenergic receptor (G_q recruitment leading to IP_3 production and calcium release from intracellular stores), and opto-β_2AR, which mimics the intracellular actions of the hamster β_2-adrenergic receptor (G_s recruitment leading to elevated cytoslic cAMP). Illumination of the opto-α_1AR-expressing nucleus accumbens when mice were in a particular region of their enclosure produced a preference for that region on the following day.

2. Targeting of Photoactivatable Molecules

Photoactivatable molecules can be introduced into the experimental preparation in many ways. For caged neurotransmitters and neuromodulators, the principal options are bath perfusion (Sect. 3.1) and focal application through a capillary tube (Sect. 3.2). Caged intracellular ligands such as calcium and second messengers must be delivered by microinjection or via patch-clamp electrode, or else made membrane permeant by using AM esters to neutralize anionic moieties, an approach that is useful for delivering calcium chelators (80, 81). Exogenous proteins such as light-gated channels require expression systems (Sect. 3.3).

2.1. Bath Application

Bath application has the advantage of relative simplicity and produces uniform concentration of the compound throughout the bath. Commonly, in vitro experiments using caged compounds are done with caged compound solution pumped in a looped perfusion (82, 83). Recirculation uses less caged compound than a gravity-driven non-recirculating perfusion system.

2.2. Focal Application Through Capillary Tubing

Local pressure ejection of a caged compound solution has a number of advantages. First, the flow of the compound can be adjusted independently of the bath flow. This is particularly useful as a control to determine whether the caged compound itself affects biological function. Second, much smaller volumes are required, thus economizing on cost. For example, in our experience, an 8-h experiment using 15 mM of MNI-glutamate ($150 for 10 mg, Tocris) will use less than 50 μl of solution applied under 0.5 psi (34 mbar) through a 50-μm inner diameter capillary tube. The resulting cost is less than $4 per experiment. In contrast, a recirculating bath of 6 ml containing 2.5 mM of MNI-glutamate would cost $72 and last less than half as long even under modest ambient light conditions.

Suitable capillary glass tubing of various diameters can be purchased from Polymicro Technologies. Tubing must be carefully scored and broken to a useful length using a ceramic cleaving stone (Polymicro) or a capillary cutter (Shortix, Scientific

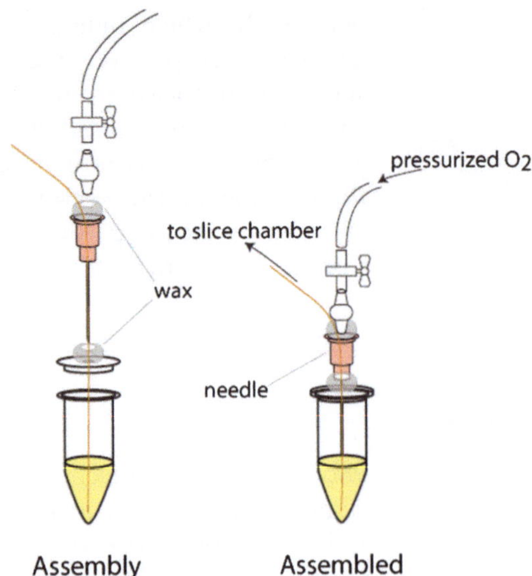

Fig. 1. Assembly of a low volume reservoir for focal elution of a caged compound solution into an experimental preparation. Bone wax is used to seal joints. The tube is covered in foil (not shown) to block room light.

Instrument Services, Inc.). A clean flat break prevents excessive turbulence of flow. A schematic for a simple pressure ejector is shown in Fig. 1.

2.3. Expression of Genetically Encodable Probes

Genetically encoded channels must be expressed within the cell of interest and then targeted to the membrane. Gene delivery methods include in utero electroporation of DNA, virus-based expression vectors, and the creation of transgenic animals.

2.3.1. In Utero Electroporation

Nucleic acids can be driven across cell membranes by strong electric fields. Several recent studies demonstrate the feasibility of electroporating transgene DNA into developing tissues (84, 85). In the brain, ChR2 has been expressed in specific layers of neocortex by prenatal electroporation to examine the functional targets of interhemispheric projections (86, 87). Electroporation can be done in utero, after which gestation can continue. One caution is that the voltages used to electroporate can damage developing tissue, leading to low viability depending on the position and geometry of the target tissue and electrodes.

2.3.2. Viral Vectors

Transgenes may be inserted into viral vectors and injected directly into tissues of interest. The simplest type of application of this method is stereotactic targeting of viral particles to a structure of interest (88) such as the mouse hypothalamus (89). In intact brain tissue, lentiviruses (90), adeno-associated viruses (91, 92), and herpes viruses (93, 94) have found wide use. Retrograde interhemispheric transport of a herpes virus expressing ChR2 has been used to identify contralaterally projecting cortical cells in the intact brain by illuminating during in vivo electrophysiological recordings (95). To enhance the specificity of virally delivered transgenes, injection of viral vectors may be performed in a cell-specific Cre-expressing mouse line (77, 96–99). In some recent applications (77, 98, 99), a double-floxed inverted strategy (100) was used to activate the transgene by inversion between two pairs of flanking *lox* sites rather than by excision of a STOP sequence. This increased the specificity of expression, and allowed tests of the causal role of dopamine neurons (98) and parvalbumin neurons (77, 99) in circuit performance and behavior. Finally, functional circuits may be infected by viruses that are able to cross synapses. References (101) and (102) provide comprehensive reviews of virus technology in neuroscience.

2.3.3. Transgenic Animals

Making a transgenic animal (103, 104) combines the spatial and temporal specificity of regulated gene expression with the possibility of having a continuous supply of experimental animals. Powerful genetic tools now allow the restriction of transgene expression by temporal and functional boundaries. For example, a transcriptional stop sequence may be "floxed" (*f*lanked by *lox*P sites) and inserted between a transgene and a strong universal promoter. The transgene may then be activated in specific tissues or cell types by the action of Cre, a site-specific DNA recombinase that excises the material between the *lox*P sites. This Cre-lox system may be used to produce particular profiles of transgene expression by crossing a floxed mouse with a cell type-specific Cre line, so that the transgene will be expressed in the tissues at the intersection of the two distributions. Reversible temporal control of transgene expression is also possible using, for example, "tet-ON" and "tet-OFF" systems (105). References (106) and (107) provide comprehensive reviews of genetic methods and their use to target neuronal subtypes.

Problems associated with the transgenic approach include the long lead time required to generate an organism successfully, uncertainty as to what cell types will express under the chosen promoter, and the possibility that expression will eventually be silenced over multiple generations by epigenetic mechanisms. In addition, at present, transgenesis is possible only in selected model organisms.

Several ChR2 transgenic animal lines are currently available. ChR2-expressing mice driven by the *thy1* promoter (108) are available from Jackson Laboratories. ChR2 and NpHR have also been introduced into the translucent nematode *C. elegans* (73, 109).

Efforts are also underway to achieve cell-type specificity for any transgene using conditional and combinatorial expression systems.

3. One-Photon Versus Two-Photon Excitation

Caged compounds are typically excited by near ultraviolet (UV) light (330–400 nm), which is strongly scattered by brain tissue, leading to limited depth penetration (110) and degradation of the focus. Also, single-photon methods do not provide a way to restrict excitation to a single *Z*-focal plane, so molecules away from the plane of interest may also be excited. The major advantages of two-photon excitation (TPE) are decreased heating by light absorption a more localized excitation spot due to reduced scattering by IR light, and an intrinsic "optical section" around the plane of focus (111).

3.1. Cost Considerations

For one-photon uncaging of glutamate, in our laboratory we use a frequency-tripled Q-switched Nd:YVO_4 laser (355 nm, DPSS Series 3501). Considerably less expensive sources of UV and near-UV light include solid-state UV lasers (e.g., Oxxius Violet), flash lamps (112, 113), and UV LEDs (available from Nichia Corporation, Japan; (114–116)).

Relative to UV, the major disadvantage of pulsed IR TPE is the cost of the excitation source. Fixed-wavelength femtosecond pulsed IR lasers suitable for two-photon uncaging are available from, e.g., Del Mar Photonics (California, USA) and Femtolasers, Inc. (Massachusetts, USA). Much more commonly used wavelength tunable IR lasers confer flexibility but are more expensive (e.g., Newport/Spectra-Physics or Coherent). The need for TPE can sometimes be obviated by careful choice of the preparation geometry, or by targeted application or expression of photoactivable molecule.

3.2. Chemical Two -Photon Uncaging

One simple way to improve the characteristics of any uncaging system is to add a second inactivating group to the molecule of interest (117). Production of active agonist then requires two photolysis events, introducing a nonlinearity by making the probability of photolysis proportional to the second power of light density. Out-of-focus uncaging is reduced since, as in the case of true two-photon uncaging, active ligand molecules will be preferentially produced in the volume where light density is maximal. This approach has been termed "chemical two-photon uncaging" (117).

In principle, the use of chemical two-photon uncaging produces an increase in spatial resolution comparable to that provided by conventional two-photon uncaging, without the significant cost of a pulsed IR light source. Instead of requiring a pulsed laser, chemical two-photon uncaging only requires the excitation events to occur within a few milliseconds, the time scale on which a molecule that has been uncaged once remains in the focal volume.

Confinement of the photolysis volume is important for extracellular uncaging of neurotransmitter in brain slices, since it limits the action of the uncaged molecules on the cells located above and below the focal volume. However, this approach still uses UV light, limiting its usefulness to the most superficial ~50 μm of brain slices (110). In addition, over repeated uncaging pulses, partial photolysis products may accumulate, though this can be reduced by the use of a local perfusion (see Sect. 3.2).

An important advantage of double caging is that it can reduce the background activity of a caged compound by making it less similar in structure to the native agonist, reducing the risk of undesired interaction with biological targets (118). Also, handling is easier since the requirement of two uncaging events makes the production of free agonist by room light or spontaneous degradation less likely. Multiple-caged compounds are generally not commercially available, but their synthesis is simplified by the fact that the design of single-caged compounds usually involves the identification of multiple caging sites. For example, the synthesis of double-caged IP_3 is achieved most simply by synthesizing triple-caged IP_3 and "photolyzing back" to the double-caged form (118).

3.3. Spot Size

A highly localized excitation spot is not always optimal. For any scanning system (as opposed to a holographic or other scanless system; see Sects. 6.4 and 6.5), a reduced spot size reduces the simultaneously excitable area. For example, an extended portion of a dendritic arbor or dendrite cannot be simultaneously illuminated with a diffraction-limited spot. In cases where spine-level spatial resolution is not necessary, a larger spot size may in fact be desirable, and may be obtained by using a flash lamp or by introducing a spatially diffuse UV laser beam, e.g., through a multimode fiber (119). For focused laser light the region of concentrated photoactivation near the focus may be expanded by using a lower numerical aperture lens or by underfilling the back aperture of a high-numerical aperture objective (120).

3.4. Absorption Spectrum of Molecular Target

A molecule's peak wavelength for two-photon absorption is often nearly twice the peak wavelength for one-photon absorption; however, this is not always the case (111), and both spectral sensitivity and absolute absorptivity must generally be determined empirically. Two-photon absorptivity (121) is usually reported in terms of the two-photon absorption cross-section, typically expressed in Göppert-Mayer units (1 GM = 10^{-50} cm^4 s/photon), or action cross-section (quantum efficiency × absorption cross-section, also often reported in units of GM). Rhodamine 6G is considered to have a large action cross-section, ~150 GM (122). The action cross-section of caged glutamate reached a usability threshold of 1 GM with the synthesis of the BHC (4) and MNI (5) caging groups, leading to the rise of two-photon glutamate uncaging (82, 123). Two-photon absorption cross-sections have not yet been measured for a large number of less commonly used cage groups, such as those which have been used to cage GABA and most of the compounds listed in Sect. 2.2.

3.5. Two-Photon Excitation of Light-Activated Channels

Unlike caged compounds, which may be synthetically designed to possess high two-photon excitabilities, the two-photon excitability of light-gated ion channels depends in part on the light-absorbing properties of the intrinsic chromophore (all-*trans* retinal). The one- versus two-photon excitabilities of light-gated proteins, as determined by the properties of the chromophore and by the photocycle dynamics described in Sect. 2.2.2, are only beginning to be explored.

One investigation has described the spectral sensitivity of ChR2 responses under TPE, using optical detection of fluorescence transients associated with a calcium-binding dye to infer photocurrent amplitudes (124). More recent electrical recordings of ChR2 currents stimulated with low-power TPE (125) give a cross-section of 260 GM and demonstrate single action potential triggering using TPE.

4. Beam Steering: Introduction

In order to activate photosensitive molecules for neurophysiological experiments, it is often necessary to have precise control over the timing and location of illumination. In the remainder of this chapter, we discuss technologies for producing patterned illumination.

To illuminate areas comparable in size to the field of view, a focused beam is unnecessary and a flash or an arc lamp may be used (see (126) for a discussion). For micron-scale spatial resolution, a focused beam must be introduced. Here we concentrate on the case in which the activating light comes from a laser and is directed through the microscope objective. An alternate approach, the use of an adjacently positioned optical fiber, removes some difficulties of power loss and chromatic aberration by the microscope objective, at the expense of some spatial resolution (126).

For some experiments, beam steering may not be required. By aligning the structure of interest with a single photolysis point, it is possible to interrogate optically a single substructural element with fine temporal resolution (38). Such an arrangement has been successfully combined with the use of a motorized stage, adjusted slowly between trials, to map responses to photorelease over a wide area, as long as time is not an issue (127–131). Where only one site of activation is needed, an alternate approach with potentially superior spatial and temporal resolution is iontophoresis of agonist (132).

In most experiments, imaging is necessary in order to locate the site of photoactivation. In the simplest case, an image can be generated using a CCD camera on a conventional microscope. However, frequently both the manipulation (photoactivation) and the measurement (imaging) require a steered beam. This situation places additional constraints on the technologies that can be used; at the very least, some technical expertise/hardware is required to coordinate the two processes.

In the final two sections of this chapter, we will examine the pros and cons of technologies for achieving sophisticated spatial and/or temporal control. We will first review technologies that have already been used to answer questions in neurophysiology at the time of this writing (Sect. 6). The issues and choices to be discussed when selecting from mature technologies are outlined in Fig. 2. In Sect. 7, we review more advanced technologies that we expect to become more accessible in the near future. In both sections, we will use the term "*XY* scanning" to refer to beam steering within a horizontal plane and "*Z* scanning" to refer to beam steering to points at varying depths in the specimen.

All beam shaping and steering technologies need to take into consideration the potential for aberration. Chromatic aberration can especially complicate photoactivation experiments since optical components are less likely to be corrected for IR or UV wavelengths than visible wavelengths. This leads to potential registration errors between photoactivation and imaging, especially in the *Z* direction (110). Many objective lenses are now corrected for commonly used wavelengths, but this must always be tested empirically. Moderate axial errors can be corrected by diverging or converging one of the wavelengths in order to displace the focus axially (133). Correction strategies for dispersive elements, which broaden femtosecond near-IR pulses, are discussed in Sect. 7.

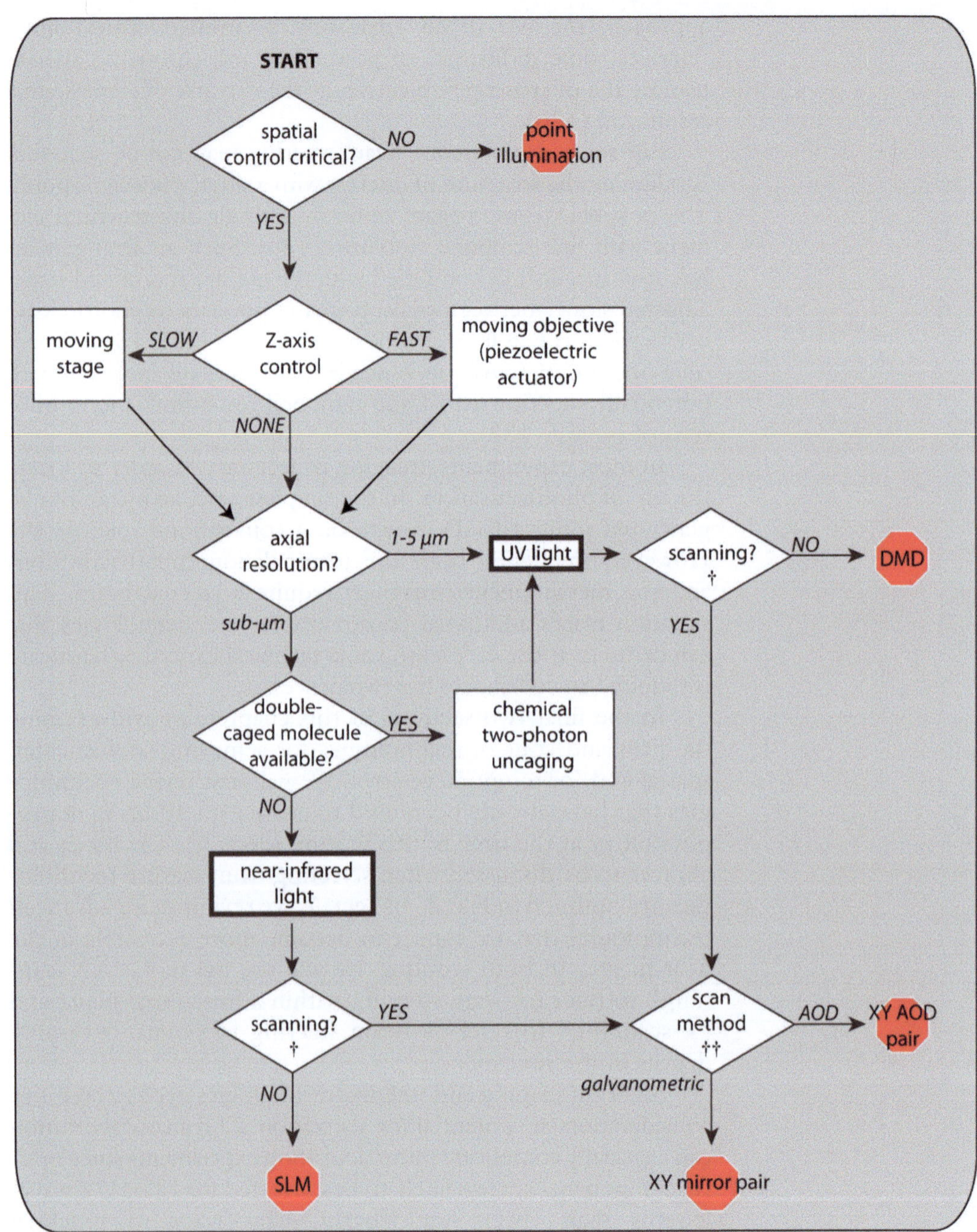

Fig. 2. Flowchart for selection of photoactivation technology. Abbreviations: *AOD* acousto-optic deflector, Sects. 6.2, 6.3, and 7.1; *DMD* digital micromirror device, Sect. 6.4; *SLM* spatial light modulator, Sect. 6.5. †Scanning (see Sect. 6.4 for an explanation of the advantages and disadvantages of scanless beam shaping). ††Scan method (see Sect. 6.3 for a comparison of galvanometric and AOD-based scanning).

5. Beam Steering: Current Technologies

5.1. XY Scanning with Mirrors Mounted on Galvanometric Scanners

Mirrors several millimeters in size, mounted on galvanometric scanners, are the standard approach to scanning microscopy (134, 135). When used to uncage neurotransmitters, galvoscanning technology has allowed groundbreaking studies of dendritic integration (123, 132, 136). Today, such technology is commercially available as components (Cambridge Technology, Massachusetts, USA), or in complete systems such as the Ultima IV from Prairie Technologies, which features one set of scanning mirrors for two-photon imaging and another set (requiring a second IR laser) for two-photon uncaging.

Newer microscope designs have gone beyond the raster scan pattern that has characterized most two-photon imaging to achieve higher visitation speeds with scanning mirrors (137). This has allowed fluorescence signal acquisition from locations separated by up to several millimeters, allowing the imaging of large networks (138). These methods are applicable to photoactivation. In principle, any scan geometry that maintains the minimum single-location exposure time necessary for sufficient photoactivation (~tens of μs) is acceptable. For example, uncaging could be done by using a spiral trajectory (X and Y sinusoids) (139) with discretization of locations achieved by blanking the beam with an electro-optic modulator (EOM) between spots. To date, most photoactivation studies have used scanning mirrors to traverse locations on a single dendrite at up to 3,000 Hz.

5.2. Acousto-Optic Deflectors

Perhaps the most rapidly developing technology for beam-steering applications is the acousto-optic deflector (AOD), which allows inertia-free scanning. The heart of an AOD is a crystal that behaves like a tunable diffraction grating with a grating constant that can be adjusted by varying the frequency of a sound wave propagating across the crystal, generated by a piezoelectric transducer interfaced with the crystal. AODs may be set up to direct most (>50%) of the input beam intensity into the first-order diffraction peak, generating a movable beam that may traverse a range of angles determined by the frequency bandwidth available to modulate the propagating sound wave. The deflected beam is mapped to a specific location in the preparation with appropriate relay optics. One AOD is required for each dimension of the coverage area.

5.2.1. The Spatial Resolution of an AOD-Based System

The number of resolvable points in the sample plane is determined by the number of resolvable angular deflection angles provided within the range of the scanning element. For a given range of angles $\Delta\theta$, a beam with an intrinsic divergence angle ϕ can occupy N resolvable angular locations, i.e.,

$$N = \frac{\Delta\theta}{\phi}. \quad (1)$$

For an AOD-based scanner, the range of deflection angles depends on the beam wavelength λ, the bandwidth of the driving acoustic frequency, Δf, and the velocity of sound in the AOD crystal medium, V:

$$\Delta\theta = \frac{\lambda \Delta f}{V}. \quad (2)$$

The divergence of the beam, ϕ, is approximated (for a uniformly illuminated aperture) by dividing the beam wavelength by the beam diameter, which for AOD-based scanners is limited by the deflector aperture size D:

$$\phi = \frac{\lambda}{D}. \quad (3)$$

Substituting Equations (2) and (3) into Equation (1) and simplifying gives the number of resolvable spots, N, in terms of the acoustic propagation time across the crystal Δt :

$$N = \Delta f \Delta t. \quad (4)$$

This quantity is sometimes referred to as the time–bandwidth product. In the case of a Gaussian beam that does not fill the aperture uniformly, N is an overestimate of the true performance. The most notable trade-offs are between spot size and scanning rate, and between bandwidth and number of resolvable spots (140). Increasing D, for example, requires less beam expansion to maintain a full back aperture and, therefore, allows more resolvable points. A useful calculator for examining design trade-offs can be found at the website of MolTech GmbH (http://www.mt-berlin.com/frames_ao/acousto_frames.htm). The ongoing development of AODs with larger apertures (e.g., 13 mm, (141)) pro-mises to increase the field of view and the number of resolvable points available with this scanning method.

TeO_2 crystals are particularly well suited for scanning applications with AODs. TeO_2 possesses a high figure of merit with a slow acoustic velocity, conferring upon it high deflation efficiency, high resolution, and a large scan angle for a relatively low RF bandwidth. TeO_2 transmits effectively over a wide spectral range including both IR and near-UV. For UV applications, screening individual crystals for custom selection is advisable for maximum efficiency (110). For pulsed IR applications, the spectral bandwidth of optical pulses leads to dispersion; pulse dispersion compensation is often used to restore original optical pulse properties (see Sect. 7.1.1). AODs are available commercially from many sources (e.g. Isomet, Brimrose Corporation, Noah Industries, and Crystal Technology Inc.).

5.2.2. XY Scanning with Two Crossed AODs

Two perpendicularly oriented AODs may be used to scan an XY plane. In this case, if the input beam fills or nearly fills the AOD aperture, an additional limit on scan angle appears due to vignetting at the second AOD aperture. To transmit all deflection angles from the initial (X) deflector through the aperture of the second (Y) deflector (allowing the maximum scan angle and hence the maximum addressable area), the two deflectors can be placed in close physical apposition (for this reason, XY pairs are often sold as a unit) or angular deflections from the first crystal's exit aperture can be optically relayed to the face of the second (142).

For patterned photoactivation with UV light in our laboratory (110), we use a Brimrose TeO_2, model 2DS-150-50-0.364, which contains two AODs with 7-mm apertures, driven by command signals from a two-channel variable frequency driver (VFE-150-50-V-B1-F2-2CH). With these parameters, we are able to visit up to 20,000 locations per second stably and reproducibly, accessing many sites in a brain slice virtually simultaneously (for example, mimicking simultaneous parallel fiber inputs to visually identified regions of a Purkinje cell dendritic arbor, dynamically adjustable from tens to hundreds of μm^2 in size). With this system, we are currently exploring the role of synchrony in branchlet-level dendritic excitation in cerebellar Purkinje neurons (Fig. 3).

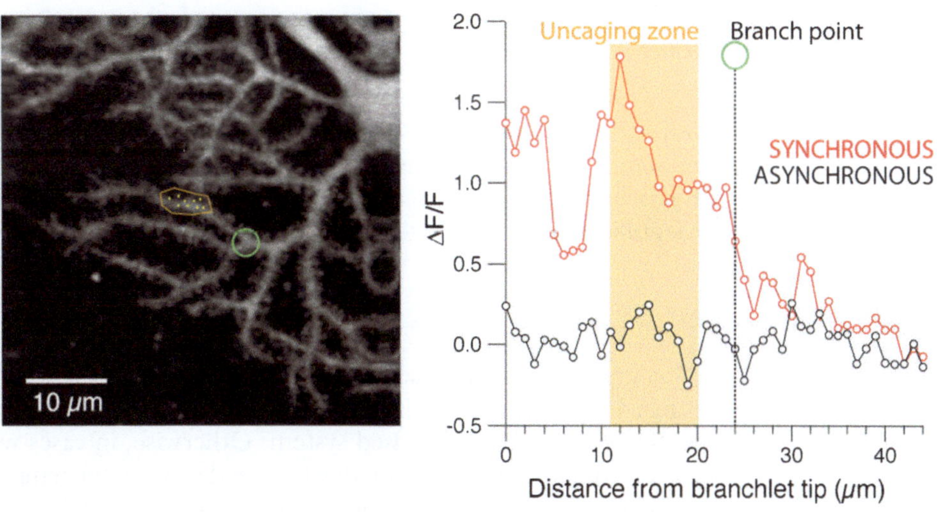

Fig. 3. Spatial structure of branchlet-level calcium transients evoked by synchronous glutamatergic input to a Purkinje cell. *Left*: Stimulus points indicated by *yellow dots*. *Right*: One-dimensional profile of calcium response to synchronous (50 μs between points) and asynchronous (10 ms between points) input (Civillico and Wang, unpublished).

5.3. XY Scanning Comparison: Galvanometric Scanners Versus AODs

The choice between AOD-based and galvanometric scanning involves a trade-off between addressing speed (where AODs are preferred) and range (mirrors). For an AOD-mediated beam deflection, the travel time between two scanned points is determined by the propagation time of sound across the beam diameter at the crystal, usually tens of microseconds. This makes any two arbitrarily spaced points within the scan area accessible within one switching time. However, AODs provide deflection over a more limited range of angles (typically ~50 mrad), while galvanometric mirrors typically span hundreds of milliradians. This disadvantage can be partially, though not completely, compensated for by choosing a lens system (see Fig. 4) that increases the angular range at the objective.

Figure 4 illustrates some of the relevant spatial relationships. In the paraxial limit, the relation between an angular beam deflection from the optical axis at the back aperture of an objective lens θ_{obj}, and the lateral displacement in the plane of focus x, is

$$x = f_{obj}\theta_{obj}, \tag{5}$$

where f_{obj} is the focal length of the lens. The focal length f_{obj} is related to the objective magnification M by

$$f_{obj} = \frac{L}{M}, \tag{6}$$

where L is a manufacturer-specific tube lens focal length (e.g. 180 mm for Olympus objectives). The deflection angle θ_{obj} is related to the deflection angle θ_{scan} at the scanner by the beam expansion factor of the focal telescope between the scanner and objective E:

$$\theta_{obj} = \frac{\theta_{scan}}{E}. \tag{7}$$

Substituting Equations (6) and (7) into Equation (5) gives the relation between scan angle θ_{scan} and lateral displacement at the focus as a function of beam expansion and objective magnification:

$$x = \frac{L}{M}\frac{\theta_{scan}}{E}. \tag{8}$$

This equation can be used to determine the available scanning range based on the maximum deflection angle produced by an AOD or galvanometric-based system. Otherwise, in cases where galvoscanning may be superior due to its wider range, the temporal/inertial limitations must be worked around, for example, by rational scan design (138).

As the technology advances, maximum available AOD aperture sizes are increasing, increasing the number of resolvable

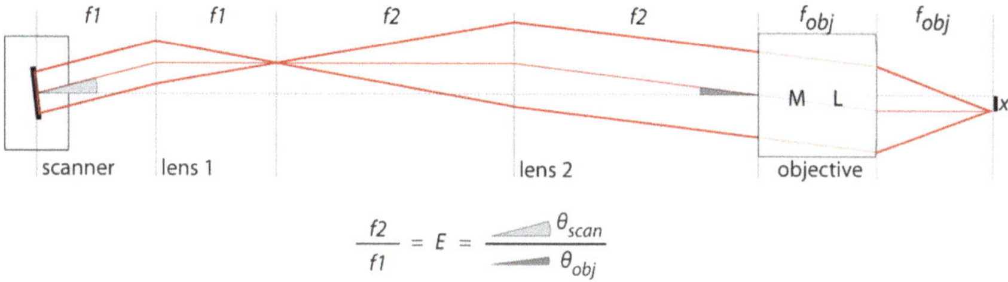

Fig. 4. Geometric optics of a one-dimensional scanning system. The scanner's pivot point is relayed to the back focal plane of the objective lens by a telescope (1:2 beam expansion in this example). The objective lens magnification (*M*) and reference tube length (*L*) determine the lateral offset *x* in the sample plane for a given value of θ_{obj}.

points and, with appropriate changes to the relay optics, enabling larger fields of view, at the cost of an increase in switching time. Note that while galvanometric scanners can produce deflections much larger than current AODs (hundreds of milliradians versus ~50 milliradians), the allowable angular range for a galvoscanning system is limited practically by the effective numerical aperture of the scan lens and by the acceptance angle of the objective (143).

5.4. Digital Multimirror Device

Methods that selectively illuminate different regions of the sample without beam-scanning deflection are said to be "scanless." The most mature scanless technology at this time is digital micromirror device (DMD) technology, a rectangular array of mirrors each approximately 100 μm^2 in size (e.g., a 10.5 × 14 mm chip, with 1,024 × 768 pixels; DMD-4000, Texas Instruments). Each mirror can be switched independently between two positions to deflect incident light selectively into the experimental preparation. Mirror movement is fast (~40 µs) and digitally controlled, allowing complex simultaneous multispot excitation patterns to be rapidly modulated. The principal disadvantage of DMD technology is that patterns are created by deflecting light out of the optical path. This limitation can be worked around by starting with a higher power light source. In an optical configuration for a neurophysiological experiment, an image of the illuminated DMD surface would be relayed to the sample plane. If coherent light is used, it must be scrambled to eliminate speckle with, for example, a vibrating mirror.

For example, compare the illumination control obtained with, e.g., a DMD-4000 to that obtained from a scanning configuration. Consider an optical setup consisting of a $\lambda = 355$ nm Gaussian beam incident upon an 8 mm wide DMD focused through a 40× objective with an NA of 0.8. The projected image is 200-µm wide in the sample plane. The light energy delivered is distributed over all the mirrors that project to this region. In contrast, in the scanning case, all of the beam power is concentrated in a spot approximately 280 nm in the lateral direction assuming a diffraction-limited focus, giving a density of approximately 500,000 times more energy per μm^2 than the DMD case. This reduced power per unit area dictates that a higher-energy light source will be required relative to a scanning configuration. Because each mirror represents a point, laser illumination is unnecessary and a flash lamp may increase the amount of light energy available.

Current DMD models from Texas Instruments are available in configurations optimized for UV or visible light in which the window in front of the mirror array is coated for high transmission at different wavelengths. In practice, the UV-optimized coating enhances UV transmission by about 10%. This technology is not ideal for TPE because of the power loss described above, as well as the fact that the manufacturing process involves coating the mirrors with aluminum rather than a dielectric coating, making them relatively poor reflectors of near-IR wavelengths. For these reasons, DMDs are most useful for one-photon excitation of molecules activated by visible or UV wavelengths.

Investigators using DMD technology report that the greatest expense, in terms of time and cost, is the development of software controls. DMD control may be simplified with third-party add-on modules such as the accessory light packages (ALPs) supplied by Vialux (Germany) that provide on-board memory and facilitate synchronization and display of mirror sequences, enabling the loading of temporal sequences onto the DMD chip which can then be delivered at the maximum display rates of 16–32 kHz, depending on the chip model. A set of tools written by Dr. Nicholas Hartell of the University of Leicester to control the ALP/DMD-3000 is available as an extension to Igor Pro (Wavemetrics, OR).

5.5. Holographic Beam Shaping with a Spatial Light Modulator

Another scanless technology for generating *XY* patterns is a liquid crystal spatial light modulator (SLM), which is a display that can be programmed to modulate the phase of the photoactivation beam to produce a desired intensity pattern at the focal plane. When the phase mask is suitably chosen, generally by an iterative algorithm (144), a beam reflected from the SLM will be shaped into a hologram, i.e., the phase of the light is modulated so that the highest intensity regions in the focal plane are arranged to produce an illumination pattern of arbitrary shape and size. A recent study (145) has demonstrated the utility of this approach for UV uncaging over subcellular regions defined "on the fly." The SLM can generate many spots with true simultaneity, and even has the ability to generate diffraction-limited spots over a 50-μm *Z*-focal range above and below the focal plane (V. Emiliani, personal communication).

In some cases, this technology is suitable for spatially patterned TPE. A recent study has demonstrated that an SLM can be used to generate arbitrarily patterned TPE and that superimposing "lens functions" on predefined phase masks confers some control over the Z-focal dimension of the excitation (146). Currently, the refresh rate for an SLM is limited to tens of Hz.

Holographic illumination patterns generated by spatial light modulators may include "ghost spots" of unwanted light. To eliminate unwanted photoactivation by these spots, an SLM may be combined with the use of a DMD as a dynamic spatial filter to increase the contrast of complex holographic patterns (C.-M. Tang, personal communication).

5.6. Z-Scanning

5.6.1. Slow Z-Scanning with a Moving Stage

Control of illumination in the Z-dimension, or axially resolved photoactivation, can be implemented most simply by controlling the distance between the microscope objective and the sample using a stepper motor coupled to the focusing knob of the objective, or a motorized stage to move the sample. Objective positioning schemes are employed in popular programs for controlling two-photon microscopes, including CfNT (R. Stepnowski, Bell Labs, and M. Müller, Max Planck Inst. Med. Res.) and ScanImage (147). While such approaches are suitable for collecting a series of images in sequence (e.g., to reconstruct cell morphology), they are not optimized for manipulating the plane of focus on neurophysiological time scales (e.g., for multisite photoactivation).

5.6.2. Faster Z-Scanning with a Piezoelectric Actuator

Rapid control of the Z-focal plane of an activating beam may be achieved by driving an oscillating piezoelectric actuator to modulate the Z-focal position of the objective. Actuators from Physik Instrumente (Germany) have found use in the imaging of neuronal and glial activity (139, 148) and could be adapted for targeted photoactivation. They may be conveniently added on to many common microscope builds, including the complete system sold by Prairie Technologies. The displacement of the objective is a function of the applied voltage; converter boxes and PC interfaces are available that allow the objective to be driven by an arbitrary waveform. In practice, inertia of a moving objective requires a smooth trajectory such as a sinusoid for fidelity to the command signal. The position of the objective can be combined with the known positions of *XY* scanning elements to reconstruct a three-dimensional scan path (139); the same approach could be used to produce a three-dimensional photoactivation trajectory.

6. Beam Steering: Emerging Technologies

Here we describe several state-of-the-art beam-steering technologies. While these methods are being developed in the context of imaging, they are amenable to use in photoactivation as well.

6.1. Pushing the Envelope with AODs

6.1.1. Two-Photon Excitation with AODs

Two-photon scanning systems using AODs for beam steering have recently become feasible. The principal optical limitation to be overcome is the dispersion of ultrashort laser pulses as they pass through an AOD. Ultrashort pulses are necessarily composed of a range of wavelengths that will travel at different speeds through typical AOD crystal materials, and be diffracted to different locations (see Equation (3) above; see (141)), resulting in pulses that are broadened temporally and spatially.

Pulse broadening may be compensated by adding dispersion to the laser pulses that is equal in magnitude and opposite in sign to that introduced by the AOD(s), for example, with a temporal pre-chirper composed of a pair of prisms (149, 150) or with a diffraction grating or acousto-optic modulator (AOM) (AA Optoelectronic, Orsay, France; (141, 150)). Spatial and temporal dispersion may be compensated simultaneously with a single tilted prism (151) placed at the correct distance from a pair of crossed AODs. The need for dispersion correction may be reduced by the use of slightly longer laser pulses.

6.1.2. Lensing with AODs

Rather than stepping the drive signal to a series of constant frequencies as for *XY* scanning (Sect. 6.2.2), lensing may be achieved by continuous modulation of the driving frequency to an AOD. Driving an AOD with a linear frequency sweep produces a continuously varying diffraction grating constant across the crystal face, resulting in a convergent or divergent beam and a fixed axial displacement of the focus that is proportional to the rate of change of the frequency. At the same time, however, the frequency modulation produces continuous lateral displacement of the focus (see Equation 2). To maintain independent control along one lateral dimension while directing the focus axially, the beam must be directed through a pair of AODs driven by counterpropagating frequency chirps (152, 153). In this configuration, the offset between the chirp center frequencies determines the lateral position, while the chirp rate determines the axial position as before. In this way, an AOD pair driven by counterpropagating frequency chirps functions as a cylindrical lens; two orthogonal pairs function as a spherical lens.

The allowable bandwidth for the driving frequencies constrains the addressable volume as follows: Because the axial displacement in Z depends on the chirp range while the lateral displacement in X or Y depends on chirp offset between the members of the X or Y pair, respectively, larger axial excursions (wider chirp range) require smaller lateral excursions (smaller distances between chirp centers) in order to fit within the allowed bandwidth. For this reason, the volume addressable by a 4-AOD scanner is octahedral, with the maximum XY span at $Z=0$.

Using four AODs, a system has been demonstrated with axial and lateral scan ranges of 50 and 200 μm, respectively, at 60× magnification (142). With this system, calcium transients could be recorded in three dimensions from hippocampal CA3 cells. The investigators suggest that axial and lateral ranges of 200 and 350 μm, respectively, are possible with this technology. The addressable volume is primarily limited by the acceptance angle of the AOD crystals; wider acceptance angles (154) will further improve this method in the near future. Acquisition rates up to tens of kilohertz are possible, giving comparable time resolution to an *XY*-only AOD-based scanning system, with the advantage of motion along the *Z*-axis. Note that the use of four AODs allows x, y, and z positioning without moving the objective. These two independent effects allow for any scan trajectory within the addressable volume, within the limited acoustic bandwidth shared by the axial and lateral displacement signals.

A recently developed system makes use of four custom-designed AODs (163). The custom-designed AODs have an optically rotated crystal orientation, and the second AOD of each *X* and *Y* pair has a narrow transducer, which results in a wide acceptance angle to the curved wavefront from the first AOD of each pair. This results in a larger overall scan volume. With a 0.8 NA, 40× objective, the AOD scanner can focus over a >100 μm range. The custom AODs are much thinner than standard scanning AODs, reducing the temporal dispersion of the AODs sufficiently to enable a prism-based pre-chirper to compensate for temporal dispersion of the ultrashort laser pulses. This enables low-noise two-photon images to be obtained at much lower powers than without a pre-chirper. These design features allow random access point measurements in three dimensions at rates up to 30 kHz throughout an octahedral volume beneath the microscope objective.

The innovative neurospy project (155) achieves three-dimensional scanning with two AODs and a custom-designed Yb: KYW laser emitting long (310 fs) pulses in the 1,030-nm range. The longer pulse duration reduces spatial and temporal dispersion by the AODs. The two AODs can be used either as a single-plane *XY* scanner as described in Sect. 6.2.2 (each position specified by one frequency pair), or to implement a three-dimensional scan path (chirped driving signals). Relative to the 4-AOD systems described above, elimination of one AOD of each pair sacrifices some flexibility in the trajectory because axial displacements are coupled to particular lateral displacements. This system can be constructed at a fraction of the cost of other scanning systems and can be run entirely from a LeCroy Waverunner 64xi oscilloscope. Documentation, instructions, and software for the construction and operation of this "open-source" system are available at http://www.neurospy.org.

6.2. Temporal Focusing

By introducing geometrical dispersion of ultrashort laser pulses outside the objective focal plane, for example, with a diffraction grating, it is possible to create a temporal focusing effect, in which a spatiotemporal light pattern derived from an ultrashort pulse is temporally as it propagates through a volume, compressed at the temporal focal plane, and dispersed again as it propagates further. Temporal focusing is the converse of the spatial focusing employed in conventional TPE; instead of a spatially restricted, temporally extended focus, there is a spatially extended, temporally restricted focus. Temporal focusing has been demonstrated for depth-resolved two-photon imaging without scanning (156), and has recently been combined with SLM-based holographic illumination to enable the placement of an arbitrarily shaped photoactivation region with 5 μm resolution at the focal plane of the objective (157).

6.3. Other Methods

Additional three-dimensional imaging methods show promise for future use in targeted photoactivation but are still in early stages of development. Rather than moving the objective, a variable-focus lens employing a fluid-filled cavity (158) might be used to adjust the focal plane dynamically. A membrane-deformable mirror could be used for this purpose ((159, 160); e.g., Flexible Optical B.V., The Netherlands). A recent study (161) provided a remarkable demonstration of three-dimensional scanning using *XY* scanning with a pair of AODs, as described in Sect. 6.2.2, to direct a beam into the ends of a matrix of single-mode optical fibers, one fiber per desired imaging spot. By micropositioning the exit ends of the fibers, imaging spots could be positioned within a three-dimensional volume as desired. Target points were defined by a three-dimensional image reconstruction that was acquired with a parallel mirror-based scanning system.

Acknowledgments

We thank Karl Deisseroth, Valentina Emiliani, Jonathan A.N. Fisher, Mark McDonald, Ashlan Reid, Angus Silver, Cha-Min Tang, Stephan Thiberge, and Dejan Vučinić for helpful discussions and comments on this chapter.

E.F.C. is supported by a Robert Leet and Clara Guthrie Patterson Postdoctoral Fellowship in Brain Circuitry. J.P.R. is supported by a National Science Foundation Graduate Research fellowship. S.S.-H.W. is a W.M. Keck Foundation Distinguished Young Investigator and is supported by National Institutes of Health grant NS045193 and the National Science Foundation.

References

1. Walker JW, McCray JA, Hess GP (1986) Photolabile protecting groups for an acetylcholine receptor ligand. Synthesis and photochemistry of a new class of *o*-nitrobenzyl derivatives and their effects on receptor function. Biochemistry 25:1799–1805
2. Milburn T, Matsubara N, Billington AP, Udgaonkar JB, Walker JW, Carpenter BK, Webb WW, Marque J, Denk W, McCray JA et al (1989) Synthesis, photochemistry, and biological activity of a caged photolabile acetylcholine receptor ligand. Biochemistry 28:49–55
3. Wilcox M, Viola RW, Johnson KW, Billington AP, Carpenter BK, McCray JA, Guzikowski AP, Hess GP (1990) Synthesis of photolabile precussors of amino acid neurotransmitters. J Org Chem 55:1585–1589
4. Furuta T, Wang SS, Dantzker JL, Dore TM, Bybee WJ, Callaway EM, Denk W, Tsien RY (1999) Brominated 7-hydroxycoumarin-4-ylmethyls: photolabile protecting groups with biologically useful cross-sections for two photon photolysis. Proc Natl Acad Sci U S A 96:1193–1200
5. Canepari M, Nelson L, Papageorgiou G, Corrie JET, Ogden D (2001) Photochemical and pharmacological evaluation of 7-nitroindolinyl-and4-methoxy-7-nitroindolinyl-amino acids as novel, fast caged neurotransmitters. J Neurosci Methods 112:29–42
6. Matsuzaki M, Ellis-Davies GCR, Nemoto T, Miyashita Y, Iino M, Kasai H (2001) Dendritic spine geometry is critical for AMPA receptor expression in hippocampal CA1 pyramidal neurons. Nat Neurosci 4:1086–1092
7. Papageorgiou G, Corrie JET (2000) Effects of aromatic substituents on the photocleavage of 1-acyl-7-nitroindolines. Tetrahedron 56:8197–8205
8. Gee KR, Niu L, Schaper K, Hess GP (1995) Caged bioactive carboxylates. Synthesis, photolysis studies, and biological characterization of a new caged N-methyl-D-aspartic acid. J Org Chem 60:4260–4263
9. Niu L, Gee KR, Schaper K, Hess GP (1996) Synthesis and photochemical properties of a kainate precursor and activation of kainate and AMPA receptor channels on a microsecond time scale. Biochemistry 35:2030–2036
10. Huang YH, Muralidharan S, Sinha SR, Kao JP, Bergles DE (2005) Ncm-D-aspartate: a novel caged D-aspartate suitable for activation of glutamate transporters and *N*-methyl-D-aspartate (NMDA) receptors in brain tissue. Neuropharmacology 49:831–842
11. Huang YH, Sinha SR, Fedoryak OD, Ellis-Davies GCR, Bergles DE (2005) Synthesis and characterization of 4-methoxy-7-nitroindolinyl-D-aspartate, a caged compound for selective activation of glutamate transporters and *N*-methyl-D-aspartate receptors in brain tissue. Biochemistry 44:3316–3326
12. Ellis-Davies GCR (2007) Caged compounds: photorelease technology for control of cellular chemistry and physiology. Nat Methods 4:619–628
13. Fino E, Araya R, Peterka DS, Salierno M, Etchenique R, Yuste R (2009) RuBi-Glutamate: two-photon and visible-light photoactivation of neurons and dendritic spines. Front Neural Circuits 3:2
14. Gee KR, Wieboldt R, Hess GP (1994) Synthesis and photochemistry of a new photolabile derivative of GABA-neurotransmitter release and receptor activation in the microsecond time region. J Am Chem Soc 116:8366–8367
15. Cürten BC, Kullmann PHMK, Bier ME, Kandler KK, Schmidt BS (2005) Synthesis, photophysical, photochemical and biological properties of caged GABA, 4-[[(2H-1-Benzopyran-2-one-7-amino-4-methoxy) carbonyl] amino] butanoic acid. Photochem Photobiol 81:641–648
16. Rial Verde EM, Zayat L, Etchenique R, Yuste R (2008) Photorelease of GABA with visible light using an inorganic caging group. Front Neural Circuits 2:2
17. Trigo FF, Papageorgiou G, Corrie JE, Ogden D (2009) Laser photolysis of DPNI-GABA, a tool for investigating the properties and distribution of GABA receptors and for silencing neurons in situ. J Neurosci Methods 181(2):159–169
18. Shembekar VR, Chen Y, Carpenter BK, Hess GP (2007) Coumarin-caged glycine that can be photolyzed within 3 microseconds by visible light. Biochemistry 46:5479–5484
19. Roitman MF, Stuber GD, Phillips PE, Wightman RM, Carelli RM (2004) Dopamine operates as a subsecond modulator of food seeking. J Neurosci 24:1265–1271
20. Sombers LA, Beyene M, Carelli RM, Wightman RM (2009) Synaptic overflow of dopamine in the nucleus accumbens arises from neuronal activity in the ventral tegmental area. J Neurosci 29:1735–1742

21. Breitinger HG, Wieboldt R, Ramesh D, Carpenter BK, Hess GP (2000) Synthesis and characterization of photolabile derivatives of serotonin for chemical kinetic investigations of the serotonin 5-HT_3 receptor. Biochemistry 39:5500–5508
22. Muralidharan S, Nerbonne JM (1995) Photolabile "caged" adrenergic receptor agonists and related model compounds. J Photochem Photobiol B 27:123–137
23. Lee TH, Gee KR, Ellinwood EH, Seidler FJ (1996) Combining "caged-dopamine" photolysis with fast-scan cyclic voltammetry to assess dopamine clearance and release autoinhibition in vitro. J Neurosci Methods 67:221–231
24. Kao JP (2006) Caged molecules: principles and practical considerations. Curr Protoc Neurosci Chapter 6:Unit 6.20
25. Kuner T, Li Y, Gee KR, Bonewald LF, Augustine GJ (2008) Photolysis of a caged peptide reveals rapid action of *N*-ethylmaleimide sensitive factor before neurotransmitter release. Proc Natl Acad Sci USA 105:347–352
26. Engels J, Schlaeger EJ (1977) Synthesis, structure, and reactivity of adenosine cyclic 3′,5′-phosphate benzyl triesters. J Med Chem 20:907–911
27. Kaplan JH, Forbush B, Hoffman JF (1978) Rapid photolytic release of adenosine 5′-triphosphate from a protected analogue: utilization by the Na:K pump of human red blood cell ghosts. Biochemistry 17: 1929–1935
28. Walker JW, Somlyo AV, Goldman YE, Somlyo AP, Trentham DR (1987) Kinetics of smooth and skeletal muscle activation by laser pulse photolysis of caged inositol 1,4,5-trisphosphate. Nature 327:249–252
29. Dantzig JA, Higuchi H, Goldman YE (1998) Studies of molecular motors using caged compounds. Meth Enzymol 291:307–348
30. Bamberg E, Clarke RJ, Fendler K (2001) Electrogenic properties of the Na^+, K^+-ATPase probed by presteady state and relaxation studies. J Bioenerg Biomembr 33:401–405
31. Adams SR, Tsien RY (1993) Controlling cell chemistry with caged compounds. Annu Rev Physiol 55:755–784
32. Kaplan JH, Ellis-Davies GCR (1988) Photolabile chelators for the rapid photorelease of divalent cations. Proc Natl Acad Sci USA 85:6571–6575
33. Ellis-Davies GCR, Kaplan JH (1994) Nitrophenyl-EGTA, a photolabile chelator that selectively binds Ca^{2+} with high affinity and releases it rapidly upon photolysis. Proc Natl Acad Sci USA 91:187–191
34. Makings LR, Tsien RY (1994) Caged nitric oxide. Stable organic molecules from which nitric oxide can be photoreleased. J Biol Chem 269:6282–6285
35. Ellis-Davies GCR (2008) Neurobiology with caged calcium. Chem Rev 108:1603–1613
36. Adams SR, Kao JPY, Tsien RY (1988) Biologically useful chelators that take up calcium (2+) upon illumination. J Am Chem Soc 111:7957–7968
37. Khodakhah K, Armstrong CM (1997) Inositol trisphosphate and ryanodine receptors share a common functional Ca^{2+} pool in cerebellar Purkinje neurons. Biophys J 73:3349–3357
38. Walker JW, Reid GP, Trentham DR (1989) Synthesis and properties of caged nucleotides. Meth Enzymol 172:288–301
39. Sarkisov DV, Wang SS-H (2008) Order-dependent coincidence detection in cerebellar Purkinje neurons at the inositol trisphosphate receptor. J Neurosci 28(1):133–142
40. Gee KR, Lee HC (1998) Characterization and application of photogeneration of calcium mobilizers cADP-ribose and nicotinic acid adenine dinucleotide phosphate from caged analogs. Meth Enzymol 291:403–415
41. Shirokova N, Niggli E (2008) Studies of RyR function in situ. Methods 46:183–193
42. Zhelyaskov VR, Godwin DW (1998) Photolytic generation of nitric oxide through a porous glass partitioning membrane. Nitric Oxide 2:454–459
43. Furuta T, Takeuchi H, Isozaki M, Takahashi Y, Kanehara M, Sugimoto M, Watanabe T, Noguchi K, Dore TM, Kurahashi T, Iwamura M, Tsien RY (2004) Bhc-cNMPs as either water-soluble or membrane-permeant photoreleasable cyclic nucleotides for both one- and two-photon excitation. Chembiochem 5:1119–1128
44. Nerbonne JM (1996) Caged compounds: tools for illuminating neuronal responses and connections. Curr Opin Neurobiol 6:379–386
45. Lester HA, Nerbonne JM (1982) Physiological and pharmacological manipulations with light flashes. Annu Rev Biophys Bioeng 11:151–175
46. Gurney AM (1994) Flash photolysis of caged compounds. In: Ogden DA (ed) Microelectrode techniques. The Plymouth workshop handbook. The company of Biologists pp 389–406
47. Sarkisov DV, Wang SS-H (2007) Combining uncaging techniques with patch-clamp recording and optical physiology. In: Walzw (ed) Patch clamp analysis: advanced techniques, vol 38, 2nd edn. Humana Press pp 149–168

48. Kramer RH, Chambers JJ, Trauner D (2005) Photochemical tools for remote control of ion channels in excitable cells. Nat Chem Biol 1:360–365
49. Szobota S, Gorostiza P, Del Bene F, Wyart C, Fortin DL, Kolstad KD, Tulyathan O, Volgraf M, Numano R, Aaron HL, Scott EK, Kramer RH, Flannery J, Baier H, Trauner D, Isacoff EY (2007) Remote control of neuronal activity with a light-gated glutamate receptor. Neuron 54:535–545
50. Gorostiza P, Volgraf M, Numano R, Szobota S, Trauner D, Isacoff EY (2007) Mechanisms of photoswitch conjugation and light activation of an ionotropic glutamate receptor. Proc Natl Acad Sci USA 104:10865–10870
51. Chambers JJ, Banghart MR, Trauner D, Kramer RH (2006) Light-induced depolarization of neurons using a modified shaker K^+ channel and a molecular photoswitch. J Neurophysiol 96:2792–2796
52. Banghart M, Borges K, Isacoff E, Trauner D, Kramer RH (2004) Light-activated ion channels for remote control of neuronal firing. Nat Neurosci 7:1381–1386
53. Fortin DL, Banghart MR, Dunn TW, Borges K, Wagenaar DA, Gaudry Q, Karakossian MH, Otis TS, Kristan WB, Trauner D, Kramer RH (2008) Photochemical control of endogenous ion channels and cellular excitability. Nat Methods 5:331–338
54. Chambers JJ, Kramer RH (2008) Light-activated ion channels for remote control of neural activity. Methods Cell Biol 90:217–232
55. Knöpfel T (2008) Expanding the toolbox for remote control of neuronal circuits. Nat Methods 5:293–295
56. Nagel G, Szellas T, Huhn W, Kateriya S, Adeishvili N, Berthold P, Ollig D, Hegemann P, Bamberg E (2003) Channelrhodopsin-2, a directly light-gated cation-selective membrane channel. Proc Natl Acad Sci USA 100: 13940–13945
57. Boyden ES, Zhang F, Bamberg E, Nagel G, Deisseroth K (2005) Millisecond-timescale, genetically targeted optical control of neural activity. Nat Neurosci 8:1263–1268
58. Zhang F, Wang LP, Boyden ES, Deisseroth K (2006) Channelrhodopsin-2 and optical control of excitable cells. Nat Methods 3:785–792
59. Ernst OP, Sánchez Murcia PA, Daldrop P, Tsunoda SP, Kateriya S, Hegemann P (2008) Photoactivation of channelrhodopsin. J Biol Chem 283:1637–1643
60. Nagel G, Brauner M, Liewald JF, Adeishvili N, Bamberg E, Gottschalk A (2005) Light activation of channelrhodopsin-2 in excitable cells of *Caenorhabditis elegans* triggers rapid behavioral responses. Curr Biol 15:2279–2284
61. Bamann C, Kirsch T, Nagel G, Bamberg E (2008) Spectral characteristics of the photocycle of channelrhodopsin-2 and its implication for channel function. J Mol Biol 375:686–694
62. Nikolic K, Grossman N, Grubb MS, Burrone J, Toumazou C, Degenaar P (2009) Photocycles of channelrhodopsin-2. Photochem Photobiol 85:400–411
63. Lin JY, Lin MZ, Steinbach P, Tsien RY (2009) Characterization of engineered channelrhodopsin variants with improved properties and kinetics. Biophys J 96: 1803–1814
64. Berndt A, Yizhar O, Gunaydin LA, Hegemann P, Deisseroth K (2009) Bi-stable neural state switches. Nat Neurosci 12:229–234
65. Zhang F, Prigge M, Beyrière F, Tsunoda SP, Mattis J, Yizhar O, Hegemann P, Deisseroth K (2008) Red-shifted optogenetic excitation: a tool for fast neural control derived from Volvox carteri. Nat Neurosci 11:631–633
66. Zhang YP, Oertner TG (2006) Optical induction of synaptic plasticity using a light-sensitive channel. Nat Methods 4:139–141
67. Ishizuka T, Kakuda M, Araki R, Yawo H (2006) Kinetic evaluation of photosensitivity in genetically engineered neurons expressing green algae light-gated channels. Neurosci Res 54:85–94
68. Bormuth V, Howard J, Schaffer E (2007) LED illumination for video-enhanced DIC imaging of single microtubules. J Microsc 226:1–5
69. Benavides JM, Webb RH (2005) Optical characterization of ultrabright LEDs. Appl Opt 44:4000–4003
70. Lanyi JK (1990) Halorhodopsin, a light-driven electrogenic chloride-transport system. Physiol Rev 70(2):319–330
71. Han X, Boyden ES (2007) Multiple-color optical activation, silencing, and desynchronization of neural activity, with single-spike temporal resolution. PLoS ONE 2(3):e299
72. Zhang F, Wang LP, Brauner M, Liewald JF, Kay K, Watzke N, Wood PG, Bamberg E, Nagel G, Gottschalk A, Deisseroth K (2007) Multimodal fast optical interrogation of neural circuitry. Nature 446:633–639
73. Zhang F, Aravanis AM, Adamantidis A, de Lecea L, Deisseroth K (2007) Circuit-breakers: optical technologies for probing neural signals and systems. Nat Rev Neurosci 8:577–581

74. Gradinaru V, Thompson KR, Deisseroth K (2008) eNpHR: a *Natronomonas* halorhodopsin enhanced for optogenetic applications. Brain Cell Biol 36:129–139
75. Chow H, Qian XQ, Boyden ES (2009) High-performance halorhodopsin variants for improved genetically-targetable optical neural silencing. Frontiers in systems neuroscience. In: Conference abstract: computations and systems neuroscience
76. Gradinaru V, Mogri M, Thompson KR, Henderson JM, Deisseroth K (2009) Optical deconstruction of parkinsonian neural circuitry. Science 324:354–359
77. Sohal VS, Zhang F, Yizhar O, Deisseroth K (2009) Parvalbumin neurons and gamma rhythms enhance cortical circuit performance. Nature 459:698–702
78. Sjulson L, Miesenböck G (2008) Photocontrol of neural activity: biophysical mechanisms and performance in vivo. Chem Rev 108(5):1588–1602
79. Airan RD, Thompson KR, Fenno LE, Bernstein H, Deisseroth K (2009) Temporally precise in vivo control of intracellular signalling. Nature 458:1025–1029
80. Tsien RY (1989) Fluorescent probes of cell signaling. Annu Rev Neurosci 12:227–253
81. Stosiek C, Garaschuk O, Holthoff K, Konnerth A (2003) In vivo two-photon calcium imaging of neuronal networks. Proc Natl Acad Sci USA 100:7319–7324
82. Harvey CD, Svoboda K (2007) Locally dynamic synaptic learning rules in pyramidal neuron dendrites. Nature 450:1195–1202
83. Nikolenko V, Poskanzer KE, Yuste R (2007) Two-photon photostimulation and imaging of neural circuits. Nat Methods 4:943–950
84. Saito T (2006) In vivo electroporation in the embryonic mouse central nervous system. Nat Protoc 1:1552–1558
85. Tabata H, Nakajima K (2008) Labeling embryonic mouse central nervous system cells by *in utero* electroporation. Dev Growth Differ 50:507–511
86. Petreanu LP, Huber DH, Sobczyk AS, Svoboda KS (2007) Channelrhodopsin-2-assisted circuit mapping of long-range callosal projections. Nat Neurosci 10:663–668
87. Petreanu L, Mao T, Sternson SM, Svoboda K (2009) The subcellular organization of neocortical excitatory connections. Nature 457:1142–1145
88. Cetin A, Komai S, Eliava M, Seeburg PH, Osten P (2006) Stereotaxic gene delivery in the rodent brain. Nat Protoc 1:3166–3173
89. Adamantidis AR, Zhang F, Aravanis AM, Deisseroth K, de Lecea L (2007) Neural substrates of awakening probed with optogenetic control of hypocretin neurons. Nature 450:420–424
90. Dittgen T, Nimmerjahn A, Komai S, Licznerski P, Waters J, Margrie TW, Helmchen F, Denk W, Brecht M, Osten P (2004) Lentivirus-based genetic manipulations of cortical neurons and their optical and electrophysiological monitoring in vivo. Proc Natl Acad Sci USA 101:18206–18211
91. Bartlett JS, Wilcher R, Samulski RJ (2000) Infectious entry pathway of adeno-associated virus and adeno-associated virus vectors. J Virol 74:2777–2785
92. Wallace DJ, zum Alten Borgloh SM, Astori S, Yang Y, Bausen M, Kügler S, Palmer AE, Tsien RY, Sprengel R, Kerr JN, Denk W, Hasan MT (2008) Single-spike detection in vitro and in vivo with a genetic Ca(2+) sensor. Nat Methods 5:797–804
93. Song CK, Enquist LW, Bartness TJ (2005) New developments in tracing neural circuits with herpesviruses. Virus Res 111:235–249
94. Ekstrand MI, Enquist LW, Pomeranz LE (2008) The alpha-herpesviruses: molecular pathfinders in nervous system circuits. Trends Mol Med 14:134–140
95. Lima SQ, Hromádka T, Znamenskiy PZ, Zador AM (2009) Photostimulation-assisted identification of neuronal populations (PINP): a new method of tagging neuronal populations for identification during in vivo electrophysiological recording. PLoS ONE 4(7):e6099
96. Kuhlman SJ, Huang ZJ (2008) High-resolution labeling and functional manipulation of specific neuron types in mouse brain by Cre-activated viral gene expression. PLoS ONE 3:e2005
97. Atasoy D, Aponte Y, Su HH, Sternson SM (2008) A FLEX switch targets Channelrhodopsin-2 to multiple cell types for imaging and long-range circuit mapping. J Neurosci 28:7025–7030
98. Tsai HC, Zhang F, Adamantidis A, Stuber GD, Bonci A, de Lecea L, Deisseroth K (2009) Phasic firing in dopaminergic neurons is sufficient for behavioral conditioning. Science 324:1080–1084
99. Cardin JA, Carlén M, Meletis K, Knoblich U, Zhang F, Deisseroth K, Tsai LH, Moore CI (2009) Driving fast-spiking cells induces gamma rhythm and controls sensory responses. Nature 459:663–667

100. Schnütgen F, Doerflinger N, Calléja C, Wendling O, Chambon P, Ghyselinck NB (2003) A directional strategy for monitoring Cre-mediated recombination at the cellular level in the mouse. Nat Biotechnol 21:562–565
101. Davidson BL, Breakefield XO (2003) Viral vectors for gene delivery to the nervous system. Nat Rev Neurosci 4:353–364
102. Callaway EM (2008) Transneuronal circuit tracing with neurotropic viruses. Curr Opin Neurobiol 18:617–623
103. Hogan BH, Constantini FC, Lacey EL (1994) Production of transgenic mice. In: Hogan B, Beddington R, Costantini F, Lacey E (eds) Manipulating the mouse embryo: a laboratory manual. Cold Spring Harbor Laboratory Press, Cold Spring Harbor, NY, pp 217–252
104. Feng G, Mellor RH, Bernstein M, Keller-Peck C, Nguyen QT, Wallace M, Nerbonne JM, Lichtman JW, Sanes JR (2000) Imaging neuronal subsets in transgenic mice expressing multiple spectral variants of GFP. Neuron 28:41–51
105. Sprengel R, Hasan MT (2007) Tetracycline-controlled genetic switches. Handb Exp Pharmacol 178:49–72
106. Dymecki SM, Kim JC (2007) Molecular neuroanatomy's "three Gs": a primer. Neuron 54:17–34
107. Luo L, Callaway EM, Svoboda K (2008) Genetic dissection of neural circuits. Neuron 57:634–660
108. Arenkiel BR, Peca J, Davison IG, Feliciano C, Deisseroth K, Augustine GJ, Ehlers MD, Feng G (2007) In vivo light-induced activation of neural circuitry in transgenic mice expressing channelrhodopsin-2. Neuron 54:205–218
109. Liewald JF, Brauner M, Stephens GJ, Bouhours M, Schultheis C, Zhen M, Gottschalk A (2008) Optogenetic analysis of synaptic function. Nat Methods 5:895–902
110. Shoham S, O'Connor DH, Sarkisov DV, Wang SS-H (2005) Rapid neurotransmitter uncaging in spatially defined patterns. Nat Methods 2:837–843
111. Zipfel WR, Williams RM, Webb WW (2003) Nonlinear magic: multiphoton microscopy in the biosciences. Nat Biotechnol 21:1369–1377
112. Ellis-Davies GCR (2000) Basics of photoactivation. Yuste R, Lanni F, Konnerth A (eds) In: Imaging neurons: a laboratory manual. Cold Spring Harbor Laboratory Press, Cold Spring Harbor, NY, pp 24.1–24.8
113. Kandler K, Givens RS, Katz LC (2000) Photostimulation with caged glutamate. Yuste R, Lanni F, Konnerth A (eds) In: Imaging neurons: a laboratory manual. Cold Spring Harbor Laboratory Press, Cold Spring Harbor, NY, pp 27.1–27.9
114. Bernardinelli Y, Haeberli C, Chatton JY (2005) Flash photolysis using a light emitting diode: an efficient, compact, and affordable solution. Cell Calcium 37:565–572
115. Venkataramani S, Davitt KM, Xu H, Zhang J, Song YK, Connors BW, Nurmikko AV (2007) Semiconductor ultra-violet light-emitting diodes for flash photolysis. J Neurosci Methods 160:5–9
116. Xu H, Zhang J, Davitt KM, Song Y, Nurmikko AV (2008) Application of blue-green and ultraviolet micro-LEDs to biological imaging and detection. J Phys D Appl Phys 41:94013
117. Pettit DL, Wang SS-H, Gee KR, Augustine GJ (1997) Chemical two-photon uncaging: a novel approach to mapping glutamate receptors. Neuron 19:465–471
118. Sarkisov DV, Gelber SE, Walker JW, Wang SS-H (2007) Synapse specificity of calcium release probed by chemical two-photon uncaging of inositol 1,4,5-trisphosphate. J Biol Chem 282:25517–25526
119. Wang SSH, Augustine GJ (1995) Confocal imaging and local photolysis of caged compounds: dual probes of synaptic function. Neuron 15:755–760
120. Matsuzaki M, Ellis-Davies GCR, Kasai H (2008) Three-dimensional mapping of unitary synaptic connections by two-photon macro photolysis of caged glutamate. J Neurophysiol 99:1535–1544
121. Göppert-Mayer M (1931) Über elementarakte mit zwei quantensprüngen. Ann Phys 9:273–294
122. Albota MA, Xu C, Webb WW (1998) Two-photon fluorescence excitation cross sections of biomolecular probes from 690 to 960 nm. Appl Opt 37:7352–7356
123. Gasparini S, Magee JC (2006) State-dependent dendritic computation in hippocampal CA1 pyramidal neurons. J Neurosci 26:2088–2100
124. Mohanty SK, Reinscheid RK, Liu X, Okamura N, Krasieva TB, Berns MW (2008) In-depth activation of ChR2 sensitized excitable cells with high spatial resolution using two-photon excitation with near-IR laser microbeam. Biophys J 95(8):3916–3926
125. Rickgauer JP, Tank DW (2009) Two-photon excitation of channelrhodopsin-2 at saturation. Proc Natl Acad Sci USA 106:15025–15030

126. Tang CM (2006) Photolysis of caged neurotransmitters: theory and procedures for light delivery. Curr Protoc Neurosci Chapter 6:Unit 6.21
127. Callaway EM, Katz LC (1993) Photostimulation using caged glutamate reveals functional circuitry in living brain slices. Proc Natl Acad Sci USA 90:7661–7665
128. Dalva MB, Katz LC (1994) Rearrangements of synaptic connections in visual cortex revealed by laser photostimulation. Science 265:255–258
129. Dantzker JL, Callaway EM (2000) Laminar sources of synaptic input to cortical inhibitory interneurons and pyramidal neurons. Nat Neurosci 3:701–707
130. Shepherd GM, Stepanyants A, Bureau I, Chklovskii D, Svoboda K (2005) Geometric and functional organization of cortical circuits. Nat Neurosci 8:782–790
131. Shepherd GM, Svoboda K (2005) Laminar and columnar organization of ascending excitatory projections to layer 2/3 pyramidal neurons in rat barrel cortex. J Neurosci 25:5670–5679
132. Major G, Polsky A, Denk W, Schiller J, Tank DW (2008) Spatiotemporally graded NMDA spike/plateau potentials in basal dendrites of neocortical pyramidal neurons. J Neurophysiol 99:2584–2601
133. Bliton AC, Lechleiter JD (1995) Optical considerations at ultraviolet wavelengths in confocal microscopy. In: Pawley JBY (ed) Handbook of biological confocal microscopy, 2nd edn. Plenum, New York, pp 431–444
134. Denk W, Delaney KR, Gelperin A, Kleinfeld D, Strowbridge BW, Tank DW, Yuste R (1994) Anatomical and functional imaging of neurons using 2-photon laser scanning microscopy. J Neurosci Methods 54:151–162
135. Tsai PS, Nishimura N, Yoder EJ, White A, Dolnick E, Kleinfeld D (2002) Principles, design and construction of a two photon scanning microscope for in vitro and in vivo studies. In: Frostig R (ed) Methods for in vivo optical imaging. CRC Press pp 113–171
136. Gasparini S, Losonczy A, Chen X, Johnston D, Magee JC (2007) Associative pairing enhances action potential back-propagation in radial oblique branches of CA1 pyramidal neurons. J Physiol 580:787–800
137. Lörincz A, Rózsa B, Katona G, Vizi ES, Tamás G (2007) Differential distribution of NCX1 contributes to spine-dendrite compartmentalization in CA1 pyramidal cells. Proc Natl Acad Sci USA 104:1033–1038
138. Lillis KP, Eng A, White JA, Mertz J (2008) Two-photon imaging of spatially extended neuronal network dynamics with high temporal resolution. J Neurosci Methods 172:178–184
139. Göbel W, Kampa BM, Helmchen F (2006) Imaging cellular network dynamics in three dimensions using fast 3D laser scanning. Nat Methods 4:73–79
140. Bullen A, Patel SS, Saggau P (1997) High-speed, random-access fluorescence microscopy I. High-resolution optical recording with voltage-sensitive dyes and ion indicators. Biophys J 73(1):477–491
141. Kremer Y, Léger JF, Lapole R, Honnorat N, Candela Y, Dieudonné S, Bourdieu L (2008) A spatio-temporally compensated acousto-optic scanner for two-photon microscopy providing large field of view. Opt Express 16:10066–10076
142. Reddy G, Kelleher K, Fink R, Saggau P (2008) Three-dimensional random access multiphoton microscopy for functional imaging of neuronal activity. Nat Neurosci 11:713–720
143. Oheim M, Beaurepaire E, Chaigneau E, Mertz J, Charpak S (2001) Two-photon microscopy in brain tissue: parameters influencing the imaging depth. J Neurosci Methods 111:29–37
144. Gerchberg RW, Saxton WO (1972) A practical algorithm for the determination of the phase from image and diffraction plane pictures. Optik 35:237–246
145. Lutz C, Otis TS, Desars V, Charpak S, Digregorio DA, Emiliani V (2008) Holographic photolysis of caged neurotransmitters. Nat Methods 5(9):821–827
146. Nikolenko V, Watson BO, Araya R, Woodruff A, Peterka DS, Yuste R (2008) SLM microscopy: scanless two-photon imaging and photostimulation with spatial light modulators. Front Neural Circuits 2:5
147. Pologruto TA, Sabatini BL, Svoboda K (2003) ScanImage: flexible software for operating laser-scanning microscopes. Biomed Eng Online 2:13
148. Hoogland TM, Kuhn B, Göbel W, Huang W, Nakai J, Helmchen F, Flint J, Wang SS (2009) Radially expanding transglial calcium waves in the intact cerebellum. Proc Natl Acad Sci USA 106:3496–3501
149. Iyer V, Losavio BE, Saggau P (2003) Compensation of spatial and temporal dispersion for acousto-optic multiphoton laser-scanning microscopy. J Biomed Opt 8:460

150. Salomé R, Kremer Y, Dieudonné S, Léger JF, Krichevsky O, Wyart C, Chatenay D, Bourdieu L (2006) Ultrafast random-access scanning in two-photon microscopy using acousto-optic deflectors. J Neurosci Methods 154:161–174

151. Zeng S, Lv X, Zhan C, Chen WR, Xiong W, Jacques SL, Luo Q (2006) Simultaneous compensation for spatial and temporal dispersion of acousto-optical deflectors for two-dimensional scanning with a single prism. Opt Lett 31:1091–1093

152. Friedman N, Kaplan A, Davidson N (2000) Acousto-optic scanning system with very fast nonlinear scans. Opt Lett 25:1762–1764

153. Reddy GD, Saggau P (2005) Fast three-dimensional laser scanning scheme using acousto-optic deflectors. J Biomed Opt 10(6):064038

154. Chang IC, Assoc A, Santa Clara CA (1994) Acousto-optic modulator with wide bandwidth and large angular aperture. Electron Lett 30:1190–1191

155. Vučinić D, Sejnowski TJ (2007) A compact multiphoton 3D imaging system for recording fast neuronal activity. PLoS ONE 2:8

156. Oron D, Silberberg Y (2005) Spatiotemporal coherent control using shaped, temporally focused pulses. Opt Express 13:9903–9908

157. Papagiakoumou E, de Sars V, Oron D, Emiliani V (2008) Patterned two-photon illumination by spatiotemporal shaping of ultrashort pulses. Opt Express 16:22039–22047

158. Oku H, Hashimoto K, Ishikawa M (2004) Variable-focus lens with 1-kHz bandwidth. Opt Express 12:2138–2149

159. Zhu L, Sun PC, Fainman Y (1999) Aberration-free dynamic focusing with a multichannel micromachined membrane deformable mirror. Appl Opt 38:5350–5354

160. Amir W, Carriles R, Hoover EE, Planchon TA, Durfee CG, Squier JA (2007) Simultaneous imaging of multiple focal planes using a two-photon scanning microscope. Opt Lett 32:1731–1733

161. Rózsa B, Katona G, Vizi ES, Várallyay Z, Sághy A, Valenta L, Maák P, Fekete J, Bányász A, Szipocs R (2007) Random access three-dimensional two-photon microscopy. Appl Opt 46:1860–1865

162. Keller M, Kao JPY, Egger M, Niggli E, (2007) Calcium waves driven by "sensitization" wavefronts. Cardiovasc Res 74:39–45

163. Kirleby PA, Naga Srinivas NKM, Silver RA (2010) A compact acousto-optic lens for 2D and 3D fentosecond based 2-photon microscopy. Opt Express 18:13721–13745

Chapter 3

Are Caged Compounds Still Useful?

Graham C.R. Ellis-Davies

Abstract

Since much of the life of cells is controlled by their chemistry, caged compounds can be used to intervene in this life in a myriad of specific ways. Organic chemists have synthesized the widest possible array of caged compounds for use by biologists. The smallest possible chemical unit (protons) to the "largest" (RNA and DNA) have been caged. Further, nonnatural products have been caged and used for blocking one aspect of cell function. Many caged compounds have been used for rapid activation of cell function, as uncaging often occurs in less than a millisecond. Studies with caged calcium and caged glutamate have proved particularly powerful in this regard. But will caged compounds continue into the second decade of the third millennium, their fourth decade? With the rise of other optical methods for control of cell function, are caged compounds still useful?

Key words: Caged molecules, Caged neurotransmitters, Caged second messengers

1. Introduction

What would Cajal think of GFP? A fluorescent Golgi stain in living animals! Like all important technical scientific advances, we now take it for granted. Discovered in 1962 but not deployed until the early 1990s, but since, then the jellyfish's gift to modern science has been mutated into a photoresponsive marker, but it is "just" a marker or sensor. The impact of genetically encoded fluorescent proteins on biology cannot be overstated; we really live in the "post-GFP era." What about the "other side of the (optical) engram," actuation? Recently, there has been a second revolution in genetically encoded optical probes that addresses the need for the photocontrol of cellular chemistry in living cells and animals. In 2002 and 2003 two novel light-activated ion channels were described, called channelrhodopsin (ChR)-1 and -2. It is ChR2, which conducts mostly sodium that is starting to have tremendous

James J. Chambers and Richard H. Kramer (eds.), *Photosensitive Molecules for Controlling Biological Function*, Neuromethods, vol. 55, DOI 10.1007/978-1-61779-031-7_3, © Springer Science+Business Media, LLC 2011

impact on neuroscience. So much so, that those of us who work on the development of other "old-fashioned" optical stimulation methods, feel under a lot of pressure from the channelrhodopsin approach: "Can't channelrhodopsin do it all?" I am often asked, "Why bother with making caged compounds these days?" Put another way: Are caged compounds still useful?

It is important to realize that optical cellular stimulation methods are not in strict competition. We do not require one industrial standard to control membrane potential. The historical competition between VHS recording formats or, more recently, HD DVD formats makes some sense, as it is certainly easier for consumers to have only one type of video recording (in a particular era). But do scientific methods have exactly the same consumer strictures? Certainly a modern computer is to be preferred to a PDP11, but patch clamp has not replaced field potential recording. Older techniques are not always completely replaced by modern ones, if the former still offer something the latter do not. Obviously, as someone who has been developing caged compounds since 1985, I am not wholly unbiased, but I hope to make the case for optical chemical probes as a still viable alternative to channelrhodopsin for the photo-stimulation of neurons and astrocytes. During the first half of 2009, the leading practitioner of the channelrhodopsin method (Karl Deisseroth) published five papers in Science and Nature (1–5), but we must not be totally seduced by such heady numbers. Since no technology is perfect, we shall see that optical methods often have complementary strengths and weaknesses, such is the case with caged compounds and channelrhodopsin.

It is important to mention here that there is a "third way." Several chapters in this book will discuss another method that uses chemical probes to control membrane function: *cis–trans* isomerisation of azobenzene chemical probes. This method, developed at Berkeley, by Isacoff, Kramer, Trauner, and co-workers is similar to caged compounds in that light controls the state of the binding of a small organic molecule to an ion channel (6, 7). However, this method is also significantly different from traditional caged compounds, in the latter relies upon unidirectional destruction of a covalent chemical bond to release the caged compound. Azobenzenes, however, are photoreversible: short wavelengths cause activation, long wavelengths inhibition. Importantly, little fatigue is seen through many duty cycles of azobenzene chemical probes, and they allow firing of action potentials at high frequencies (as much as 50 Hz) without suffering desensitization (7). Initially, they required tethering to a mutated ion channel, then covalent cross-linking, now potassium channels can be modulated by simple solution application of an azobenzene probe in solution (8). Azobenzenes have been used in cultured neurons, brain slices, and zebra fish, but no reports in living mice have appeared.

The azobenzene method appeared about 1 year before the ChR2 was used in neurons (9) but has yet to capture the imagination of the neuroscience community in the same way that the channelrhodopsin method has, probably because the probes are not yet commercially available. Thus, this "third way" has tremendous potential, but I will not discuss it in any more detail here, except to say it ought to be a significant part of the neuroscientist's optical toolbox (10).

2. A Short History of Uncaging

2.1. Why Are Caged Compounds Useful?

The first real biological experiment with a caged compound illustrates many of the strengths and weaknesses of the uncaging method. Much of the life of the cell is internal, or intracellular. If one wanted temporally and chemically precise activation of an enzyme inside a cell how could this be accomplished? One obvious answer is to stimulate a process on the outside that is coupled to something on the inside, but what happens if such coupling is not present? In 1976, patch clamp technology barely existed, so how could one study the sodium pump with high temporal resolution? The answer was to develop caged ATP (11). The sodium pump hydrolyzes ATP inside cells to create the ionic gradients essential for neuronal functional and cell health. Red blood cells can by opened under certain conditions, then re-sealed "right side out" or "inside out," and using this preparation much of the biochemistry of active transport was understood in the 1960s and 1970s. But since such manipulations of cells take a substantial amount of time, they do not allow one to define certain partial reactions with high temporal precision. Consequently, Kaplan et al. realized that blocking the gamma phosphate of ATP with a photoremovable protecting group would allow them to load "pro-ATP" (in pharmacology parlance) into a re-sealed red blood cell and thus photochemistry to initiate sodium efflux in a well-defined manner. This simple experiment proved revolutionary for many areas of biology and still exemplifies many of the strengths of the uncaging approach, namely (1) the concentration of substrates in normally inaccessible intracellular compartments can be controlled; (2) the time point of initiation of a biological process only requires light; (3) upstream metabolic cellular demands can be dissociated from the downstream consequences of those demands, thus uncaging allows one to parse complex signaling cascades; (4) global, synchronous activation of intracellular processes can be effected; and (5) the life of a cell can be controlled by selective activation of a one process, as if one flips a light switch, it can be "turned on" noninvasively. Shining light on cells is often spoken of as being "noninvasive," as low light levels are relatively benign to cell health and light passes through cells, allowing one to excite intracellular probes, often without physically touching cells at all. In 1971, Fork used light to fire action potentials without the addition of any exogenous chromophore (12). The exact mechanism was not well understood, and might involve temporary plasma membrane rupture. For this reason, this method has not been much used.

2.2. The Correct Order of Things

Caged ATP was next utilized by Goldman and Trentham in a series of classic experiments on the biophysics of skeletal muscle cross-bridge cycle. Rapid photolysis of ATP, ADP, or P_i with a frequency-doubled ruby laser (pulse-width 35–50 ns) enabled the detailed study of the partial reactions of muscle contraction (13–16). To set the scene for these time-resolved studies, the rate of uncaging of Kaplan's caged ATP was determined first (17). It turned out that NPE-ATP released ATP slightly too slowly for true kinetic resolution of some of the muscle contraction steps. Nevertheless, these studies set the standard for all subsequent studies using caged compounds as: (1) the rate of substrate release was determined before application to rapid kinetic studies and (2) the amount of uncaged nucleotide was quantified. Many faster caged ATP probes were made in the next 30 years, but they have only given access to handful of biological experiments not performed with NPE-ATP.

2.3. Inside and Out

Even though the experiments by Kaplan et al. in 1978 are probably the seminal ones for caged compounds, strictly speaking Engels and Schlaeger made caged cyclic-AMP a full year before caged ATP (18). However, I think it is fair to say they did not fully appreciate the importance of this compound, as it was one in a series of many esters they made for delivery into cells, the others being nonphotolabile, were hydrolyzed by intracellular esterases. Nevertheless, the *idea* of photochemically uncaging a second messenger was presented in their work. With the synthesis and use of caged cGMP in 1984 (19) and caged IP3 in 1987 (20), intracellular uncaging of a signaling molecule, as opposed to an energy source (ATP), was firmly established. cGMP is membrane permeable as the caging chromophore neutralizes the single negative charge, allowing the caged probe to be "loaded" into a cell by simple diffusion through the plasma membrane. Whereas NPE-IP3 bears five negative changes, so was loaded into permeabilized smooth muscle fibers. In both cases the experiments required maintaining extracellular presence of the caged compound. Nevertheless, the kinetics of release was orders of magnitude faster than traditional means of switching intracellular solutions in such biological preparations. The caged IP3 experiment was the first definitive proof that smooth muscle contraction was mediated by IP3, and thus it was a landmark biological study. It exemplifies the reductionistic power of caged compounds to focus in one element of a signaling pathway. This is often called being "necessary and sufficient," though strictly nothing is absolutely sufficient: at the very least IP3 needs its calcium channel to be useful.

With the development of caged cAMP and ATP, it was only a matter of time before a full range of biological important organic molecules was caged (21, 22). This has turned out to be more or less true. Caging cations was, however, another matter. Since ions like Ca^{2+} do not form covalent bonds, they are not amenable to caging like ATP or cAMP, therefore a new strategy had to be devised (23). Two similar approaches were developed independently in the late 1980s, both of which turned out to be quite useful. See Fig. 1 for a timeline.

2.4. Caged Calcium

The study of Ca^{2+} signaling was revolutionized in 1980 when Roger Tsien made quin2 (24). Not satisfied with mere passive observation of [Ca^{2+}], Tsien realized that a full understanding of Ca^{2+}-signaling cascades could only be achieved by development of caged calcium probes. He introduced the first Ca^{2+}-selective cage in 1986, called nitr-2 (25), shortly after his lab had made fura-2 and indo-1 (26). However, nitr-2 was deficient in some of its properties, therefore the Tsien lab made several other Ca^{2+} cages, the most widely used of which is called nitr-5 (27). All these probes use the caged ATP/cAMP photochemistry (21), which causes the calcium buffering capacity of BAPTA to be reduced by about 40-fold. Independent of the Tsien group, Jack Kaplan and I took a different approach to caging Ca^{2+} and synthesized photolabile derivatives of EGTA and EDTA. The idea was to cut the chelators in to two, so completely disrupting the cation coordination sphere (28). Since tetracarboxylic chelators were known to have high affinities (EDTA 32 nM, EGTA 150 nM at pH 7.2), and dicarboxylates have low affinities (ca. 1 mM), most of the bound calcium would be photoreleased. The first EGTA derivativeI made had a disappointing low calcium affinity (28), and thus was a useless probe. However, the EDTA derivative (DM-nitrophen, (29)) at pH 7.2 had a very high affinity for Ca^{2+} of 5 nM and irradiation reduced the affinity 600,000-fold (a 15,000-fold larger change than nitr-5), to 3 mM, and has turned out to be much more useful than we could have possibly imagined. (I have recently reviewed the design and development of caged calcium in detail (23).) So in 1988 there were two versions of caged calcium: one which was calcium-selective but photochemically very inefficient, and the other was extremely chemically and photochemically effective but relatively divalent cation nonselective. Since both have proved very effective for many biological experiments, it goes to show you do not need to be perfect to be useful.

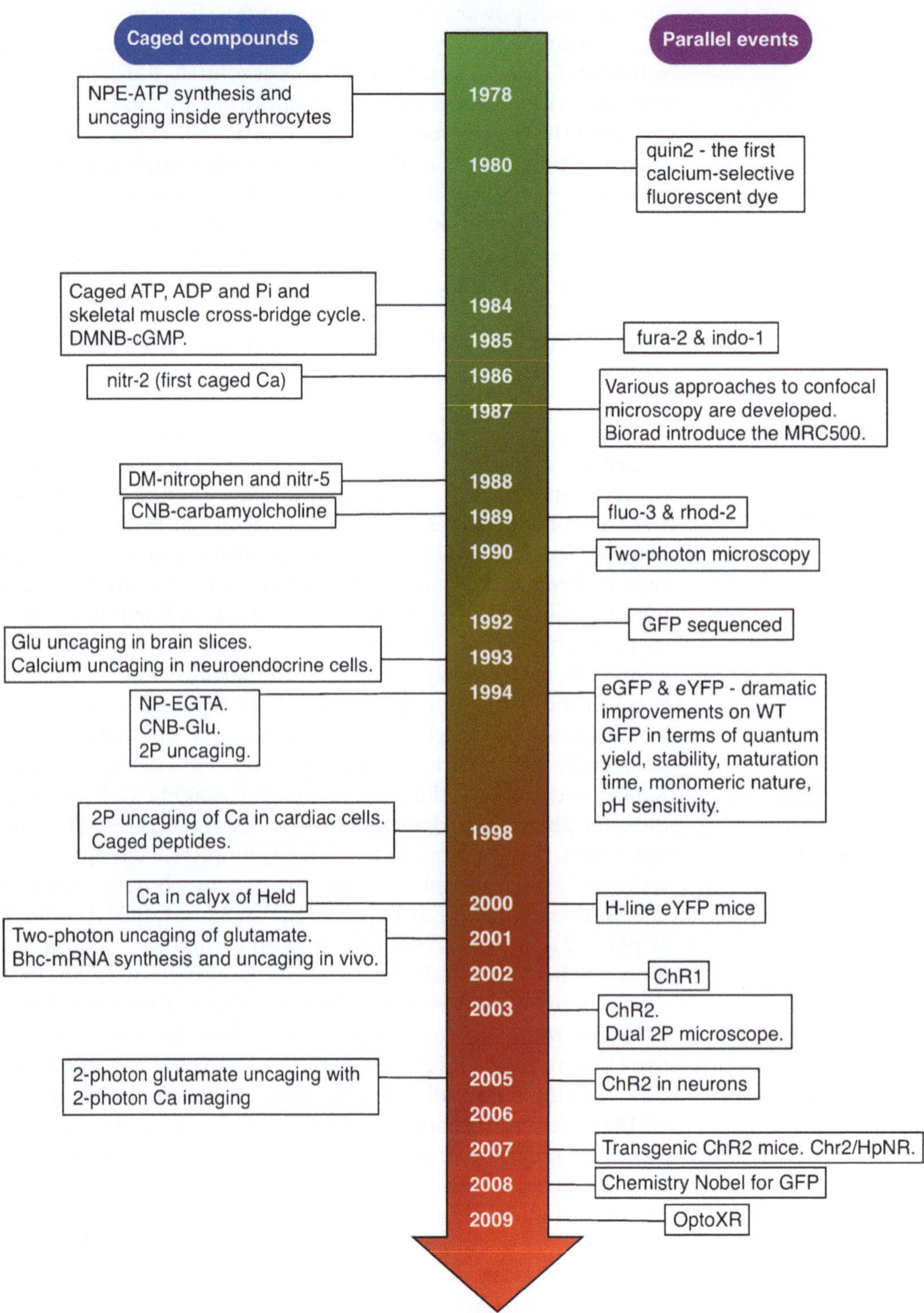

Fig. 1. Timeline for important developments in the field of caged compounds, with significant parallel developments of other optical techniques useful to biologists.

2.5. Never Let Your Own Lack of Imagination Be a Limit for Others

Since DM-nitrophen is only 500-fold more selective for calcium than magnesium, under "standard intracellular physiological conditions" (i.e. 100 nm calcium and 1 mM magnesium), DM-nitrophen is essentially caged magnesium (30). However, if you were desperate, you could leave magnesium and ATP out of your intracellular pipette solution, and viola, you get a really good caged calcium probe. This is what Thomas and Almers did in 1991: through a patch pipette, DM-nitrophen was dialyzed into melanotrophs with fura-2; whole-cell uncaging produced rapid, globally even step increases in calcium concentration (31). This allowed measurement of distinct stages of the rapid secretory (capacitance) events in these pituitary cells to be correlated with quantitative calcium measurements for the first time. In 1993, landmark studies of secretory events in chromaffin cells and melanotrophs appeared which exemplify many of the strengths of calcium probes for the study of calcium signaling in neuroendocrine cells (32, 33). Uncaging of calcium produces a temporally rapid "space clamp" of [Ca^{2+}] that can be quantified using calcium microfluorometry on the millisecond time scale. Subsequently, DM-nitrophen has been used in many studies of neuroendocrine cells and neurons (34, 35), one personal favorite, and shows caged calcium can be strikingly useful.

Since the Ca^{2+} hypothesis of neurotransmitter secretion was advanced by Sir Bernard Katz in 1965 (36), a "Holy Grail" experiment for synaptic physiologists was to define the quantitative relationship between the concentration of Ca^{2+} experienced locally by individual synaptic vesicles, and the postsynaptic response evoked by secretion of the neurotransmitter from such a vesicle. Well-studied synapses from the mammalian CNS such as pyramidal neurons are too small and delicate for such detailed study. One specialized synapse has proved sufficiently large and robust for detailed examination with double patch clamp techniques, namely the calyx of Held (37). The calyx of Held is an excitatory glutamatergic synapse arising from globular bushy cells in the anterior ventral cochlear nucleus onto a principal cell in the medial nucleus of the trapezoid body. This synapse has a diameter of >10 μm and is amenable to whole-cell patch clamp recording from pre- and postsynaptic cells in situ (i.e. in acutely isolate brain slices that preserve intact the complex architecture of neuronal cells in the mammalian CNS). Photolysis of DM-nitrophen rapidly released Ca^{2+} to varying concentrations (2–25 mM), depending on the irradiation power, throughout the presynaptic terminal. [Ca^{2+}]i was quantified in the same way as in neuroendocrine cells. This time the photoreleased calcium was correlated not with capacitance changes but with the evoked postsynaptic currents. These experiments showed that each synaptic vesicle experiences on average a very rapid pulse of Ca^{2+} in less than 1 ms after the action potential, with a size in the range of 10–25 μM. Similar to chromaffin cells, the secretory complex bound 4–5 Ca^{2+} ions (37).

In 1987, we almost did not make DM-nitrophen as we "knew" it would not be any good as a caged calcium probe, as its affinity for magnesium was too close to that of calcium to make truly selective and therefore useful. I always say to our incoming graduate students: never let your own lack of imagination limit others' creativity.

2.6. A Personal Anecdote – Only 16 Years

In 1992, I finally made caged calcium. I will never forget the day. I had many abortive attempts at the synthesis of a photolabile derivative of EGTA. However, in 1992 I had finally made what we now call NP-EGTA. With much trembling, I did the first calcium titration and it was obvious it had the long sought after high affinity for calcium (less than 100 nM at pH 7.2). Adding 1 mM magnesium to a 1 mM solution of NP-EGTA with 50% calcium had no effect on free calcium as shown by fluo-3, laser (a single pulse from a frequency-doubled ruby) irradiation saturated the indicator. When we finally published the report of this work in 1994 (38), Kaplan remarked that he had the idea for caged calcium in 1978, and it had only taken 16 years to bring it to fruition. The principle behind NP-EGTA was the same as its EDTA-based cousin, so it did not add much that was conceptually new to the field of caged compounds, except that one could now release large amounts of calcium in the presence of normal [Mg] and [MgATP]. In 1996, we finally published the rate of release of calcium by NP-EGTA and DM-nitrophen (39).

2.7. Speed Matters

Since it is physically impossible to accomplish rapid solution exchange inside a cell, the power of intracellularly uncaging a second messenger is rather obvious. But when it comes to neurotransmitters, one would think that rapid flow techniques could affect fast increases in glutamate concentration, say, in order to activate AMPA receptors with a similar time course to synaptic activation. After all, the target receptor is sitting there, on the surface, willing and waiting. For intact cells, this turns out *not* to be true, as the experiments with the first really good caged glutamate (called CNB-Glu, (40)) showed vividly. Since the rise time of AMPA receptors is a few hundred microseconds, rapid flow application of a solution of glutamate simply takes too long to equilibrate at the cell surface *before densitization takes over*, blunting the evoked response. Caged transmitters solve this problem. Photolysis of CNB-Glu produced 20-fold larger peak current when compared to rapid solution exchange (Fig. 2 of (40)). Furthermore, the rise time after laser uncaging was 1 ms compared to 14 ms for rapid flow. Flash photolysis experiments showed that CNB-Glu uncaged with a rate of about 48,000/s. CNB-GABA was introduced the same year and both caged compounds have been commercially available since then, allowing for many synaptic studies (review: (41)). Hess and collaborators went on to apply CNB caging chromophore to almost every other neurotransmitter (kainate, serotonin, dopamine, NMDA, etc.) but compared to CNB-Glu, these caged compounds received relatively little use.

Even though CNB-Glu satisfies most of the design criteria for a good caged glutamate, this has not prevented organic chemists making many more photolabile glutamates. I know of at least eight caged glutamate probes that are photochemically more efficient than CNB-Glu (as measured by the product of their quantum yield and extinction coefficient), but these probes have not enabled a single significant neurological study. Thus, the rationale for this work has not been clear. Admittedly, CNB-Glu is a little hydrolytically unstable in solution at pH 7.4, but even this problem has not really hindered its use by many laboratories for UV uncaging on cultured neurons and acute brain slices (41). CNB-Glu does have one important photochemical limitation: it is not sensitive to two-photon excitation. Since UV uncaging releases glutamate throughout the light path, this method does not readily allow highly localized uncaging in three dimensions. Nonlinear excitation, using the two-photon microscope, was designed to address this need. Having said this, it would be remiss of me not to mention the seminal glutamate uncaging paper appeared in 1993 (42). UV photolysis of a relatively slow caged glutamate in acute brain slices provided the first intimations of how powerful caged glutamate could be for neuroscience.

2.8. Two Photons Are Better Than One?

When Webb and co-workers introduced the first laser-scanning two-photon microscope in 1990 (43), there were no caged neurotransmitters that had been designed for effective two-photon excitation. In spite of this, Denk outlined many of the possibilities, in his 1994 paper entitled "Two-photon scanning photochemical microscopy: mapping ligand-gated ion channel distributions," for two-photon photolysis of neurotransmitters in the extracellular milieu (44). Principally, Denk showed that diffraction-limited two-photon excitation of a caged neurotransmitter (CNB-carbamoylcholine (45)) could produce exquisite axial localization of receptor activation, and that pseudo-random scanning is the preferred mode of functional receptor mapping when compared to sequential pixel uncaging. It took several more years for this technique to be "perfected." During that period, hardware and probes both improved dramatically, making two-photon microscopy the "turn-key" technique it is today. The importance of the development of solid-state "two-photon laser" cannot be underestimated. Denk used a colliding pulse, mode-locked dye laser at 640 nm, which required different dyes for tuning. Now broadband Ti:sapphire lasers are effortlessly tuned by computer over a wide range. The necessary probe development was done in academia.

Using an early version of the Ti:sapphire laser (output at 705 nm), Lipp and Niggli were able to use DM-nitrophen to create highly localized intracellular calcium transients that initiated global calcium waves inside cardiac myocytes or localized calcium release events (sparks), depending upon the magnitude of the incident laser power (46). This seminal work, provided me with the vital clue as to how to improve CNB-Glu and NP-EGTA so that these caged compounds would undergo effective two-photon excitation in the 705–725 nm range (this is the lowest tuning window for Ti:sapphire lasers). I synthesized "dimethoxy (DM) versions" of these caged compounds (i.e. DMCNB-Glu and DMNPE-4); the electron donating methoxy groups apparently make the nitroaromatic chromophore more likely to absorb two IR photons (47). Two-photon uncaging of calcium has not attracted much interest subsequently, but localized photorelease of glutamate has proved very useful.

The presence of the dimethoxy groups on DMCNB-Glu made it even more hydrolytically labile than its parent, but it effectiveness convinced us that a similar, more stable probe could be very useful indeed. Thus, swapping the rather electronically neutral 5-methoxycarbonylmethyl substituent of "NI"-Glu (48) with a methoxy group *para* to the nitro substituent created a stable caged glutamate (MNI-Glu) that proved to be reasonably sensitive to two-photon excitation (49). [Curiously, a second methoxy group at the 5-position hindered the photochemistry (50), but electron withdrawal at this position dramatically improved it, making two-photon uncaging of neurotransmitters in brain slices even more useful (51)].

So are two photons better than one? Certainly when it comes to axial confinement of excitation, diffraction-limited, two-photon uncaging of glutamate produces a much smaller volume of glutamate release than regular one-photon uncaging such that visually selected, single synapses in acute brain slices can be selectively stimulated with astonishing subcellular precision (52–71). Significantly, two-photon uncaging of glutamate allows one to tune the stimulation intensity such that quantal release can be mimicked (49) and can fire action potentials (72, 73). Using these techniques certain aspects of the biochemistry inside spine heads and the nature of dendritic summation have begun to be investigated in a rational way for the first time. Thus, two photons are better than one, in some instances.

2.9. Size Does not Matter

The history so far has been approximately linear and focused on small organic molecules that activate proteins. But the latter are, in fact, simply large organic molecules. In principle, these also could be caged, if detailed structure–activity relationships are known. Using site-directed, unnatural amino acid mutagenesis, Schultz and co-workers synthesized caged T4 lysozyme as proof-of-principle that an enzyme could be caged using these techniques (74). An alternative approach to such selective caging is "shot gun caging" where by several amino acid residues (e.g. lysines) are covalently modified with a highly reactive probe. G-actin was caged using this approach in 1994 (75). The most significant large molecules to be caged are RNA. Beginning with the RNA for GFP (76), Okamoto and colleagues have developed a method for caged RNA for use in zebra fish that allowed elegant experiments to be performed in vivo (77, 78). Several other proteins have been caged, showing there is no real limit to the size of a molecule that can be used with the technique (79, 80).

2.10. Boutique Caged Compounds

I have been deliberately selective, as there are many comprehensive reviews that can be consulted. In my brief historical survey of caged compounds, I have tried to give a basic overview of the very broad range of compounds that can be photomanipulated using uncaging technology: nucleotides, intracellular second messengers, cations, amino acids and neurotransmitters, proteins and peptides, enzymes, RNA, etc. Most of these probes have been used in multiple studies by many laboratories around the world; however, many caged compounds have not gone much further than the original proof-of-principle first publication. Perhaps, the reason for this is there are really no pressing biological questions waiting to be answered using such technology, or the synthesis of (more) the cage is so difficult and time-consuming, or it is rather unstable, that it is hard to justify making more compound. I will end my survey with a somewhat arbitrary selection of "boutique caged compounds." These probes illustrate further the strength of the technique:

1. *Caged agonists.* Miesenbock and co-workers developed a caged capsaicin for activation of "alien receptors" which they genetically encode into neurons. Capsaicin is the natural product that opens TRPV1 ion channels. This channel is not part of the CNS, so expression of TRPV1 in cultured hippocampal neurons enabled photolysis of caged capsaicin to evoke action potentials (81). Obviously, photocontrol of nonnatural agonists can only be accomplished with caged compounds.
2. *Caged gases.* Nitric oxide is one of the most important second messengers in mammals, as it regulates the tone of smooth muscle. In the CNS, it is a retrograde second messenger that plays an important part in LTD (82, 83).
3. *Caged peptides.* Protein–protein interactions can be blocked by a synthetic version of the binding domain of one partner. Just like other signaling cascades, normal means of adding such inhibitory peptides has little temporal or spatial control. In contrast, microinjection of a caged inhibitory peptide into white blood cells introduces a latent blocker of protein–protein interaction (84). The interaction between calmodulin

and myosin light chain kinase controls cell motility (crawling). Several other caged peptides have been synthesized in the subsequent 10 years (80).

4. *Caged protons. Escherichia coli* have chemoreceptors that respond to certain chemical gradients, including protons. Rapid photolytic release of protons allowed for the control of bacterial motility with great temporal precision (85).

3. An Even Shorter History of Optogenetics

Expression of ChR2 in cultured hippocampal neurons allowed repetitive firing of action potentials with short pulses of light, with low jitter and high pulse fidelity. Much to everyone's surprise, cultured mammalian neurons required no additional co-factors (9). Pairing of ChR2 with the well-known chloride pump, halorhodopsin (HpHR), allowed bidirectional dual wavelength control of neuronal membrane potential (86, 87). Transgenic mice are now commercially available with ChR2 expressed in subsets of neurons, just like a few GFP mice lines (in fact ChR2 is tagged with YFP). Since the visual system converts the absorption of light into increases of intracellular [cGMP], channelrhodopsins can be manipulated to control other G-proteins such a PLC and adenylate cyclase (3, 88). Blue light-sensing photoreceptors use flavins to control the latter and this membrane protein can also be expressed in vivo to control [cAMP] (89). Finally, vertebrate rhodopsin 4 allows light to be used to modulate inwardly rectifying potassium channels (90). There are several more 7-transmembrane receptors systems that could also be modulated by light, with molecular tinkering (e.g., phosphodiesterases, Rho, GIRK, PI3K, MAP kinases, GABA-B receptors, mGluR, etc.). Phosphopdiesterases linked to HpHR would offer the possibility of partnering with the recently developed opto-β_2AR (3), to yield bidirectional control of [cAMP] with yellow and blue light. I will be out on a limb: we are only at the beginning of the "channelrhodopsin revolution."

So what are the advantages and disadvantages of various optical methods for controlling cell function? First, let us consider the ChR family. Just like GFP, the biggest single strength of ChR is that it is encoded by DNA. Thus, this method has enabled the development of transgenic mice that stably express the ChR2 light-activated ion channel. In principle, using genetics, one can target selected subsets of neurons in any part of the CNS. Illumination of ChR2 allows high rates of action potential firing to be imposed at will upon neurons in vivo or ex vivo. But what are the disadvantages of the ChR approach? (1) It requires DNA (95, 96), (2) single-channel currents are low, (3) the channel can desensitize or reach fatigue, (4) ChR is not readily activated by two-photon illumination, so highly localized excitation is not possible (91), (5) protein processing can be faulty, with much protein being sequestered into inclusion bodies, and (6) native membrane receptors are not targeted by ChR.

Several recent reports have sought to address issues 2–5. For example, mutations have prolonged the channel open time over a wide range, by delaying the closing time, allowing much more current to flow for every photon absorbed. The single channel current per se was not improved much by these mutants. Note, this was done at the expense of slowing the rise time (92). Other mutations have improved other aspects of the kinetics of ChR2, allowing faster firing rates before desensitization (93). A greatly improved new chloride pump has been made (eNpHR), which is mainly targeted to the plasma membrane in vivo, unlike the parent NpHR (94). It is a little mysterious as why ChR2 is not very sensitive to 2-photon excitation. The criterion for excitation is a measured cellular current; perhaps the 2-photon *action cross-section* per unit area of plasma membrane is low for ChR2. The first and last points are important issues in comparison with caged compounds, so will be discussed in more detail.

First, ChR requires DNA. How can the advantage suddenly be called a disadvantage? If one wants to study a transgenic mouse model of a neuronal disease, having to apply DNA through a virus to the mice is a serious, time-consuming extra step. Making a double transgenic mouse is even more labor intensive. In contrast, caged probes can be directly applied to a mouse model, because they are exogenous probes. Thus, spine head physiology or neuronal circuits can be studied simply by making brain slices of *any* mouse (rat, ferret, cat) by caged compound photolysis. A second, practical issue is that since protein expression is inevitably variable, ChR does easily enable the production of stereotypical photo-stimulation in the way caged compounds do. Having said all this, there is no doubt that many laboratories will be adding channelrhodopsin probes to their transgenic mouse models of various diseases.

A second, and arguably fundamental, problem with ChR2 is that it does not target native membrane receptors. In contrast, glutamate uncaging does activate native AMPA, NMDA, and mGlu receptors. Recently, the Deisseroth lab has made new members of the channelrhodopsin family that are not light-activated ion channels, but light-activated enzymes (called optoXR) like those in the visual system (3). Two other labs have also taken similar approaches to using light-regulated receptors to control intracellular signaling (89, 90). None of these recent innovations target *directly* ionotropic channels in neurons. In contrast, caged probes use native receptors as their targets.

4. Caged Compounds Are Still Useful

Caged compounds still offer biologists several unique and/or powerful features for optical stimulation of cell function:

1. An impressively wide array of receptors can be stimulated. Much effort from organic chemists has lead to the development of caged compounds for almost every type of signaling molecule, not just those in the CNS.
2. Caged drugs allow for receptors subtype activation or blockade. Since caged compounds are not restricted to natural products, they offer the possibility of photomanipulation of any organic molecule.
3. Native receptors are directly targeted by photolysis of caged compounds. Probes such as caged calcium and glutamate are used to activate directly normal cellular machinery.
4. Uncaging can be used to release substrates either inside or outside cells, and can occur very rapidly so as to "switch on" a biological process. Since many of the photochemical process used for uncaging occur in the submicrosecond time domain, caged compounds initiate an essentially instantaneous change in substrate concentration.
5. Uncaging allows spatially selective substrate release. Since exogenous probes are used and can be applied to a large volume, substrate release can effect in large or small voxels, depending on the nature of the excitation paradigm. In particular, two-photon uncaging allows highly localized stimulation of receptors.
6. Uncaging may be graded in a very fine manner. Release of caged substrates can be quantified so that activation can be "titrated" in situ, allowing for subthreshold or supra-threshold activation that is very precise.

References

1. Tsai HC, Zhang F, Adamantidis A et al (2009) Phasic firing in dopaminergic neurons is sufficient for behavioral conditioning. Science 324(5930):1080–1084
2. Gradinaru V, Mogri M, Thompson KR, Henderson JM, Deisseroth K (2009) Optical deconstruction of parkinsonian neural circuitry. Science 324(5925):354–359
3. Airan RD, Thompson KR, Fenno LE, Bernstein H, Deisseroth K (2009) Temporally precise in vivo control of intracellular signalling. Nature 458(7241):1025–1029
4. Sohal VS, Zhang F, Yizhar O, Deisseroth K (2009) Parvalbumin neurons and gamma rhythms enhance cortical circuit performance. Nature 459(7247):698–702
5. Cardin JA, Carlen M, Meletis K et al (2009) Driving fast-spiking cells induces gamma rhythm and controls sensory responses. Nature 459(7247):663–667
6. Banghart M, Borges K, Isacoff E, Trauner D, Kramer RH (2004) Light-activated ion channels for remote control of neuronal firing. Nat Neurosci 7(12):1381–1386

7. Volgraf M, Gorostiza P, Numano R, Kramer RH, Isacoff EY, Trauner D (2006) Allosteric control of an ionotropic glutamate receptor with an optical switch. Nat Chem Biol 2(1):47–52
8. Fortin DL, Banghart MR, Dunn TW et al (2008) Photochemical control of endogenous ion channels and cellular excitability. Nat Meth 5(4):331–338
9. Boyden ES, Zhang F, Bamberg E, Nagel G, Deisseroth K (2005) Millisecond-timescale, genetically targeted optical control of neural activity. Nat Neurosci 8(9):1263–1268
10. Kramer RH, Chambers JJ, Trauner D (2005) Photochemical tools for remote control of ion channels in excitable cells. Nat Chem Biol 1(7):360–365
11. Kaplan JH, Forbush B III, Hoffman JF (1978) Rapid photolytic release of adenosine 5'-triphosphate from a protected analogue: utilization by the Na:K pump of human red blood cell ghosts. Biochemistry 17(10): 1929–1935
12. Fork RL (1971) Laser stimulation of nerve cells in Aplysia. Science 171(974):907–908
13. Goldman YE, Hibberd MG, McCray JA, Trentham DR (1982) Relaxation of muscle fibres by photolysis of caged ATP. Nature 300(5894):701–705
14. Goldman YE, Hibberd MG, Trentham DR (1984) Relaxation of rabbit psoas muscle fibres from rigor by photochemical generation of adenosine-5'-triphosphate. J Physiol 354:577–604
15. Goldman YE, Hibberd MG, Trentham DR (1984) Initiation of active contraction by photogeneration of adenosine-5'-triphosphate in rabbit psoas muscle fibres. J Physiol 354:605–624
16. Hibberd MG, Dantzig JA, Trentham DR, Goldman YE (1985) Phosphate release and force generation in skeletal muscle fibers. Science 228(4705):1317–1319
17. McCray JA, Herbette L, Kihara T, Trentham DR (1980) A new approach to time-resolved studies of ATP-requiring biological systems; laser flash photolysis of caged ATP. Proc Natl Acad Sci USA 77(12):7237–7241
18. Engels J, Schlaeger EJ (1977) Synthesis, structure, and reactivity of adenosine cyclic 3', 5'-phosphate benzyl triesters. J Med Chem 20(7):907–911
19. Nerbonne JM, Richard S, Nargeot J, Lester HA (1984) New photoactivatable cyclic nucleotides produce intracellular jumps in cyclic AMP and cyclic GMP concentrations. Nature 310(5972):74–76
20. Walker JW, Somlyo AV, Goldman YE, Somlyo AP, Trentham DR (1987) Kinetics of smooth and skeletal muscle activation by laser pulse photolysis of caged inositol 1, 4, 5-trisphosphate. Nature 327(6119):249–252
21. Ellis-Davies GC (2007) Caged compounds: photorelease technology for control of cellular chemistry and physiology. Nat Meth 4(8):619–628
22. Mayer G, Heckel A (2006) Biologically active molecules with a "light switch". Angew Chem Int Ed Engl 45(30):4900–4921
23. Ellis-Davies GC (2008) Neurobiology with caged calcium. Chem Rev 108(5): 1603–1613
24. Tsien RY (1980) New calcium indicators and buffers with high selectivity against magnesium and protons: design, synthesis, and properties of prototype structures. Biochemistry 19(11):2396–2404
25. Tsien RY, Zucker RS (1986) Control of cytoplasmic calcium with photolabile tetracarboxylate 2-nitrobenzhydrol chelators. Biophys J 50(5):843–853
26. Grynkiewicz G, Poenie M, Tsien RY (1985) A new generation of Ca2+ indicators with greatly improved fluorescence properties. J Biol Chem 260(6):3440–3450
27. Adams SR, Kao JPY, Grynkiewicz G, Minta A, Tsien RY (1988) Biologically useful chelators that release ca-2+ upon illumination. J Am Chem Soc 110(10):3212–3220
28. Ellisdavies GCR, Kaplan JH (1988) A new class of photolabile chelators for the rapid release of divalent-cations – generation of caged-ca and caged-mg. J Org Chem 53(9): 1966–1969
29. Kaplan JH, Ellis-Davies GC (1988) Photolabile chelators for the rapid photorelease of divalent cations. Proc Natl Acad Sci USA 85(17):6571–6575
30. Ellis-Davies GC (2006) DM-nitrophen AM is caged magnesium. Cell Calcium 39(6): 471–473
31. Thomas P, Almers W (1991) Simultaneous measurements of increased Ca and secretion in single rat melanotrophs evoked by photolytic release of caged Ca. Biophys J 59:597a
32. Thomas P, Wong JG, Almers W (1993) Millisecond studies of secretion in single rat pituitary cells stimulated by flash photolysis of caged Ca2+. EMBO J 12(1):303–306
33. Neher E, Zucker RS (1993) Multiple calcium-dependent processes related to secretion in bovine chromaffin cells. Neuron 10(1): 21–30

34. Becherer U, Rettig J (2006) Vesicle pools, docking, priming, and release. Cell Tissue Res 326(2):393–407
35. Sorensen JB (2004) Formation, stabilisation and fusion of the readily releasable pool of secretory vesicles. Pflugers Arch 448(4): 347–362
36. Katz B, Miledi R (1965) The effect of calcium on acetylcholine release from motor nerve terminals. Proc R Soc Lond B Biol Sci 161:496–503
37. Schneggenburger R, Neher E (2000) Intracellular calcium dependence of transmitter release rates at a fast central synapse. Nature 406(6798):889–893
38. Ellis-Davies GC, Kaplan JH (1994) Nitrophenyl-EGTA, a photolabile chelator that selectively binds Ca2+ with high affinity and releases it rapidly upon photolysis. Proc Natl Acad Sci USA 91(1):187–191
39. Ellis-Davies GC, Kaplan JH, Barsotti RJ (1996) Laser photolysis of caged calcium: rates of calcium release by nitrophenyl-EGTA and DM-nitrophen. Biophys J 70(2): 1006–1016
40. Wieboldt R, Gee KR, Niu L, Ramesh D, Carpenter BK, Hess GP (1994) Photolabile precursors of glutamate: synthesis, photochemical properties, and activation of glutamate receptors on a microsecond time scale. Proc Natl Acad Sci USA 91(19): 8752–8756
41. Eder M, Zieglgansberger W, Dodt HU (2004) Shining light on neurons – elucidation of neuronal functions by photostimulation. Rev Neurosci 15(3):167–183
42. Callaway EM, Katz LC (1993) Photostimulation using caged glutamate reveals functional circuitry in living brain slices. Proc Natl Acad Sci USA 90(16):7661–7665
43. Denk W, Strickler JH, Webb WW (1990) Two-photon laser scanning fluorescence microscopy. Science 248(4951):73–76
44. Denk W (1994) Two-photon scanning photochemical microscopy: mapping ligand-gated ion channel distributions. Proc Natl Acad Sci USA 91(14):6629–6633
45. Milburn T, Matsubara N, Billington AP et al (1989) Synthesis, photochemistry, and biological-activity of a caged photolabile acetylcholine-receptor ligand. Biochemistry 28(1):49–55
46. Lipp P, Niggli E (1998) Fundamental calcium release events revealed by two-photon excitation photolysis of caged calcium in Guinea-pig cardiac myocytes. J Physiol 508(Pt 3): 801–809
47. Ellis-Davies GCR (1999) Localized photolysis of caged compounds. J Gen Physiol 114:1a
48. Papageorgiou G, Ogden DC, Barth A, Corrie JET (1999) Photorelease of carboxylic acids from 1-acyl-7-nitroindolines in aqueous solution: rapid and efficient photorelease of L-glutamate. J Am Chem Soc 121(27): 6503–6504
49. Matsuzaki M, Ellis-Davies GC, Nemoto T, Miyashita Y, Iino M, Kasai H (2001) Dendritic spine geometry is critical for AMPA receptor expression in hippocampal CA1 pyramidal neurons. Nat Neurosci 4(11):1086–1092
50. Papageorgiou G, Ogden D, Kelly G, Corrie JET (2005) Synthetic and photochemical studies of substituted 1-acyl-7-nitroindolines. Photochem Photobiol Sci 4(11):887–896
51. Ellis-Davies GC, Matsuzaki M, Paukert M, Kasai H, Bergles DE (2007) 4-Carboxymethoxy-5, 7-dinitroindolinyl-Glu: an improved caged glutamate for expeditious ultraviolet and two-photon photolysis in brain slices. J Neurosci 27(25):6601–6604
52. Carter AG, Sabatini BL (2004) State-dependent calcium signaling in dendritic spines of striatal medium spiny neurons. Neuron 44(3):483–493
53. Smith MA, Ellis-Davies GC, Magee JC (2003) Mechanism of the distance-dependent scaling of Schaffer collateral synapses in rat CA1 pyramidal neurons. J Physiol 548(Pt 1):245–258
54. Araya R, Jiang J, Eisenthal KB, Yuste R (2006) The spine neck filters membrane potentials. Proc Natl Acad Sci USA 103(47):17961–17966
55. Bloodgood BL, Sabatini BL (2005) Neuronal activity regulates diffusion across the neck of dendritic spines. Science 310(5749): 866–869
56. Bloodgood BL, Sabatini BL (2007) Nonlinear regulation of unitary synaptic signals by CaV(2.3) voltage-sensitive calcium channels located in dendritic spines. Neuron 53(2):249–260
57. Noguchi J, Matsuzaki M, Ellis-Davies GC, Kasai H (2005) Spine-neck geometry determines NMDA receptor-dependent Ca2+ signaling in dendrites. Neuron 46(4):609–622
58. Ngo-Anh TJ, Bloodgood BL, Lin M, Sabatini BL, Maylie J, Adelman JP (2005) SK channels and NMDA receptors form a Ca2+ -mediated feedback loop in dendritic spines. Nat Neurosci 8(5):642–649
59. Tanaka J, Horiike Y, Matsuzaki M, Miyazaki T, Ellis-Davies GC, Kasai H (2008) Protein synthesis and neurotrophin-dependent structural plasticity of single dendritic spines. Science 319(5870):1683–1687

60. Araya R, Nikolenko V, Eisenthal KB, Yuste R (2007) Sodium channels amplify spine potentials. Proc Natl Acad Sci USA 104(30):12347–12352
61. Losonczy A, Magee JC (2006) Integrative properties of radial oblique dendrites in hippocampal CA1 pyramidal neurons. Neuron 50(2):291–307
62. Losonczy A, Makara JK, Magee JC (2008) Compartmentalized dendritic plasticity and input feature storage in neurons. Nature 452(7186):436–441
63. Remy S, Csicsvari J, Beck H (2009) Activity-dependent control of neuronal output by local and global dendritic spike attenuation. Neuron 61(6):906–916
64. Lee SJ, Escobedo-Lozoya Y, Szatmari EM, Yasuda R (2009) Activation of CaMKII in single dendritic spines during long-term potentiation. Nature 458(7236):299–304
65. Sobczyk A, Scheuss V, Svoboda K (2005) NMDA receptor subunit-dependent [Ca2+] signaling in individual hippocampal dendritic spines. J Neurosci 25(26):6037–6046
66. Harvey CD, Yasuda R, Zhong H, Svoboda K (2008) The spread of Ras activity triggered by activation of a single dendritic spine. Science 321(5885):136–140
67. Harvey CD, Svoboda K (2007) Locally dynamic synaptic learning rules in pyramidal neuron dendrites. Nature 450(7173): 1195–1200
68. Sobczyk A, Svoboda K (2007) Activity-dependent plasticity of the NMDA-receptor fractional Ca2+ current. Neuron 53(1): 17–24
69. Gasparini S, Magee JC (2006) State-dependent dendritic computation in hippocampal CA1 pyramidal neurons. J Neurosci 26(7):2088–2100
70. Matsuzaki M, Honkura N, Ellis-Davies GC, Kasai H (2004) Structural basis of long-term potentiation in single dendritic spines. Nature 429(6993):761 766
71. Honkura N, Matsuzaki M, Noguchi J, Ellis-Davies GC, Kasai H (2008) The subspine organization of actin fibers regulates the structure and plasticity of dendritic spines. Neuron 57(5):719–729
72. Matsuzaki M, Ellis-Davies GC, Kasai H (2008) Three-dimensional mapping of unitary synaptic connections by two-photon macro photolysis of caged glutamate. J Neurophysiol 99(3):1535–1544
73. Nikolenko V, Poskanzer KE, Yuste R (2007) Two-photon photostimulation and imaging of neural circuits. Nat Meth 4(11):943–950
74. Mendel D, Ellman JA, Schultz PG (1991) Construction of a light-activated protein by unnatural amino-acid mutagenesis. J Am Chem Soc 113(7):2758–2760
75. Marriott G (1994) Caged protein conjugates and light-directed generation of protein-activity – preparation, photoactivation, and spectroscopic characterization of caged g-actin conjugates. Biochemistry 33(31):9092–9097
76. Ando H, Furuta T, Tsien RY, Okamoto H (2001) Photo-mediated gene activation using caged RNA/DNA in zebrafish embryos. Nat Genet 28(4):317–325
77. Okamoto H, Hirate Y, Ando H (2004) Systematic identification of factors in zebrafish regulating the early midbrain and cerebellar development by odered differential display and caged mRNA technology. Front Biosci 9:93–99
78. Ando H, Kobayashi M, Tsubokawa T, Uyemura K, Furuta T, Okamoto H (2005) Lhx2 mediates the activity of Six3 in zebrafish forebrain growth. Dev Biol 287(2):456–468
79. Shigeri Y, Tatsu Y, Yumoto N (2001) Synthesis and application of caged peptides and proteins. Pharmacol Ther 91(2):85–92
80. Lee HM, Larson DR, Lawrence DS (2009) Illuminating the chemistry of life: design, synthesis, and applications of "caged" and related photoresponsive compounds. ACS Chem Biol 4(6):409–427
81. Zemelman BV, Nesnas N, Lee GA, Miesenbock G (2003) Photochemical gating of heterologous ion channels: remote control over genetically designated populations of neurons. Proc Natl Acad Sci USA 100(3):1352–1357
82. Lev-Ram V, Jiang T, Wood J, Lawrence DS, Tsien RY (1997) Synergies and coincidence requirements between NO, cGMP, and Ca2+ in the induction of cerebellar long-term depression. Neuron 18(6):1025–1038
83. Makings LR, Tsien RY (1994) Caged nitric oxide. Stable organic molecules from which nitric oxide can be photoreleased. J Biol Chem 269(9):6282–6285
84. Walker JW, Gilbert SH, Drummond RM et al (1998) Signaling pathways underlying eosinophil cell motility revealed by using caged peptides. Proc Natl Acad Sci USA 95(4): 1568–1573
85. Khan S, Castellano F, Spudich JL et al (1993) Excitatory signaling in bacterial probed by caged chemoeffectors. Biophys J 65(6):2368–2382
86. Han X, Boyden ES (2007) Multiple-color optical activation, silencing, and desynchronization of neural activity, with single-spike temporal resolution. PLoS ONE 2(3):e299

87. Zhang F, Wang LP, Brauner M et al (2007) Multimodal fast optical interrogation of neural circuitry. Nature 446(7136):633–639
88. Kim JM, Hwa J, Garriga P, Reeves PJ, RajBhandary UL, Khorana HG (2005) Light-driven activation of beta 2-adrenergic receptor signaling by a chimeric rhodopsin containing the beta 2-adrenergic receptor cytoplasmic loops. Biochemistry 44(7):2284–2292
89. Schroder-Lang S, Schwarzel M, Seifert R et al (2007) Fast manipulation of cellular cAMP level by light in vivo. Nat Meth 4(1):39–42
90. Li X, Gutierrez DV, Hanson MG et al (2005) Fast noninvasive activation and inhibition of neural and network activity by vertebrate rhodopsin and green algae channelrhodopsin. Proc Natl Acad Sci USA 102(49):17816–17821
91. Wang H, Peca J, Matsuzaki M et al (2007) High-speed mapping of synaptic connectivity using photostimulation in Channelrhodopsin-2 transgenic mice. Proc Natl Acad Sci USA 104(19):8143–8148
92. Berndt A, Yizhar O, Gunaydin LA, Hegemann P, Deisseroth K (2009) Bi-stable neural state switches. Nat Neurosci 12(2):229–234
93. Lin JY, Lin MZ, Steinbach P, Tsien RY (2009) Characterization of engineered channelrhodopsin variants with improved properties and kinetics. Biophys J 96(5):1803–1814
94. Gradinaru V, Thompson KR, Deisseroth K (2008) eNpHR: a Natronomonas halorhodopsin enhanced for optogenetic applications. Brain Cell Biol 36(1–4):129–139
95. Bertalan K. Andrasfalvy, Boris V. Zemelman, Jianyong Tang, and Alipasha Vaziri (2010) Two-photon single-cell optogenetic control of neuronal activity by sculpted light 107(26):11981–11986
96. Eirini Papagiakoumou, Francesca Anselmi, Aurélien Bègue, Vincent de Sars, Jesper Glückstad, Ehud Y Isacoff & Valentina Emiliani (2010) Nature methods 7(10):848–854

Chapter 4

Chromophores for the Delivery of Bioactive Molecules with Two-Photon Excitation

Timothy M. Dore and Hunter C. Wilson

Abstract

The localized release of bioactive molecules from "caged compounds" through two-photon excitation (2PE) is an emerging technology for the study of biological processes in cell and tissue culture and whole animals. Several advantages are realized when 2PE drives the activation of the biological effector: (1) excitation is tightly localized to femtoliter-sized volumes; (2) there is less photodamage to biological tissues; and (3) deeper penetration into the sample is achieved. A barrier to widespread use and an expansion of applications for the pinpoint three-dimensional delivery of biological effectors are the small number of available caging groups and phototriggers with sufficient sensitivity to 2PE, appropriate photolysis kinetics, and necessary physiological compatibility. Chromophores based on nitrobenzyl, nitroindoline, coumarin, *ortho*-hydroxycinnamic acid, quinoline, and other structural motifs have been designed to regulate the action of biologically active compounds with 2PE. Design principles from structure–property relationships elucidated for two-photon absorbing materials can be applied to the design of caging groups and phototriggers for high efficiency 2PE-mediated release of bioeffectors. The conjugation size, symmetry, and the strength of donor and acceptor groups impact the overall sensitivity to 2PE, but these factors must be balanced with the need for biocompatibility and the ability to drive photochemical reactions with rapid kinetics.

Key words: 2-photon excitation, Caged compounds, Photoactivation, Photoremovable protecting groups, Phototriggers, Photochemistry

1. Introduction

Over the past few decades, "caged compounds" have emerged as important tools for the study of biological systems (for recent reviews, see (1–5)). A caged compound is a substance in which a photoremovable protecting group (PPG), or "caging group," is attached to a biologically active molecule to inactivate it. Upon exposure to light, the PPG releases, or "uncages," the substrate in its active form. The utility of these PPGs is derived from their

James J. Chambers and Richard H. Kramer (eds.), *Photosensitive Molecules for Controlling Biological Function*, Neuromethods, vol. 55, DOI 10.1007/978-1-61779-031-7_4, © Springer Science+Business Media, LLC 2011

ability to effect large jumps "instantaneously" in the concentration of the bioactive molecules in cell or tissue culture or whole animals. It is important to note that the term cage does not refer to a physical encapsulation of the bioactive effector molecule; rather, a covalent bond between the effector and the cage is sufficient to inactivate the effector. Upon irradiation, photons provide the energy to cleave the covalent bond, and the effector molecule regains its activity, enabling the system to be studied under the conditions of a sudden influx of the bioactive molecules.

When two-photon excitation (2PE) drives the photolysis reaction, the excitation volume is restricted to the focus of a pulsed laser beam passed through a microscope objective. Three main advantages over conventional single-photon excitation (1PE) processes are realized: (1) excitation is tightly localized to a volume on the order of a femtoliter (1 fL is about the volume of an *E. coli* bacterium and much smaller than a mammalian cell), so the uncaging event can be limited to a specific cell or a cellular region within a complex tissue preparation; (2) there is less photodamage to biological tissues; and (3) deeper penetration into the sample is achieved. Since its invention in 1990, two-photon microscopy has held the promise of performing three-dimensionally resolved photoactivation of bioactive molecules (6, 7). The initial objective was to mimic the release of a neurotransmitter from the synaptic vesicle at a single synapse within intact brain tissue. Some success toward this specific goal has been achieved (for reviews, see (8–11)), but it is clear that a barrier to widespread use and an expansion of applications for the pinpoint three-dimensional delivery of biological effectors are the small number of available caging groups with sufficient sensitivity to 2PE, appropriate photolysis kinetics, and necessary physiological compatibility.

2. Two-Photon Excitation

2.1. Theory

Maria Göppert-Mayer conceived the idea that an atom could become excited through the simultaneous absorption of two photons (12, 13), but it was not until 30 years later in 1961 that the phenomenon was observed (14). Denk, Strickler, and Webb exploited the three-dimensional localization of excitation and reported in 1990 the observation of fluorogenic indicators within biological samples using a laser-scanning two-photon microscope that they designed and built (6). Later, Denk released carbamoylcholine (an acetylcholine receptor activator) from a caged version using 2PE to map the distribution of the acetylcholine-gated ion channels in cell culture (7), which ushered in research efforts to further exploit 2PE to mediate the photoactivation of caged compounds in a three-dimensionally resolved manner.

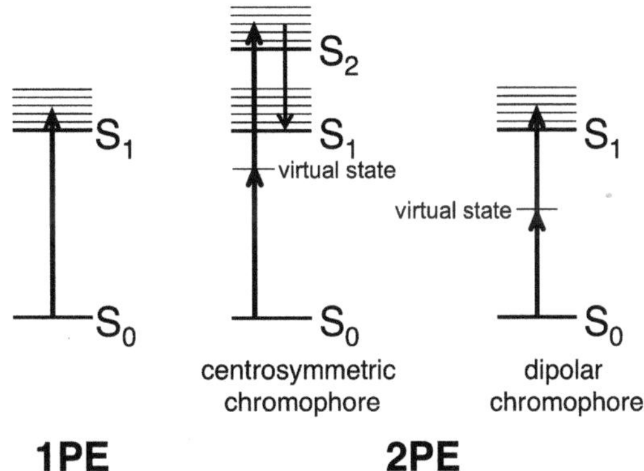

Fig. 1. 1PE versus 2PE of a chromophore. Jablonski diagrams illustrate the allowed electronic transitions for 1PE and for 2PE processes in centrosymmetric and non-centrosymmetric dipolar chromophores.

To effect 2PE, a single molecule absorbs two nonresonant photons of the same wavelength nearly simultaneously, generating an electronically excited state that can undergo the same photophysical and photochemical processes as the excited state generated by 1PE, but uses light that is twice the wavelength required to achieve 1PE (Fig. 1). The 2PE process exploits a short-lived (<5 fs) virtual excited state that is a mixture of the ground and excited states, and only exists while the molecule interacts with the first photon. Absorption of a second photon while the chromophore is in the virtual state enables the transition to the electronically excited state from which photochemistry can occur. For centrosymmetric chromophores, an electronic transition that is allowed for the absorption of two photons is forbidden for a 1PE process; therefore, 2PE leads to a different excited state (S_2) than would be achieved with 1PE (S_1). It also means that the optimal two-photon absorption wavelength is not exactly twice the 1PE absorbance maximum. Because the chromophores used to make caged compounds are relatively large and have many overlapping vibrational levels, enabling internal conversion to the lowest excited state (S_1), the subsequent photochemistry will proceed from the same excited state attained by 1PE (Kasha's rule). For non-centrosymmetric dipolar chromophores, there is no reversal in the selection rules for 1PE and 2PE. The transition to the excited state is allowed by 1PE and 2PE (although depending on the chromophore's symmetry, some excited states might not be allowed for 1PE, 2PE, or both), and the maximum 2PE absorption is typically twice the maximum 1PE absorption.

In contrast to 1PE, 2PE depends on the simultaneous interaction of two photons, so the probability of absorption increases with the square of the light intensity. It is a nonlinear process, whereas 1PE depends linearly on the light intensity. To generate

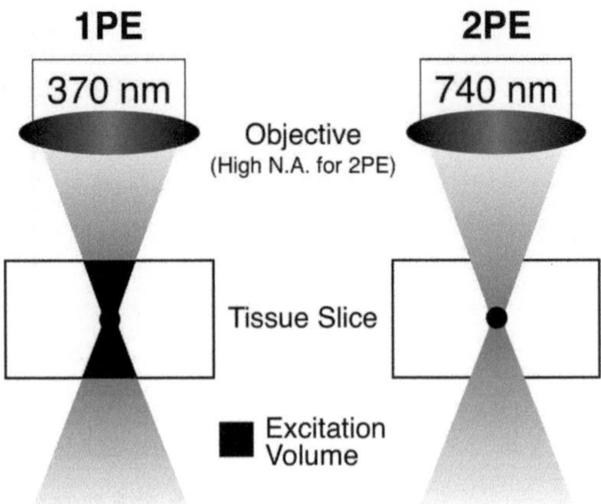

Fig. 2. Spatial selectivity of excitation with 2PE. 1PE lacks three-dimensional selectivity of excitation; all photons have sufficient energy to excite the chromophore. 2PE is restricted to the focal volume because the probability of excitation is proportional to z^{-4}, where z is the distance from the focus of the beam.

the large density of photons necessary to effect 2PE requires a focused and pulsed laser. The most commonly used light source in modern two-photon microscopes is a mode-locked, fs-pulsed, Ti:sapphire laser with tuning ranges of 700–1,000 nm, <140-fs pulses, 90-MHz repetition rates, and average power of up to 1 W. The nonlinear power dependence of excitation for 2PE means that excitation only occurs at the focus of the beam (Fig. 2), called the focal volume, and if a high numerical aperture (NA) objective is used, this volume can be as small as a femtoliter. In contrast to 1PE, no excitation occurs outside of the focal volume with 2PE, since the intensity of a focused laser beam decreases as the square of the distance from the focus (z) and the probability of 2PE scales quadratically with light intensity. The number of molecules excited by 2PE is proportional to z^{-4}; therefore, excitation probabilities outside of the focus are negligible.

Aside from the spatial restriction of excitation imposed by 2PE, there are two other important advantages that are relevant to releasing bioactive molecules from their caged conjugates: reduced photodamage and increased depth penetration. First, less photodamage occurs with 2PE than with 1PE, because photobleaching cannot occur outside of the focal volume, since the wavelengths used to initiate 2PE fall in the near-IR (700–1,100 nm, or the "phototherapeutic window") region of the electromagnetic spectrum. These wavelengths are not generally absorbed by native biomolecules in cells and tissues. Unless the sample contains hemoglobin, melanin, or chlorophyll, only water is the principle absorber in the IR, and its absorption coefficient in the wavelengths relevant to 2PE in a biological system is very low.

Out-of-focus excitation is negligible because the probability of two photons scattering to the same location at the same time and being absorbed by a chromophore is practically zero. Damage within the focal volume can be a problem if the average laser power used for uncaging is larger than 10 mW, which has been shown to be the threshold for collateral destruction of biological structures (15). As a second advantage, 2PE enables a greater depth of penetration in biological samples because longer IR wavelengths of light are not as readily scattered or absorbed by native chromophores as the shorter UV wavelengths used in conventional 1PE uncaging experiments.

2.2. Measuring Two-Photon Excitation Susceptibility

The measurement of the susceptibility of a PPG to a photoreaction induced by 2PE is known as its two-photon uncaging action cross-section, δ_u, which is quantified by a unit called the Göppert-Mayer (GM, 10^{-50} (cm^4 s)/photon), named in honor of Maria Göppert-Mayer, who first hypothesized the phenomenon of 2PE. It is the product of the two-photon absorbance cross-section, δ_a, and the quantum yield of the photolysis reaction, Q_u. Since δ_a represents the probability that two photons will be absorbed by the PPG and Q_u indicates the efficiency of the reaction (that is, what portion of the photons absorbed by the chromophore will lead to release of the substrate), δ_u is a measure of the probability that a PPG will absorb two photons and undergo productive photocleavage from its effector. Because photochemical reactions proceed from their lowest triplet or singlet state regardless of the excitation method, the quantum efficiency for a 1PE reaction is assumed to be the same as that for a 2PE reaction. Roger Tsien proposed that a PPG relevant to the study of biological systems should have a δ_u value exceeding 0.1 GM (16), though others have calculated that values must be on the order of 3.84–30.8 GM, depending on the rates of diffusion through the biological medium (15).

There are several methods of measuring δ_u that are practical and do not require detailed knowledge of the laser beam parameters. For a qualitative estimate of whether one caging group is more sensitive to 2PE than another, one can compare the reaction progress of one caged compound irradiated with a Ti:sapphire laser with that of another one with a known δ_u in the same setup (17). Reaction progress can be measured by HPLC or with the readout of a fluorescent indicator of the reaction product. Estimates of the relative sensitivity of the caged compound to 2PE are obtained.

A method developed by Tsien (16) and based on the work of Webb (18, 19) enables measurement of δ_u of a PPG using a simple apparatus and fluorescein as an external standard with a known fluorescence quantum yield (20) and δ_a (18, 19, 21). This method has the advantage of being general; it can be used for any caged compound, even if the products are nonfluorescent.

A mode-locked, fs-pulsed Ti-sapphire laser is focused onto the center of a sample or fluorescein reference solution in a 1 × 1 × 10 mm cuvette with a side window. A radiometer is used to measure the fluorescence output of the fluorescein standard, then 20-μL aliquots of the sample are successively irradiated for different lengths of time, and the progress of the reaction assessed by HPLC. The value of δ_u is determined by comparison of the known photochemical parameters of the chromophore fluorescein with the decay of the cage-bound effector using the following equation:

$$\delta_u = \frac{N_p \phi Q_{fF} \delta_{aF} C_F}{< F(t) > C_S}$$

where N_p represents the number of molecules that undergo photoreaction (as determined by HPLC and calculated from the initial rate of reaction), ϕ is the detector collection efficiency for measuring the fluorescence of fluorescein at a right angle to the excitation laser beam, Q_{fF} is the fluorescence quantum yield of fluorescein (20), δ_{aF} is the fluorescein two-photon absorbance cross-section (18, 19, 21), $<F(t)>$ represents the time-averaged fluorescent photon flux from fluorescein, and C_F and C_S are the respective concentrations of the fluorescein standard and the caged substrate under evaluation. This method gives reproducible measurements of δ_u that enable reliable comparison of the 2PE sensitivities of different caging groups to one another. Absolute δ_u values are probably accurate within a factor of 2.

For caging groups that release a fluorophore or a substrate that can be detected by a fluorescent indicator, δ_u can be measured by quantifying the fluorescence generated after the 2PE uncaging event has been triggered (22). A brief pulse (8.5 μs) from a high-intensity Ti:sapphire laser is used to initiate 2PE uncaging, then the laser intensity is immediately reduced by an electro-optical modulator (EOM) to elicit fluorescence from the reaction product or chelator pair, which has an amplitude that is proportional to δ_u. The decay of this signal as the fluorescent species diffuses out of the excitation volume is measured and can be related to δ_u if a kinetic model for the release reaction is known. For example, δ_u for Ca^{2+} caged by azid-1 can be measured using fluo-3 as the fluorescent indicator and fitting a plot of the fluorescence ratio ($\Delta F(\tau)$) versus time (τ) with the following equation:

$$\Delta F(\tau) = \left(\frac{2}{\pi}\right)^{3/2} \frac{\delta_u \gamma^2 F_0^2 R^{1/2} \Delta t}{\delta_f E w_r^3 (1+\tau)(1+R\tau)^{1/2}} \frac{[\mathrm{cCa}^{2+}]_0}{[\mathrm{fCa}^{2+}]_0^2}$$

where γ is the ratio of the average powers of the uncaging and monitoring lasers, F_0 is the average fluorescence of the equilibrium concentration of Ca^{2+}-bound fluo-3, R is the square of the ratio of the two beam waists, Δt is the duration of the uncaging pulse train of the laser, δ_f is the fluorescence action cross-section of Ca^{2+}-bound fluo-3, E is the efficiency of the detection system, w_r is the radial beam waist, and $[cCa^{2+}]_0$ and $[fCa_{2+}]_0$ are the concentrations of caged and fluo-2-bound Ca^{2+}, respectively. If a dye is used, it is important to take into account the kinetics of chelation as they compare to the rate at which species diffuse out of the irradiation volume. Errors in measurements arise from estimating the illumination volume and determining the efficiency of the detector. Both are meaningless when measuring relative values of δ_u on the same instrument, but absolute values are correct within a factor of 2.

Fluorescence correlation spectroscopy (FCS) after 2PE can be used to calculate δ_u (23). FCS is a single-molecule detection technique that measures fluctuations in fluorescence intensity in a small (fL) sample volume. For the purposes of extracting δ_u from FCS data, the fluctuations are a result of the initial photolysis reaction that occurs after excitation with 2PE. In the experiment, fluctuations in the fluorescence emission in the focal volume of a Ti:sapphire laser of a solution of a caged fluorescent dye are measured just after 2PE. Autocorrelation curves ($G(\tau)$ versus τ) are recorded at different laser powers and each are fitted to the equation for the autocorrelation coefficient $G(\tau)$:

$$G(\tau) = \frac{N_{tot} + \alpha^2 N_2}{(N_{tot} + \alpha N_2)^2} \frac{1}{1 + \tau / \tau_D}$$

where N_{tot} is the total number of molecules in the illuminated spot, N_2 is the number of product molecules, α is the brightness of the product relative to the caged fluorophore, and τ_D is the diffusion time of the caged fluorophore and product through the beam waist. From the fit, values of $G(0)$ and τ_D are calculated, then $1/G(0)$ is plotted as a function of the laser power (P) and fit to the following equation:

$$\frac{1}{G(0)} = N_{tot} \frac{(1 + \alpha\beta P^2)^2}{1 + \alpha^2 \beta P^2}$$

which enables the calculation of the value of β, from which δ_u is derived. Measurement of δ_u of DMNB (0.02 GM at 740 nm) using the FCS after 2PE method is in good agreement with the values obtained from other methods (0.03 and 0.01 GM (16, 22)). Because a fluorescent signal is required, the method is not general to all caged compounds. $G(\tau)$ contains kinetic assumptions that are tailored to the behavior of different compounds; therefore, employment of FCS in the measurement of δ_u requires a firm understanding of the kinetic processes that govern the uncaging of the PPG studied.

3. Design Considerations for PPGs with Sensitivity to 2PE

The design of a PPG used to release a bioactive molecule with 2PE depends a great deal on the application, but there are some general considerations. To be effective as a tool for biological study, the ideal PPG for 2PE must meet many of the criteria for being a good conventional caging group for 1PE. Covalently linking the PPG to the bioactive molecule must render the effector biologically inactive, but upon excitation with light at wavelengths that are not detrimental to the biological preparation, the PPG should release the bioactive molecule efficiently and in high yield. The uncaging must be on a timescale shorter than the process under investigation, which for fast signal transduction events, such as neurotransmission, means less than 1 ms. For example, the AMPA-type glutamate channel GluR1Q_{flip} opens and closes with rate constants of 29,000 s^{-1} (τ=34 μs) and 2,100 s^{-1} (τ=480 μs), respectively (24), so the rate of glutamate release must be quite rapid to study the physiology of these receptors effectively. Photochemical side reactions that do not release the active effector should be minimized. The caged compound and any remnant photoproducts of the PPG must not affect the system under study or interfere with the measurement of the biological activity during the experiment. It must be stable in the dark and adequately soluble in physiological buffers of high ionic strength. If the caged compound is to be used intracellularly, then it will ideally cross the cell membrane passively and not require invasive procedures such as microinjection. The caged compound should be synthetically accessible and able to be purified to a high standard.

In addition to the criteria for being a good caging group for 1PE, the ideal PPG that mediates the release of a biological effector through 2PE should have a large δ_u at wavelengths in the physiological optical window that are easily accessed by the Ti:sapphire lasers that typically drive modern two-photon microscopes (690–1,100 nm). A large cross-section is required to enable the use of low average laser power (<10 mW), which keeps tissue damage within the focal volume to a minimum. For exploiting the tight localization of 2PE, release must occur within the focal volume of the laser beam. To ensure this, the timescale for discharging the bioactive molecule must be faster than the diffusional escape time from the focal volume, which is estimated at ~300 μs from a volume produced by a 1.25 NA lens illuminated at 700 nm (22), and is generally in the range of τ=113–900 μs (25). A slower release time will limit the spatial resolution, because the bioactive substrate will have diffused away from the focal volume before it has become activated (Fig. 3). Additionally, each

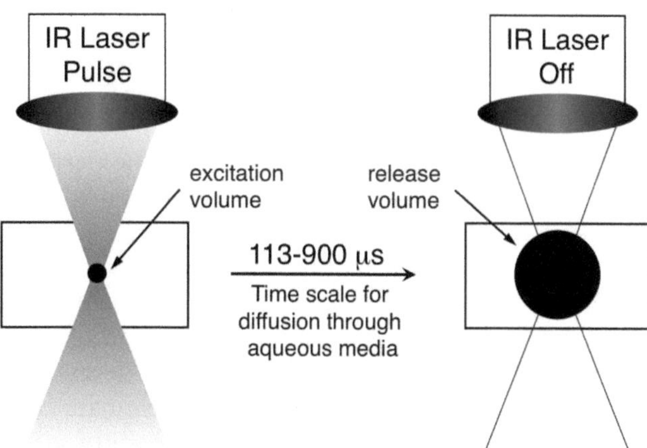

Fig. 3. Slow release time increases release volume relative to the excitation volume.

pulse from the laser must initiate the release of enough bioactive substrate to have the desired physiological effect. This means that the excitation and release must be efficient and rapid enough to overcome diffusion, which depletes the focal volume of a bioactive substrate as it is produced by the photolysis reaction. Tsien has suggested a minimum value of $\delta_u = 0.1$ GM (16), but Ogden has calculated that if one assumes a low diffusion coefficient ($D = 100\ \mu m^2/s$) typical of an extracellular environment, δ_u needs to be at least 3.84 GM to deplete the concentration of a caged compound in the focal volume by half when using a laser with an average power of 5 mW (15). Larger diffusion coefficients require greater values of δ_u. The limitations imposed by diffusion out of the focal volume and the chromophore's sensitivity to 2PE have tremendous design implications for the applications requiring effector release in volumes of 1 fL, such as simulating the release of neurotransmitters into the synaptic cleft from vesicles in the presynaptic neuron.

During the last decade, a great deal has been learned about optimizing the δ_a of new materials, and structure–property relationships have been elucidated for a number of classes of materials (for recent reviews, see (26–31)). Some of these principles can be applied to the design of PPGs with large values of δ_u for biological use. Recall that $\delta_u = \delta_a \times Q_u$, so increasing δ_a will also increase δ_u. Typically, molecules with large values of δ_a have extended conjugation, polarizability, and high extinction coefficients for π–π^* transitions (32, 33). Increasing the size of the conjugation dramatically increases δ_a. Conjugation can be increased by using larger chromophores or by introducing conformational rigidity in the molecule, which forces coplanarity of the π-system and increases conjugation lengths. Nevertheless, extensive conjugation has limitations for caging groups.

Large delocalized π-systems tend to have low energy excited states; therefore, they might not be able to collect enough energy to drive the photochemistry of substrate release. Also, increasing chromophore size can negatively impact the biocompatibility, solubility, and membrane permeability of the caged compound. Symmetry plays a role; centrosymmetric molecules typically have larger δ_a than dipolar ones. For driving photochemical reactions, this might have drawbacks because the large δ_a is derived from resonance with an intermediate state (S_1, Fig. 1) that is lower in energy than the final state accessed by 2PE (S_2). Through internal conversion, S_2 will decay rapidly to S_1, which is where the photochemistry will originate (Kasha's rule). When a centrosymmetric dye absorbs two photons, a large portion of the energy of one photon that might be needed to drive the subsequent photochemistry is lost as heat. In contrast, the energy of both photons absorbed by a dipolar chromophore can be used to drive the photochemistry. Donor and acceptor groups at the ends and in the center of the molecule enhance δ_a; in particular, terminal electron-donating groups, such as amines, are better than electron-withdrawing groups, which improve δ_a when they are placed near the center. The structural factors influencing δ_a are interrelated, and optimizing δ_a of the caging group must be balanced with the influence the modifications have on the photochemical reaction that releases the bioactive molecule.

4. Caging Groups with Sensitivity to Two-Photon Excitation

Most of the well-studied caging groups used with great success in conventional 1PE applications have inadequate sensitivity to 2PE, which has spawned the introduction of novel caging groups that have been built with high 2PE susceptibility as a primary objective. The structures of a selected number of these chromophores are illustrated in Fig. 4.

4.1. Chromophores Based on Nitrobenzyl Protecting Groups

Some of the most widely employed PPGs are based upon the well-studied 2-nitrobenzyl (NB) group (34, 35). The caging and release of the neurotransmitter and energetically important biomolecule ATP by Kaplan (36) using NB helped in great part to usher in photochemistry as a tool for the dynamic study of biological systems. Numerous variations of NB, such as DMNB, NPE, and CNB (Fig. 4), have been created to modify the core characteristics of the chromophore, including wavelength of maximum absorption, molar absorptivity, and solubility (Table 1). NB derivatives have the advantage of protecting a range of biologically relevant functional groups, including carboxylates,

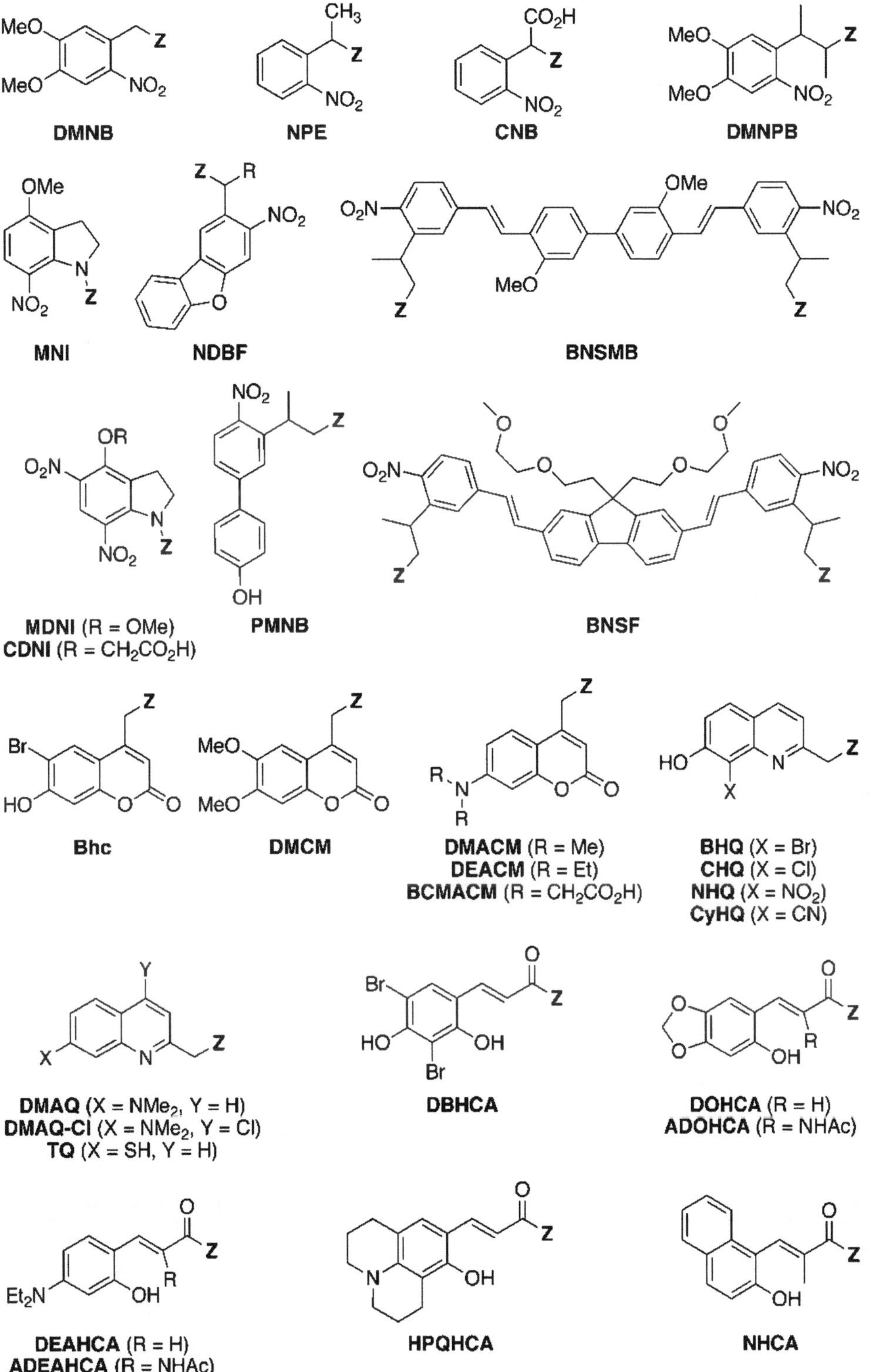

Fig. 4. Structures of selected chromophores used as caging groups.

Table 1
Photophysical and photochemical properties of selected caging groups

Protecting group	Released group[a]	λ_{max} (nm)	ε (M^{-1} cm^{-1})	Q_u	$\varepsilon \times Q_u$	δ_u (GM (nm))	τ_{rel} (s)	References
DMNB	OAc	346	6,100	0.005	31	0.03 (740)	5×10^{-3}	(16, 35)
NPE	OAc	265	4,200	0.52	2,184	0.019 (640)	3.2×10^{-1}	(15, 35, 36, 140)
CNB	γ-Glu	262	5,100	0.14	714	–	3.0×10^{-5}	(141–143)
DMNPB	γ-Glu	350	4,500	0.26	1,170	0.17 (740)	7.6×10^{-4}	(51)
NDBF	Ca^{2+}	330	18,400	0.7	12,880	0.6 (720)	5.3×10^{-4}	(53)
PMNB	γ-Glu	317	9,900	0.1	990	0.45	4.9×10^{-4}	(54)
BNSMB	γ-Glu	400	39,340	0.30	11,800	0.9 (800)	–	(55)
BNSF	γ-Glu	415	63,960	0.25	16,000	5.0 (800)	–	(55)
MNI	γ-Glu	347	4,330	0.085	368	0.06 (720)	2×10^{-7}[b]	(17, 59, 62)
MDNI	γ-Glu	350	8,600	[c]	–	0.06 (730)	–	(60, 67)
CDNI	γ-Glu	–	6,200[d]	[e]	–	– (720)	–	(68)
Bhc	OAc	370	15,000	0.037	555	0.72 (740)	$<1 \times 10^{-6}$	(8, 16)
DMCM	$OP(O)(OEt)_2$	346	11,400	0.08	912	– (728)	$\sim 2 \times 10^{-9}$	(94)
DMACM	ATP	385	15,300	0.086	1,316	– (750)[f]	6.3×10^{-10}	(94, 95)
DEACM	γ-Glu	385	10,000	0.11	1,100	–	7.1×10^{-10}[g]	(96, 99)
BCMACM	γ-Glu	379	13,100	0.10	1,310	0.4 (740)[h]	–	(101–103)
DBHCA	OEt	369	25,000	0.05	1,250	1.6 (750)	4.5×10^{2}	(108, 109)
DOHCA	OEt	360	16,000	0.03	480	3.8 (750)	3.1×10^{2}	(109)
ADOHCA	OBu	356	17,000	0.07	1,190	2.0 (750)	7.7×10^{0}	(110)

DEAHCA	OEt	378	28,000	0.02	560	2.0 (750)	2.7×10^2	(109)
ADEAHCA	OBu	380	28,000	0.05	1,400	0.3 (750)	3.3×10^1	(110)
HPQHCA	OEt	394	20,000	0.03	600	4.7 (750)	1.1×10^2	(109)
NHCA	OEt	339	6,000	0.25	1,500	2.5 (750)	5×10^1	(109)
BHQ	OAc	369	2,600	0.29	754	0.59 (740)	$<1 \times 10^{-6}$	(111–113)
CHQ	OAc	370	2,800	0.10	280	0.12 (740)	–	(117)
CyHQ	OAc	364	7,700	0.31	2,387	0.32 (740)	–	(117)
NHQ	OAc	350	6,500	0.00	0	0.00 (740)	–	(117)
TQ	OAc	369	5,200	0.063	328	0.42 (740)	–	(117)
DMAQ	OAc	368	4,600	0.046	211	0.13 (740)	–	(117)
DMAQ-Cl	OAc	386	3,300	0.090	234	0.47 (740)	–	(117)

(–) not reported

[a]γ-Glu is the γ-carboxylate of glutamate; ATP is the γ-phosphate of adenosine triphosphate

[b]Measured for methyl 2-(1-acetyl-7-nitroindolin-5-yl)acetate (59), but no kinetic studies on MNI have been reported

[c]This value is controversial. Ellis-Davies reports $Q_u = 0.47$ (67) and Corrie reports $Q_u = 0.14$ (60). Corrie also points out other discrepancies between his progressive photolysis experiments and those Ellis-Davies reports

[d]Measured at 325 nm

[e]The controversy over the value of Q_u for MDNI indicates that the value of $Q_u = \sim 0.5$ Ellis-Davies reports for CDNI should be independently checked

[f]Sensitivity to 2PE indicated for $OP(O)(OEt)_2$

[g]Measured for 8-bromo-adenosine-3′,5′-cyclic monophosphate (8-Br-cAMP)

[h]δ_u measured for phenyl acetate

phosphates, alcohols, and amines. Although DMNB, NPE, and CNB are by far the most widely used PPGs, these NB derivatives, and many others like them (23), are insensitive to 2PE. For example, DMNB has a δ_u value of only 0.03 GM at 740 nm (16, 22, 23, 37), well below the lower limits proposed for physiological use. NB chromophores exhibit slow release kinetics; rate constants are in the 1–1,000 s^{-1} range. Additionally, NB photoproducts possess a nitroso group, which is potentially toxic to biological preparations, though little evidence of widespread problems arising from by-product interference has been reported.

The photochemistry of the NB PPG proceeds through singlet and triplet excited states (38, 39), but the subsequent dark reaction pathways (34, 40), which are strongly influenced by solvent, pH, and buffer concentrations, are most important for understanding the release of the bioactive molecule (Fig. 5). Transfer of the benzylic proton of **1** to an oxygen of the nitro group generates the *Z*-nitronic acid **2**, which is in equilibrium with the aci-nitro anion **3** and the *E*-nitronic acid **4** (41, 42). Cyclization of **3** or **4** generates 1,3-dihydrobenz[*c*]isoxazol-1-ol derivative **5** (43), which decomposes to the arylnitroso 7 and the released effector species with hemiacetal **6** serving as a likely intermediate under physiological conditions (43, 44). Early studies suggest that aci-nitro anion decomposition to the nitroso by-product is the rate-determining step leading to effector release (45, 46), but more recent experiments show that hemiacetal **6** decays much more slowly to 7 than the aci-nitro anion **3**, making it the rate-limiting intermediate under physiological conditions (43). The decay of **5** can also be slower than that of the aci-nitro anion (43). This illustrates an important point: Using the decay of the easily observable aci-nitro anion as an indicator of the release rate can be misleading. Actual release rates for NB-based

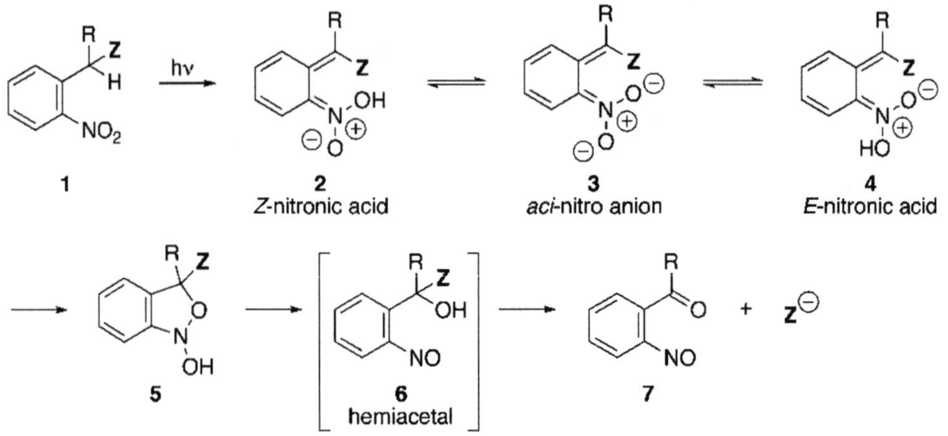

Fig. 5. The mechanism of the photolysis of NB-based caging groups.

PPGs can be much slower in some cases, so it is important to investigate the photochemistry of any new PPG fully.

There are examples of traditional NB-derived caging groups exhibiting larger than normal uncaging action cross-sections. These observations are due to antenna effects, where a chromophore more sensitive to 2PE assists with the absorption of light and transfer of the excitation energy to the NB-group after excitation. NPE-HCCC1/Me and NPE-HCC1 (Fig. 6) are caged fluorophores that release a fluorescent coumarin after photolysis at 740 nm with $\delta_u = 0.37$ and 0.68 GM, respectively (47, 48). Compared to NPE-caged acetate, whose δ_u is negligible at 740 nm, this represents a tremendous increase in the sensitivity. This system has been adapted to a FRET-based fluorophore that emits a green color (the coumarin is blue) (49), and it has been applied to the study of cell–cell gap junctions (50). In another example, a DMNB-caged vanilloid, DMNB-VNA (Fig. 6), could release *N*-vanillyl-nonanoylamide (VNA) and evoke a vanilloid receptor-mediated Ca^{2+} response upon 2PE (720 nm) at relatively low laser power. No data quantifying δ_u have been reported for the

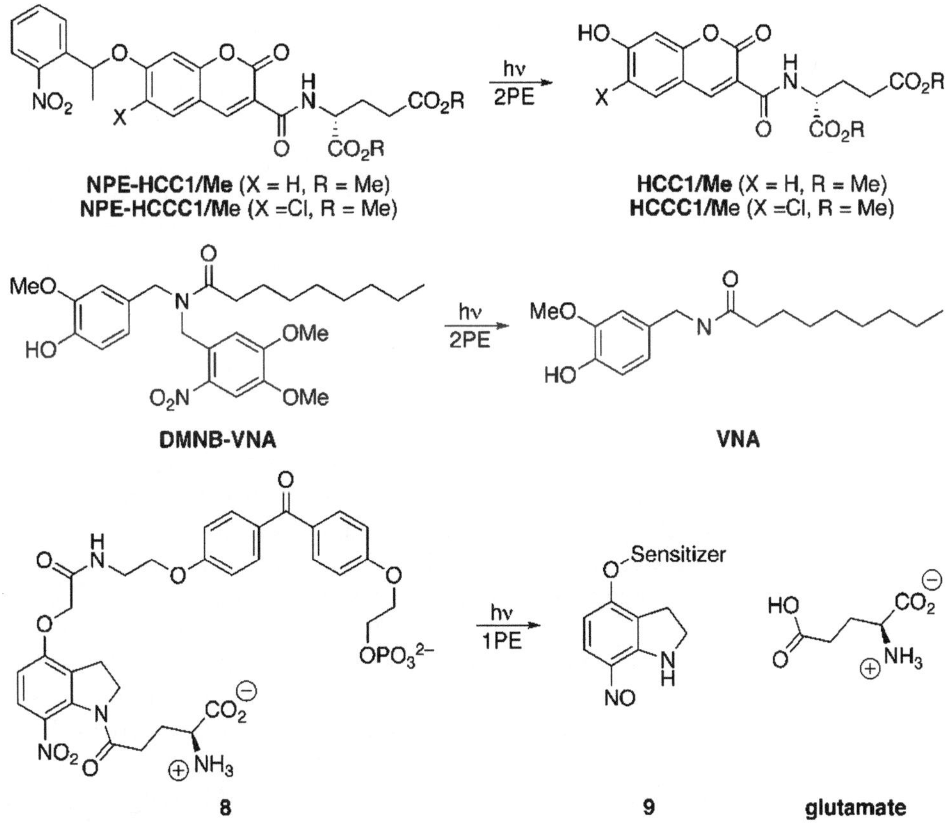

NPE-HCC1/Me (X = H, R = Me)
NPE-HCCC1/Me (X =Cl, R = Me)

HCC1/Me (X = H, R = Me)
HCCC1/Me (X =Cl, R = Me)

DMNB-VNA

VNA

8 **9** **glutamate**

Fig. 6. The antenna effect in enhancing susceptibility to 2PE.

DMNB-VNA compound. These examples illustrate an important point about caged compounds; the bioactive molecule can greatly influence the photophysical and photochemical properties of the PPG that cages it.

Several NB-based PPGs that have improved sensitivity to 2PE have been introduced in the last few years. The 3-(4,5-dimethoxy-2-nitrophenyl)-2-butyl (DMNPB) PPG has a δ_u value of 0.17 GM at 720 nm and has been used to release glutamate efficiently (51). Despite the modest improvement in δ_u over other NB-based PPGs, the decay of the aci-nitro intermediate occurs with bi-exponential kinetics at the rate of 1,400 and 75 s^{-1}, which is too slow to achieve tight localization of glutamate release. In a variation on the caged fluorophores developed by Li (47, 48), DMNPB was used to create a photoactivatable fluorescent label for His-tagged proteins (52), though, surprisingly, the coumarin did not drastically increase δ_u of DMNPB (0.21 GM) as it did for the NPE protecting group.

The principles derived from the optimization of chromophores for TPA (26–31) suggest that lengthening the conjugated system and increasing the electron donating or withdrawing strength of side groups on the chromophore improves the TPA properties. Ellis-Davies and colleagues have explored the chromophore nitrodibenzofuran (NDBF, Fig. 4), which has shown favorable photolysis through both 1PE ($Q_u = 0.7$) and 2PE ($\delta_u = 0.6$ GM) to release calcium ions (53). The extended conjugation likely contributes to the improved δ_u. NDBFs sensitivity to 1PE ($\varepsilon \times Q_u$) is among the highest known, owing to a large molar absorptivity (Table 1). Two-photon photolysis of NDBF–EGTA, a PPG for calcium ions, induced localized increases in Ca^{2+} concentration, which induced the release of more Ca^{2+} from the sarcoplasmic reticulum of intact cardiac myocytes. Flash photolysis experiments measured the Ca^{2+} release rate to be 20,000 s^{-1}, which is rapid enough to study many Ca^{2+}-activated processes. Bolze and coworkers reported that the 3-(2-propyl)-4-methoxy-4-nitrobiphenyl (PMNB) group, which has extended conjugation, releases glutamate with $\delta_u = 0.45$ GM at 800 nm (54), although aqueous solubility might prove problematic at higher concentrations. Like other NB-derived PPGs, the kinetics might be too slow for releasing fast-acting signaling molecules. The compounds 4,4′-bis-{8-[4-nitro-3-(2-propyl)-styryl]}-3,3′-di-methoxybiphenyl (BNSMB) and 2,7-bis-{4-nitro-8-[3-(2-propyl)-styryl]}-9,9-bis-[1-(3,6-dioxaheptyl)]-fluorene (BNSF) extend the conjugated system and in the case of BNSMB introduce strong electron-donating side groups to a centrosymmetric system (55). The δ_u values of BNSMB and BNSF are 0.9 and 5.0 GM, respectively, based upon the photorelease of glutamate at 800 nm. These PPGs have the ability to cage two effectors each and have good water solubility. The δ_u value of 5.0 GM for BNSF is one of the largest measured for a PPG. The kinetics of release of these chromophores have not been reported yet, but mechanistically, these are likely to be slow NB-based caging groups.

4.2. Chromophores Based on Nitroindoline Protecting Groups

1-Acyl-7-nitroindoline (NI) derivatives release carboxylates rapidly and efficiently in aqueous solution (56) and have been used to cage glutamate, GABA, and glycine (57, 58). NI-protected carboxylates are hydrolytically stable and release carboxylates on the sub-microsecond timescale. The photolysis proceeds from the triplet excited state of the nitroindoline and involves transfer of the acyl group to one of the oxygen atoms of the nitro group to generate the key intermediate **13** (Fig. 7), whose decomposition is the rate-limiting step of the pathway and occurs with an observed rate constant of $5 \times 10^6\ s^{-1}$ in water (34, 59). The mechanistic pathway followed depends on the solvent and the substituents on the aromatic ring of the indoline. In low water content solutions, the nitronic anhydride **13** undergoes nucleophilic substitution to generate nitroindoline **14** and the carboxylic acid, whereas in aqueous media, **13** decomposes through deprotonation adjacent to the indoline nitrogen atom and elimination of the carboxylate to form nitroso-3*H*-indole **15**, which tautomerizes to 7-nitrosoindole **16**. An exception to this is the case of 1-acyl-5,7-dinitroindoline derivatives, which yield both **14** and **16** in aqueous media (60). A time-resolved IR study on 1-acetyl-5,7-dinitroindoline in acetonitrile revealed that intermediate **13**

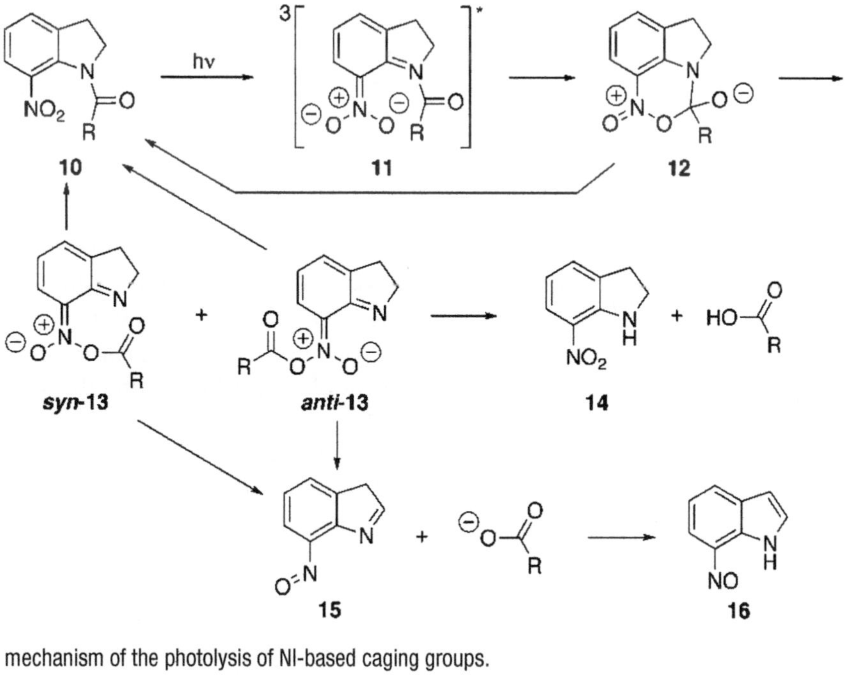

Fig. 7. The mechanism of the photolysis of NI-based caging groups.

exists as two non-interconverting stereoisomers (61). The *syn*-**13** isomer decays rapidly to **10**, while the *anti*-**13** isomer has a longer lifetime and decomposes to **15** in the presence of butylamine.

Corrie introduced the 4-methoxy-7-nitroindoline (MNI, Fig. 4) as a protecting group for glutamate (62), showing that it improves the photolysis efficiency about twofold over the 2-(7-nitroindolin-5-yl)acetate (56). Kasai used MNI-protected γ-glutamate to map functional glutamate receptors using 2PE even though it has an uncaging action cross-section that is less than 0.1 GM at 720 nm (Table 1) (17). The kinetics of carboxylate release have not been reported for MNI, but related nitroindolines release carboxylates on the sub-microsecond timescale, which is shorter than the diffusional escape time from the focus of the laser, so release is confined to the focal volume. MNI-γ-Glu has been used in several biological applications (for examples, see (63–65)), though millimolar concentrations must be used to evoke a glutamatergic response upon 2PE. There is an MNI-caged D-aspartate derivative (66), but its performance with 2PE has not yet been reported.

The PPG 4-methoxy-5,7-dinitroindolinyl (MDNI) was introduced independently by Ellis-Davies (67) and Corrie (60). There is significant disagreement between the two papers with respect to the photochemical properties of MDNI. Both groups report an improvement in quantum efficiency over MNI, but of differing magnitude: Ellis-Davies reports $Q_u = 0.47$, whereas Corrie reports $Q_u = 0.14$. Corrie also points out other discrepancies between his progressive photolysis experiments and the observations Ellis-Davies reports. Nevertheless, the sensitivity of MDNI-γ-Glu to 2PE is similar to that measured for MNI-γ-Glu (67), so adding the 5-nitro substituent to the indoline does not improve two-photon absorptivity. The related caging group 4-carboxymethoxy-5,7-dinitro-indolinyl (CDNI) cleanly releases glutamate upon photolysis, but no δ_u value is reported (68). The AMPA-receptor currents evoked from 2PE-mediated (720 nm) photorelease of glutamate on hippocampal neurons by CDNI are four- to fivefold stronger than those evoked by MNI. A comparative NMR study of the photolysis of CDNI-γ-Glu and MNI-γ-Glu in D_2O suggests that $Q_u = \sim 0.5$ for CDNI-γ-Glu, but the controversy surrounding the photochemical properties of MDNI, a structurally similar chromophore, indicates that this value should be independently verified.

One way to improve the sensitivity of the nitroindolines to 2PE would be to attach an antenna triplet sensitizer that has a robust TPA cross-section and appropriate triplet-state energy for efficient transfer to the indoline. Corrie attached a benzophenone derivative to a nitroindoline to generate a caged glutamate **8** (Fig. 6), which has improved efficiency of photorelease (69–71). Swapping the benzophenone for a chromophore that is more sensitive to 2PE would improve δ_u for the 1-acyl-7-nitroindoline PPG. One concern for physiological use is toxicity due to the generation of singlet oxygen, which competes with triplet energy transfer to the nitroindoline.

4.3. Chromophores Based on Coumarin Protecting Groups

A number of coumarin-based cages (for a review, see (72)) have demonstrated susceptibility to 2PE and have the additional advantage of rapid release kinetics. The photolysis proceeds from the singlet excited state through a solvent-assisted photoheterolysis reaction mechanism (Fig. 8) (73–76). After excitation to **18**, the excited state can decay through fluorescence or non-radiative processes to **17** or undergo heterolytic bond cleavage to form the singlet ion pair **19**. Recombination regenerates **17**, or the components of **19** can escape from the solvent cage to form the solvent separated ions **20** and the biological effector **Z**. Experiments in ^{18}O-labeled water (73) revealed that the cation **20** is trapped by water and, following rapid deprotonation, yields the uncaging remnant **21**, a proton, and **Z**. Formation of **19** is quite rapid, occurring with rate constants of $k_1 = 0.18–52 \times 10^9$ s^{-1} (76), but the rate constant for decay of **19** to alcohol **21** and release of **Z** is more difficult to measure. Time-resolved flash photolysis experiments suggest that release of carboxylate occurs at least on the sub-microsecond timescale (8) and might approach 4 ns (75), which places the coumarins among the fastest known phototriggers. The pK_a of the leaving group influences the rate of photolysis; **Z** must be a good leaving group (p$K_a \leq 10$) for photolysis to occur (73, 76). Coumarins can cage carboxylates, phosphates, sulfates, diols, ketones and aldehydes, alcohols (as the carbonate), amines (as the carbamate), and thiols (as the thiocarbonate). The carbamate or carbonate linkages required for caging amines or alcohols with coumarin derivatives pose problems for release within the excitation volume. After photolysis, the resultant carbamate

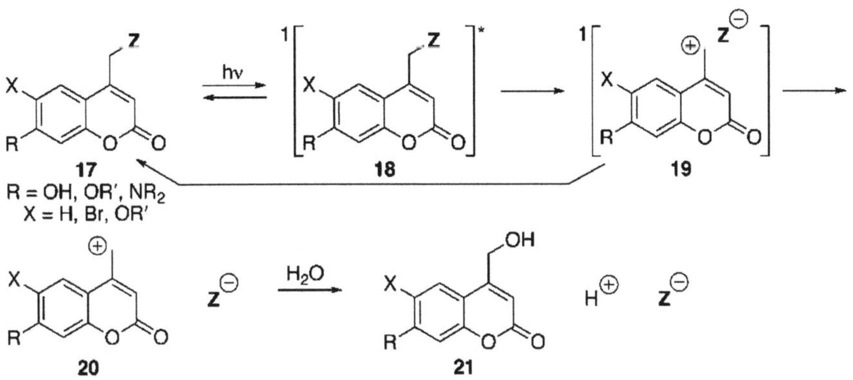

Fig. 8. The mechanism of the photolysis of coumarin-based caging groups.

or carbonate decomposes, releasing CO_2, to the corresponding amine or alcohol on timescales of milliseconds to seconds (77, 78), much slower than diffusional escape time from both the focal volume and rapid biological signal transduction processes.

6-Bromo-7-hydroxycoumarin (Bhc, Fig. 4) was the first caging group introduced that had been explicitly designed for 2PE-mediated release of neurotransmitters (16, 79). It has demonstrated application to cell and tissue culture and zebrafish, and is suitable for the protection and photorelease of many biologically relevant compounds with 1PE and 2PE, including glutamate (16), cyclic nucleotides (80), DNA and RNA (81), a phosphopeptide for photoactivating kinases (82), diols (83), 1,2-dioctanoylglycerol (84, 85), the iNOS inhibitor 1,400 W (86, 87), the protein synthesis inhibitor anisomycin (88), and nucleobases (89). Bhc has also been modified into a photocleavable linker for peptides and proteins (90) and used for three-dimensional micropatterning of agarose gels (91). A derivative called Bhc-diol protects aldehydes and ketones (92), and has been used to cage progesterone (93). The sensitivity of Bhc to 2PE photolysis varies according to the attached substrate, but in general $\delta_u = {\sim}1$ GM at 740 nm. Release of acetate occurs with $\delta_u = 0.72$ at 740 nm in aqueous buffer at pH 7.2. Photolysis of Bhc-caged carboxylates is rapid with a rate constant of at least $1.5 \times 10^6\ s^{-1}$ (8). The bromine serves to lower the pK_a of the phenol to ensure that at neutral pH, Bhc exists mostly in its phenolate form, which has a larger ε and λ_{max} than the phenolic form. The increase in absorption wavelength makes Bhc sensitive to 2PE in the 740–800 nm range at physiological pH. Some drawbacks of Bhc as a PPG include low levels of water solubility and bright fluorescence, which might limit the compound's use in certain biological studies that rely upon fluorescent indicators. A similar compound, DMCM (Fig. 4), has been reported to have sensitivity to 2PE, but δ_u has not been quantified (94).

The 7-dialkylamino-coumarin-based PPGs [7-(dimethylamino) coumarin-4-yl]methyl (DMACM) (94–96) and [7-(diethylamino) coumarin-4-yl]methyl (DEACM) (96–98) (Fig. 4) have been used to cage nucleotides and cyclic nucleotides, and DEACM is an effective caging group for the neurotransmitters glutamate (99) and glycine (100). The sensitivity of DMACM and DEACM to 2PE has not been quantified. Interestingly, the amino group increases λ_{max} by 15 and 39 nm relative to Bhc and DMCM, respectively (Table 1). Since the wavelength typically used for 2PE is approximately twice that used for 1PE, the optimum wavelength for photolysis with 2PE is expected to be longer by 30 and 78 nm. Caging groups sensitive to 2PE at disparate wavelengths offer the potential for using two caged compounds on a single biological preparation and independently releasing their substrates.

Another coumarin-based PPG, {7-[bis(carboxymethyl)amino] coumarin-4-yl}methyl (BCMACM) (101–103), has been shown to possess favorable 2PE sensitivities, with δ_u values ranging from 0.4 to 2.6 GM for cyclic nucleotides and various carbonates, carbamates, and thiocarbonates, illustrating the principle that the covalently attached substrate is capable of modulating the photochemical properties of the caging group. The two carboxylate groups confer much better water solubility on the BCMACM chromophore than the dialkylamino-coumarin derivatives.

Design principles suggest that lengthening conjugation of a chromophore will increase δ_a; therefore, one expects that oxobenzo[*f*]benzopyrans would have larger values of δ_u than the corresponding coumarins. Indeed, extended coumarins **22** and **23** (Fig. 9) have been shown to release neurotransmitter amino acids with 1PE at wavelengths near 350 nm (104–106). In our own work, we found that the tribromonapthocoumarin **24** (**Z** = OAc, λ_{max} = 412 nm) could release acetate upon 2PE at 800 nm, although **24** has poor aqueous solubility at neutral pH (107).

4.4. Chromophores Based on the Ortho-Hydroxycinnamic Acid Protecting Groups

The *ortho*-hydroxycinnamic acid (*o*-HCA) platform releases alcohols directly in a process called two-photon uncaging with fluorescence reporting (108–110). Irradiation of the *o*-HCA causes isomerization of the alkene **25** (Fig. 10), which is reversible under the reaction conditions, and subsequent cyclization to

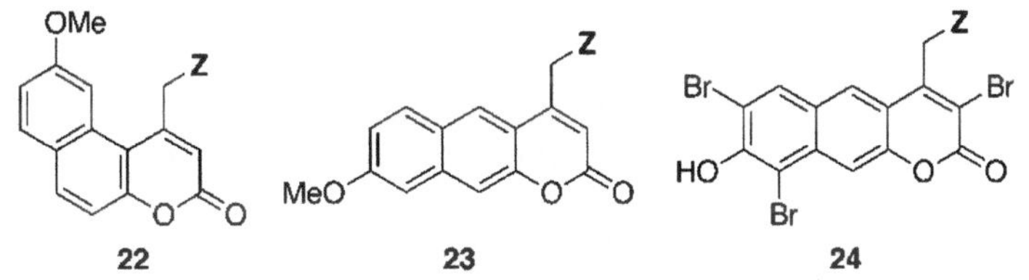

Fig. 9. Some extended coumarins that are potentially sensitive to 2PE.

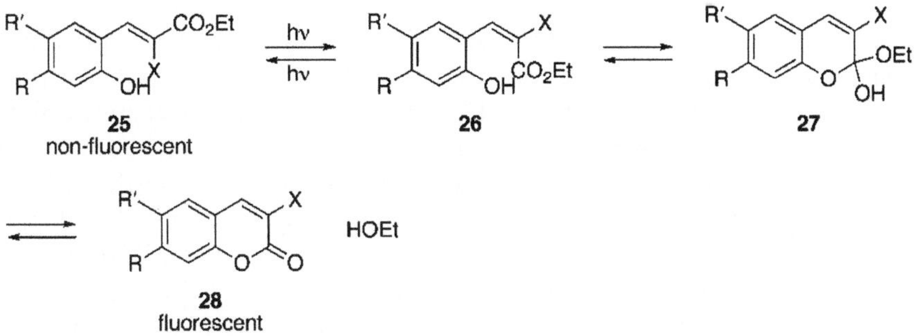

Fig. 10. The mechanism of the photolysis of *ortho*-hydroxycinnamic acid caging groups.

the fluorescent coumarin **28** with concomitant release of an alcohol. The ability to monitor the progress of uncaging through fluorescence is useful for applications that require precise titration of the release of a bioactive compound. To show proof of principle, a zebrafish embryo was incubated in a solution of DBHCA (**Z** = OEt), and a small area of the caudal fin was then illuminated with 750-nm light. Fluorescence output from the irradiated area measured by two-photon microscopy indicated that the concentration of caging remnant **28** (and ethanol, assuming a quantitative chemical yield) was ~10 μM (108).

Measurements of the two-photon action cross-section for the *E*/*Z* isomerization (though this is not the rate-determining step of the reaction and does not release alcohol) using the FCS after 2PE method reveal δ_u values greater than 1 GM at 750 nm. Comparing δ_u values of several *o*-HCA derivatives (Table 1) reveals design principles for optimizing δ_u that are similar to those shown to increase δ_a in other materials. Increasing the number of electron-donating groups on the aromatic system, as in the case of DOHCA (Fig. 4), more than doubles the cross-section relative to DBHCA. Increasing the donor strength by switching from oxygen to nitrogen as in the case of DEAHCA and HPQHCA also increases δ_u, and the conformational rigidity imposed by the cyclohexane rings of HPQHCA ensures that conjugation of the lone pair of electrons on nitrogen has a large effect. Extending the conjugated system (NHCA) improves the 2PE sensitivity relative to DBHCA. An acetamide substituent on the alkene (ADOHCA and ADEACA) has a negative effect on the cross-section.

Although the values of δ_u for the *o*-HCA system are high, it is not suitable for releasing alcohols within the focal volume. The rate constants for the cyclization of the (*Z*)-cinnamates are on the order of 10^{-2} s^{-1} (109, 110), much slower than the rate of diffusional escape from the focal volume. Nevertheless, these caging groups would have application for restricting the release of bioactive substrates to a single cell or other physically confined volumes.

4.5. Chromophores Based on Quinoline Protecting Groups

The quinoline-based PPG 8-bromo-7-hydroxyquinoline (BHQ, Fig. 4) mediates the photorelease of carboxylates, phosphates, and diols (111, 112), and owing to its mechanistic similarity to coumarins, it should also work with carbonyls, amines, and alcohols (as the carbamates and carbonates), and thiols (as the thiocarbonate). Values of δ_u range from 0.4 to 0.9 GM, depending on the substrate, and chemical yields are ~70%. BHQ has the added advantages of high quantum efficiency (Q_u = 0.29), good aqueous solubility, and low fluorescence output, which can be advantageous in applications that use fluorescent indicators to monitor biological activity.

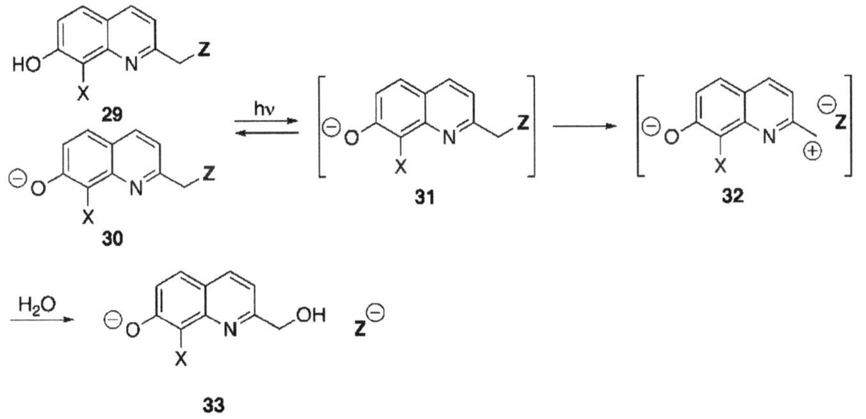

Fig. 11. The mechanism of the photolysis of quinoline-based caging groups.

The photolysis of BHQ proceeds by a solvent-assisted photoheterolysis (S_N1) mechanism through a short-lived excited state (112) similar to the mechanism for coumarins (Fig. 11). Resonance Raman spectroscopy revealed that at neutral pH, BHQ exists as a mixture of neutral (**29**) and anionic forms (**30**) (113), which correlates with $pK_a = 6.8$ measured for the phenolic proton (111). Excitation of either species generates a short-lived excited state intermediate **31** that is probably in the anionic form, because hydroxyquinolines in the excited state rapidly deprotonate (114–116). Heterolytic cleavage generates the bioactive substrate and the zwitterion-like intermediate **32**, which is trapped by water to furnish the alcohol by-product **33**. Stern–Volmer quenching studies suggest that the triplet excited state is not involved in the reaction. This result is supported by time-resolved IR (TRIR) studies with a triplet quencher and a sensitizer (112). The triplet state observed by TRIR decays at a rate of $1.7 \times 10^6\ s^{-1}$, which suggests that the release of substrate is on a sub-microsecond timescale or shorter, because the singlet excited state is expected to have a much shorter lifetime than the triplet state. An ^{18}O-labeling study confirms that the hydroxy group on **33** comes from solvent.

The bromine lowers the pK_a of the hydroxyl group, which increases the phenolate to phenol ratio at physiological pH. The phenolate absorbs at a longer wavelength than the phenolic form. Replacing the bromine with stronger electron-withdrawing groups, such as cyano (CyHQ, Fig. 4) and nitro (NHQ), dramatically increases the amount of phenolate form at physiological pH and improves hydrolytic stability in the dark, but does not confer greater susceptibility to 2PE (Table 1) (117). Whereas CyHQ has excellent sensitivity to 1PE, δ_u is a little more than half the value of BHQ. NHQ is photoinert, probably because it nonradiatively decays to the ground state through a low-lying triplet state accessed by the nitro group. As expected, replacement of the 7-hydroxyl group with the stronger electron-withdrawing dimethylamino group or a sulfhydryl increases δ_a from 2.0 GM for BHQ to 2.8, 5.2, and 6.7 GM for DMAQ, DMAQ-Cl, and TQ, respectively, but because DMAQ, DMAQ-Cl, and TQ have significantly lower quantum efficiencies relative to BHQ, δ_u is lower. This illustrates the complications associated with manipulating the structure of the chromophore to improve one desired property; it can come at the expense of another.

4.6. Other Chromophores

A number of other chromophores have been created that are specialized for the release of a specific biologically active substrate, but do not fall into the category of a classic PPG. Nevertheless, the chromophores in these systems demonstrate sensitivity to 2PE and photochemistry that is potentially useful for physiological studies, and can be instructive for designing new chromophores.

Azid-1 is a cage for calcium ions with 100% quantum efficiency and $\delta_u = 1.4$ GM at 700 nm (Fig. 12) (22, 118). Photolysis lowers the Ca^{2+} affinity of the ligand from 230 nM to 120 μM for structure **34**, which is not a dynamic range as large as that for other ligands for releasing Ca^{2+}. Nevertheless, azid-1 is extremely sensitive to light and releases Ca^{2+} rapidly enough for most applications. The extended and polar nature of the π-system of the chromophore probably contributes to its relatively large δ_u values

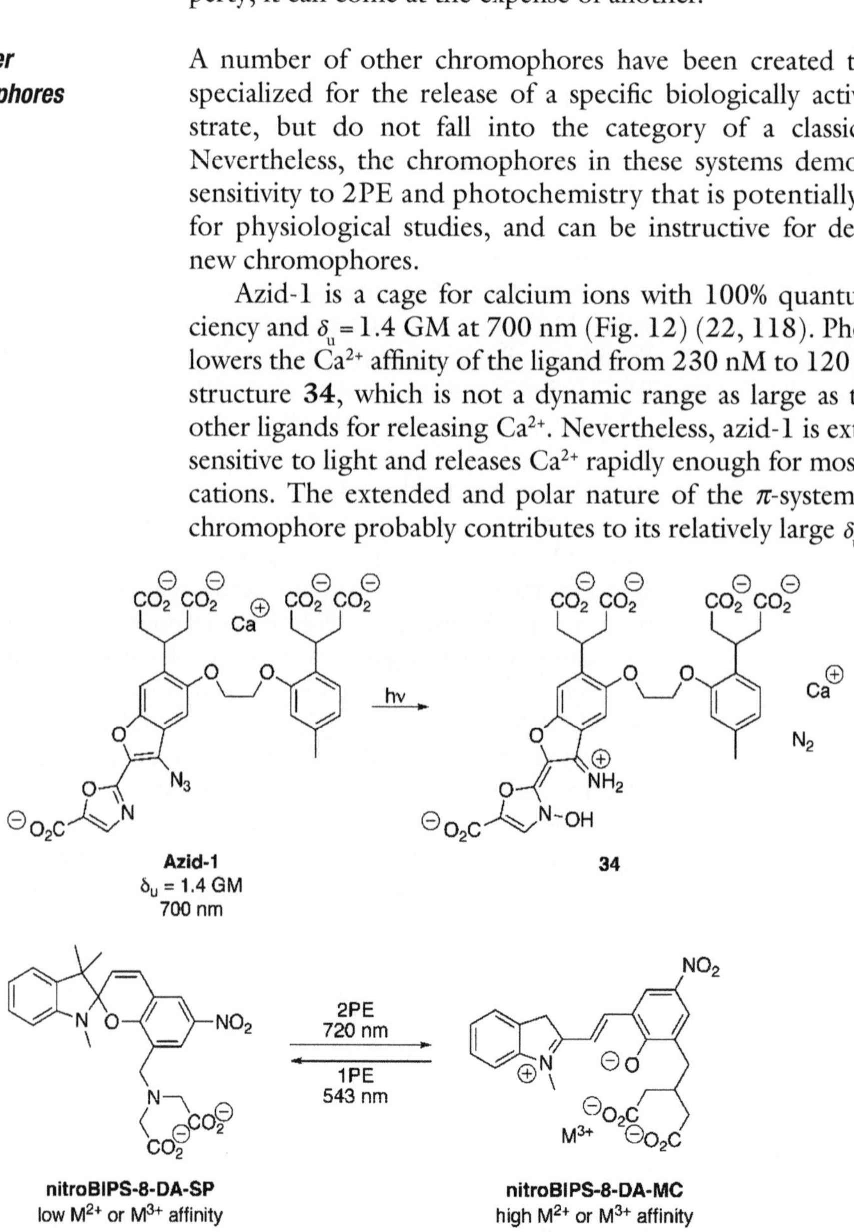

Fig. 12. Two strategies for releasing metals through 2PE. Azid-1 irreversibly releases Ca^{2+}, whereas nitroBIPS-8-DA is a reversible chelator of Gd^{3+}.

compared to other caged Ca^{2+}. Exploiting the photochemical behavior of the azid-1 chromophore for releasing other bioactive compounds presents challenges that have not yet been overcome. A promising method for manipulating cellular Ca^{2+} and other metals, including the magnetic resonance imaging (MRI) probe Gd^{3+}, is to use a chelator that is photoswitchable between high and low affinity states. The nitrobenzospiropyran iminodiacetate (nitroBIPS-8-DA, Fig. 12) has a colorless spiro state (SP) that has low affinity for a number of metals, and a fluorescent merocyanine state (MC) that binds Gd^{3+} ($K_d = 5.2$ μM in water) (119). The SP state rapidly converts to the MC state upon 2PE with 720 nm, but exposure to 543 nm (1PE) drives the system back to the nonbinding SP state. The cycle can be repeated at least ten times with high fidelity. The sensitivity of the optical switching to 2PE has not been quantified and no structure–property studies are available, but the potential to generate metal ligands that are optically switchable through 2PE is promising.

Examples already described in this chapter have demonstrated that linking an antenna sensitive to 2PE to the PPG can greatly increase its susceptibility to 2PE through the transfer of the energy needed for photolysis of the PPG. The iron nitrosyl cluster $Fe_2(\mu\text{-}SR)_2(NO)_4$ (Roussin's red salt ester, RSE) is a PPG for nitric oxide (NO), which plays a number of roles in physiology (blood pressure regulation, tumor growth inhibition, and immune defense). Ford and coworkers have shown that creating a complex of RSEs and 2PE-sensitive chromophores yields caged NO that can be released with 2PE (120). The first generation of photoreleasable NO used protoporphyrin IX (PPIX), for which $\delta_a = 2$ GM, as the antenna. The supramolecular complex [μ-S, μ-S′-protoporphyrin-IX-bis(2-thioethyl)diester]tetranitrosyl-diiron (PPIX-RSE, Fig. 13) releases NO through 2PE at 810 nm in THF (121), which is a completely irrelevant solvent system for physiology, but was used because of solubility issues with PPIX-RSE. Fluorescein has a much larger absorption cross-section than PPIX ($\delta_a = 38$ GM at 782 nm in pH 11 buffer) and better solubility in aqueous media. For Fluor-RSE, $\delta_a = 63$ GM at 800 nm in 50:50 acetonitrile/phosphate buffer pH 7.4, and since $Q_u = 0.01$ for NO release, $\delta_u = 0.63$ GM (122). Diphenylaminofluorene (AF) chromophores have enhanced absorption cross-sections, owing to the large, rigid π-system and strong donor and acceptor groups on either end of the conjugation. The sensitivity of AFX-RSE to 2PE is increased nearly fourfold ($\delta_a = 246$ GM at 800 nm in THF) relative to Fluor-RSE; therefore, $\delta_u = 4.9$ GM (123). To overcome the water solubility and biocompatibility issues associated with PPIX-RSE, Fluor-RSE, and AFX-RSE, Prasad and coworkers engineered a caged NO (2P-M, Fig. 13) from a complex of RSE and an oligo-phenylene vinylene (OPV) backbone modified with three groups of tetra(oligo(ethylene glycol)monomethyl) ether

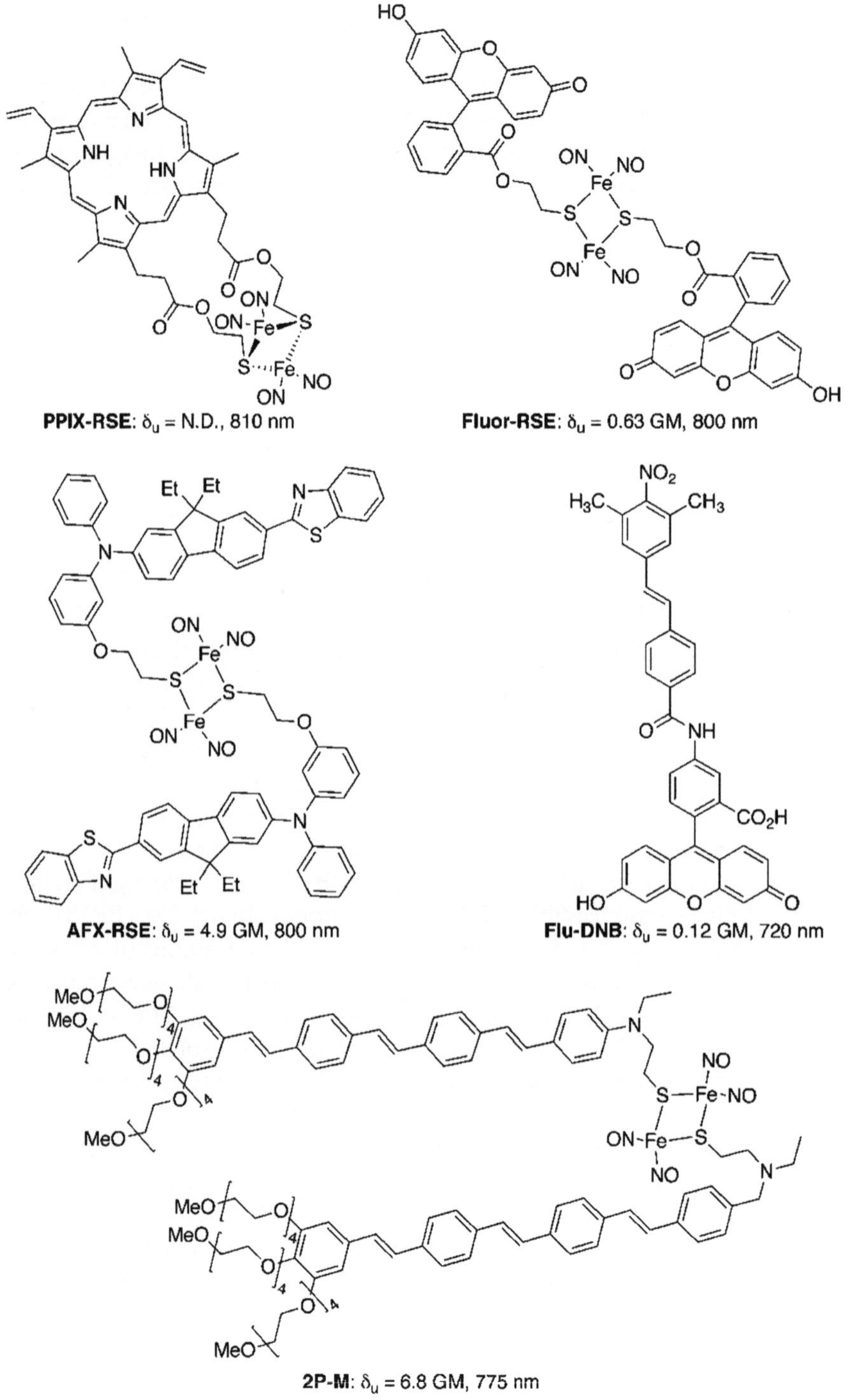

PPIX-RSE: δ_u = N.D., 810 nm

Fluor-RSE: δ_u = 0.63 GM, 800 nm

AFX-RSE: δ_u = 4.9 GM, 800 nm

Flu-DNB: δ_u = 0.12 GM, 720 nm

2P-M: δ_u = 6.8 GM, 775 nm

Fig. 13. Caging groups for the release of NO through 2PE.

to impart water solubility (124). The OPV backbone's extended conjugation combines with the strong amine and ether donor groups on the ends to make it a sensitive two-photon absorber. At 775 nm, δ_a = 682 GM for 2P-M; therefore, δ_u for release of NO is approximately 6.8 GM, based on an estimation of the quantum efficiency for NO release of 0.01 for RSE. Release of NO from 2P-M (40 μM) with 2PE was cytotoxic to HeLa cells, but excitement over this result is tempered by the fact that a 2P-M concentration of ≥8 μM considerably reduces cell viability in the absence of light. To reduce toxicity potentially caused by the nitrosyl-chelated metal ions, Nakagawa, Miyata, and coworkers designed an entirely organic NO donor (Flu-DNB, Fig. 13) that is based on 2,6-dinitrobenzene and incorporates fluorescein as a 2PE-sensitive antenna (125). The release of NO from Flu-DNB through 2PE ($\delta_u = 0.12$ GM at 720 nm) is not as efficient as that from the RSE derivatives, but if toxicity studies demonstrate biocompatibility, then it will be an improvement over the cytotoxic RSE compounds. No studies of Flu-DNB in a biological system have been reported yet.

An inorganic compound based on a ruthenium bipyridyl complex, $[Ru(bpy)_2(4\text{-aminopyridine})_2]Cl_2$ (Ru4AP), releases the neurochemical 4-aminopyridine, a potassium channel blocker, upon 2PE ($\delta_u = 0.01$–0.1 GM) (126). Replacing the bipyridyls with ligands that have greater sensitivity to 2PE might improve δ_u of the complex, but studies of this nature have not yet been reported.

The enediyne antibiotic dynemicin A is an antineoplastic agent whose biological activity depends on a light-induced Bergman cyclization that produces a *para*-benzyne diradical, which initiates DNA cleavage. Several caged enediynes have been prepared (127–129), but none have demonstrated sensitivity to 2PE. An alternative approach is to photochemically generate the enediyne in situ, which then undergoes the Bergman cyclization and DNA cleavage. Because the decarbonylation of cyclopropenones to give alkynes can be driven by 2PE (130), Popik and Poloukhtine reasoned that this reaction could be used to generate an enediyne (131). Quinone **35** (Fig. 14) undergoes 2PE at 800 nm ($\delta_u = 0.22$ GM) to generate enediyne **36**, which at physiological temperatures successively undergoes a Bergman cyclization to the diradical **37** and abstracts hydrogens from 1,3-cyclohexadiene to provide the product **38**. The study was not carried out in aqueous buffer, and this method is not suitable for control of the abstraction process within the excitation volume because the Bergman cyclization is incredibly slow ($\tau = 88.3$ h). Nevertheless, the concept is promising as a tool to manipulate biological systems, because the reaction is initiated by light in the phototherapeutic window, so phototoxicity to biological samples is minimized, and the reaction can be initiated deeper in tissues than can be achieved with UV wavelengths and 1PE.

The Wolff rearrangement of diazonaphthoquinone (DNQ) can be driven by 2PE (132, 133), and Fréchet and coworkers have described an amphiphilic molecule **39** (Fig. 14) that incorporates the hydrophobic DNQ and the hydrophilic linear

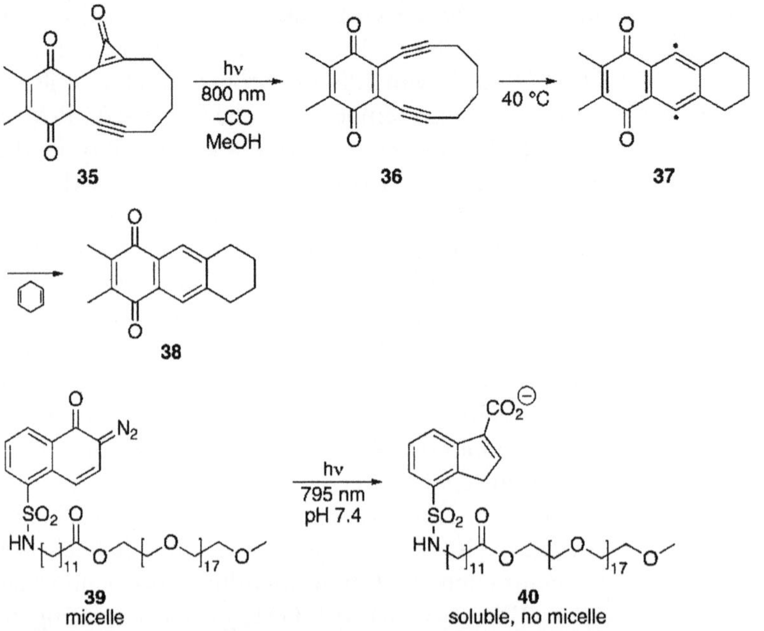

Fig. 14. Photochemical reactions driven by 2PE. The decarbonylation of cyclopropene generates an enediyne, which cyclizes to a DNA-cleaving diradical. The photo-Wolff rearrangement changes the solubility of the amphiphile, disrupting the micelles.

poly(ethylene glycol) to form micelles, which can be destroyed by a 2PE-initiated Wolff rearrangement to the carboxylate **40** that solubilizes the chain and releases the fluorescent dye Nile Red from the micelle (134). A high critical micelle concentration (cmc) and cytoxicity led to the development of a poly(ethylene oxide)-dendritic polyester copolymer hybrid (135), which has a lower cmc and is noncytotoxic even at 1 mg/mL. These and a similar strategy based on DEACM (136) have the potential to be developed into drug delivery platforms.

Cycloaddition (137) and cycloreversion (138, 139) reactions can be driven by 2PE, but like many photochemical reactions, they have not yet been exploited to direct the activity of a biologically active compound. The exploration of photochemical reactions for their sensitivity to 2PE to release or generate biologically active molecules presents an open area for research.

5. Conclusions

Caging groups and phototriggers with increased sensitivity to 2PE are emerging as invaluable tools for the study of biological systems, and they are in many ways an improvement upon the conventional caging groups activated by 1PE. When 2PE drives the activation process and the photoreaction is faster than the rate of diffusional escape from the focal volume, the release can be localized to a femtoliter. Compared to uncaging with 1PE, 2PE provides a less destructive method of delivering bioactive molecules to cell and tissue culture and whole animals, and the transparency of biological tissues to near-IR wavelengths enables the release of biological effectors deeper into the tissue than possible with UV light. New advances in theory and an emergent understanding of the roles conjugation, symmetry, and electron donors and acceptors play in enhancing 2PE susceptibility are making it easier to engineer PPGs with large cross-sections at appropriate wavelengths, but these factors must be balanced with the need for rapid photoreaction kinetics, low toxicity, aqueous solubility, dark stability, synthetic accessibility, and any specific requirements of the biological system under study, such as membrane permeability. Sometimes enhancement of one property comes at the expense of another. Chromophores based on nitrobenzyl, nitroindoline, coumarin, *ortho*-hydroxycinnamic acid, quinoline, and other structural motifs have been designed to regulate the action of biologically active compounds with 2PE. Derivatives of these have proven to be valuable tools in applications to map local signaling events. The synthesis and characterization of new chromophores with enhanced 2PE, kinetics, and other photochemical properties will contribute to a further understanding of design criteria for caging groups and phototriggers with sensitivity to 2PE and provide high precision tools for the study of physiological systems.

Acknowledgments

Thanks to Kyle T. Harris and Steven D. Flynn for helpful comments and suggestions. This work was supported in part by the National Science Foundation.

References

1. Goeldner M, Givens RS (eds) (2005) Dynamic studies in biology: phototriggers, photoswitches, and caged biomolecules. Wiley-VCH, Weinheim, Germany
2. Mayer G, Heckel A (2006) Biologically active molecules with a "light switch." Angew Chem Int Ed 45:4900–4921
3. Young DD, Deiters A (2007) Photochemical control of biological processes. Org Biomol Chem 5:999–1005
4. Ellis-Davies GCR (2007) Caged compounds: photorelease technology for control of cellular chemistry and physiology. Nat Methods 4:619–628
5. Lee H-M, Larson DR, Lawrence DS (2009) Illuminating the chemistry of life: design, synthesis, and applications of "caged" and related photoresponsive compounds. ACS Chem Biol 4:409–427
6. Denk W, Strickler JH, Webb WW (1990) Two-photon laser scanning fluorescence microscopy. Science 248:73–76
7. Denk W (1994) Two-photon scanning photochemical microscopy: mapping ligand-gated

ion channel distributions. Proc Natl Acad Sci USA 91:6629–6633

8. Dore TM (2005) Multiphoton phototriggers for exploring cell physiology. In: Goeldner M, Givens RS (eds) Dynamic studies in biology: phototriggers, photoswitches, and caged biomolecules. Wiley-VCH, Weinheim, Germany, pp 435–459
9. Benninger RKP, Hao M, Piston DW (2008) Multi-photon excitation imaging of dynamic processes in living cells and tissues. Rev Physiol Biochem Pharmacol 160:71–92
10. Rubart M (2004) Two-photon microscopy of cells and tissue. Circ Res 95:1154–1166
11. Svoboda K, Yasuda R (2006) Principles of two-photon excitation microscopy and its applications to neuroscience. Neuron 50: 823–839
12. Göppert M (1929) Über die Wahrscheinlichkeit des Zusammenwirkens zweier Lichtquanten in einem Elementarakt. Naturwissenschaften 17:932
13. Göppert-Mayer M (1931) Über Elementarakte mit zwei Quantensprüngen. Ann Phys 9:273–294
14. Kaiser W, Garrett CGB (1961) Two-photon excitation in CaF_2:Eu^{2+}. Phys Rev Lett 7:229–231
15. Kiskin NI, Chillingworth R, McCray JA, Piston D, Ogden D (2002) The efficiency of two-photon photolysis of a "caged" fluorophore, *o*-1-(2-nitrophenyl)ethylpyranine, in relation to photodamage of synaptic terminals. Eur Biophys J 30:588–604
16. Furuta T, Wang SSH, Dantzker JL, Dore TM, Bybee WJ, Callaway EM, Denk W, Tsien RY (1999) Brominated 7-hydroxycoumarin-4-ylmethyls: photolabile protecting groups with biologically useful cross-sections for two photon photolysis. Proc Natl Acad Sci USA 96:1193–1200
17. Matsuzaki M, Ellis-Davies GCR, Nemoto T, Miyashita Y, Iino M, Kasai H (2001) Dendritic spine geometry is critical for AMPA receptor expression in hippocampal CA1 pyramidal neurons. Nat Neurosci 4:1086–1092
18. Xu C, Guild J, Webb WW, Denk W (1995) Determination of absolute two-photon excitation cross sections by in situ second-order autocorrelation. Opt Lett 20:2372–2374
19. Xu C, Webb WW (1996) Measurement of two-photon excitation cross sections of molecular fluorophores with data from 690 to 1050 nm. J Opt Soc Am B 13:481–491
20. Demas JN, Crosby GA (1971) Measurement of photoluminescence quantum yields. A review. J Phys Chem 75:991–1024
21. Makarov NS, Drobizhev M, Rebane A (2008) Two-photon absorption standards in the 550–1600 nm excitation wavelength range. Opt Express 16:4029–4047
22. Brown EB, Shear JB, Adams SR, Tsien RY, Webb WW (1999) Photolysis of caged calcium in femtoliter volumes using two-photon excitation. Biophys J 76:489–499
23. Aujard I, Benbrahim C, Gouget M, Ruel O, Baudin J-B, Neveu P, Jullien L (2006) *o*-Nitrobenzyl photolabile protecting groups with red-shifted absorption: syntheses and uncaging cross-sections for one- and two-photon excitation. Chem Eur J 12:6865–6879
24. Li G, Niu L (2004) How fast does the GluR1Q_{flip} channel open? J Biol Chem 279:3990–3997
25. Kiskin NI, Ogden D (2002) Two-photon excitation and photolysis by pulsed laser illumination modelled by spatially non-uniform reactions with simultaneous diffusion. Eur Biophys J 30:571–587
26. Strehmel B, Strehmel V (2007) Two-photon physical organic, and polymer chemistry: theory, techniques, chromophore design, and applications. Adv Photochem 29:111–354
27. Rumi M, Barlow S, Wang J, Perry JW, Marder SR (2008) Two-photon absorbing materials and two-photon-induced chemistry. Adv Polym Sci 213:1–95
28. Terenziani F, Katan C, Badaeva E, Tretiak S, Blanchard-Desce M (2008) Enhanced two-photon absorption of organic chromophores: theoretical and experimental assessments. Adv Mater 20:4641–4678
29. He GS, Tan L-S, Zheng Q, Prasad PN (2008) Multiphoton absorbing materials: molecular designs, characterizations, and applications. Chem Rev 108:1245–1330
30. Pawlicki M, Collins HA, Denning RG, Anderson HL (2009) Two-photon absorption and the design of two-photon dyes. Angew Chem Int Ed 48:3244–3266
31. Kim HM, Cho BR (2009) Two-photon materials with large two-photon cross sections. Structure–property relationship. Chem Commun (Camb) 153–164
32. Albota M, Beljonne D, Bredas J-L, Ehrlich JE, Fu J-Y, Heikal AA, Hess SE, Kogej T, Levin MD, Marder SR, McCord-Maughon D, Perry JW, Rockel H, Rumi M, Subramaniam G, Webb WW, Wu X-L, Xu C (1998) Design of organic molecules with large two-photon absorption cross sections. Science 281: 1653–1656
33. Reinhardt BA, Brott LL, Clarson SJ, Dillard AG, Bhatt JC, Kannan R, Yuan L, He GS,

Prasad PN (1998) Highly active two-photon dyes: design, synthesis, and characterization toward application. Chem Mater 10: 1863–1874

34. Corrie JET (2005) Photoremovable protecting groups used for the caging of biomolecules: 2-nitrobenzyl and 7-nitorindoline derivatives. In: Goeldner M, Givens RS (eds) Dynamic studies in biology: phototriggers, photoswitches, and caged biomolecules. Wiley-VCH, Weinheim, Germany, pp 1–28
35. Corrie JET, Trentham DR (1993) Caged nucleotides and neurotransmitters. In: Morrison H (ed) Biological applications of photochemical switches. Wiley, New York, pp 243–305
36. Kaplan JH, Forbush B III, Hoffman JF (1978) Rapid photolytic release of adenosine 5′-triphosphate from a protected analogue: utilization by the Na:K pump of human red blood cell ghosts. Biochemistry 17: 1929–1935
37. Neveu P, Aujard I, Benbrahim C, Le Saux T, Allemand J-F, Vriz S, Bensimon D, Jullien L (2008) A caged retinoic acid for one- and two-photon excitation in zebrafish embryos. Angew Chem Int Ed 47:3744–3746
38. Yip RW, Wen YX, Gravel D, Giasson R, Sharma DK (1991) Photochemistry of the *o*-nitrobenzyl system in solution: identification of the biradical intermediate in the intramolecular rearrangement. J Phys Chem 95:6078–6081
39. Yip RW, Sharma DK, Giasson R, Gravel D (1985) Photochemistry of the *o*-nitrobenzyl system in solution: evidence for singlet-state intramolecular hydrogen abstraction. J Phys Chem 89:5328–5330
40. Givens RS, Kotala MB, Lee J-I (2005) Mechanistic overview of phototriggers and cage release. In: Goeldner M, Givens RS (eds) Dynamic studies in biology: phototriggers, photoswitches, and caged biomolecules. Wiley-VCH, Weinheim, Germany, pp 95–129
41. Schwörer M, Wirz J (2001) Photochemical reaction mechanisms of 2-nitrobenzyl compounds in solution I. 2-nitrotoluene: thermodynamic and kinetic parameters of the aci-nitro tautomer. Helv Chim Acta 84:1441–1458
42. Il'ichev YV, Wirz J (2000) Rearrangements of 2-nitrobenzyl compounds. 1. Potential energy surface of 2-nitrotoluene and its isomers explored with ab initio and density functional theory methods. J Phys Chem A 104:7856–7870
43. Il'ichev YV, Schwoerer MA, Wirz J (2004) Photochemical reaction mechanisms of 2-nitrobenzyl compounds: methyl ethers and caged ATP. J Am Chem Soc 126: 4581–4595
44. Corrie JET, Barth A, Munasinghe VRN, Trentham DR, Hutter MC (2003) Photolytic cleavage of 1-(2-nitrophenyl)ethyl ethers involves two parallel pathways and product release is rate-limited by decomposition of a common hemiacetal intermediate. J Am Chem Soc 125:8546–8554
45. Walker JW, Reid GP, McCray JA, Trentham DR (1988) Photolabile 1-(2-nitrophenyl)ethyl phosphate esters of adenine nucleotide analogs. Synthesis and mechanism of photolysis. J Am Chem Soc 110:7170–7177
46. Barth A, Corrie JET, Gradwell MJ, Maeda Y, Maentele W, Meier T, Trentham DR (1997) Time-resolved infrared spectroscopy of intermediates and products from photolysis of 1-(2-nitrophenyl)ethyl phosphates: reaction of the 2-nitrosoacetophenone byproduct with thiols. J Am Chem Soc 119:4149–4159
47. Zhao Y, Zheng Q, Dakin K, Xu K, Martinez ML, Li W-H (2004) New caged coumarin fluorophores with extraordinary uncaging cross sections suitable for biological imaging applications. J Am Chem Soc 126: 4653–4663
48. Dakin K, W-h L (2006) Infrared-LAMP: two-photon uncaging and imaging of gap junctional communication in three dimensions. Nat Methods 3:959
49. Zheng G, Guo Y-M, Li W-H (2007) Photoactivatable and water soluble FRET dyes with high uncaging cross section. J Am Chem Soc 129:10616–10617
50. Guo Y-M, Chen S, Shetty P, Zheng G, Lin R, W-h L (2008) Imaging dynamic cell–cell junctional coupling in vivo using Trojan-LAMP. Nat Methods 5:835–841
51. Specht A, Thomann J-S, Alarcon K, Wittayanan W, Ogden D, Furuta T, Kurakawa Y, Goeldner M (2006) New photoremovable protecting groups for carboxylic acids with high photolytic efficiencies at near-UV irradiation. Application to the photocontrolled release of L-glutamate. Chembiochem 7:1690–1695
52. Orange C, Specht A, Puliti D, Sakr E, Furuta T, Winsor B, Goeldner M (2008) Synthesis and photochemical properties of a light-activated fluorophore to label His-tagged proteins. Chem Commun (Camb) 1217–1219
53. Momotake A, Lindegger N, Niggli E, Barsotti RJ, Ellis-Davies GCR (2006) The nitrodibenzofuran chromophore: a new

caging group for ultra-efficient photolysis in living cells. Nat Methods 3:35–40

54. Gug S, Charon S, Specht A, Alarcon K, Ogden D, Zietz B, Leonard J, Haacke S, Bolze F, Nicoud J-F, Goeldner M (2008) Photolabile glutamate protecting group with high one- and two-photon uncaging efficiencies. Chembiochem 9:1303–1307
55. Gug S, Bolze F, Specht A, Bourgogne C, Goeldner M, Nicoud J-F (2008) Molecular engineering of photoremovable protecting groups for two-photon uncaging. Angew Chem Int Ed 47:9525–9529
56. Papageorgiou G, Ogden DC, Barth A, Corrie JET (1999) Photorelease of carboxylic acids from 1-acyl-7-nitroindolines in aqueous solution: rapid and efficient photorelease of L-glutamate. J Am Chem Soc 121: 6503–6504
57. Canepari M, Nelson L, Papageorgiou G, Corrie JET, Ogden D (2001) Photochemical and pharmacological evaluation of 7-nitroindolinyl- and 4-methoxy-7-nitroindolinyl-amino acids as novel, fast caged neurotransmitters. J Neurosci Methods 112:29–42
58. Maier W, Corrie JET, Papageorgiou G, Laube B, Grewer C (2005) Comparative analysis of inhibitory effects of caged ligands for the NMDA receptor. J Neurosci Methods 142:1–9
59. Morrison J, Wan P, Corrie JET, Papageorgiou G (2002) Mechanisms of photorelease of carboxylic acids from 1-acyl-7-nitroindolines in solutions of varying water content. Photochem Photobiol Sci 1:960–969
60. Papageorgiou G, Ogden D, Kelly G, Corrie JET (2005) Synthetic and photochemical studies of substituted 1-acyl-7-nitroindolines. Photochem Photobiol Sci 4:887–896
61. Cohen AD, Helgen C, Bochet CG, Toscano JP (2005) The mechanism of photoinduced acylation of amines by *N*-Acyl-5, 7-dinitroindoline as determined by time-resolved infrared spectroscopy. Org Lett 7:2845–2848
62. Papageorgiou G, Corrie JET (2000) Effects of aromatic substituents on the photocleavage of 1-acyl-7-nitroindolines. Tetrahedron 56:8197–8205
63. Zhang Y-P, Holbro N, Oertner TG (2008) Optical induction of plasticity at single synapses reveals input-specific accumulation of alpha CaMKII. Proc Natl Acad Sci USA 105:12039–12044
64. Nikolenko V, Poskanzer KE, Yuste R (2007) Two-photon photostimulation and imaging of neural circuits. Nat Methods 4:943–950
65. Beique J-C, Lin D-T, Kang M-G, Aizawa H, Takamiya K, Huganir RL (2006) Synapse-specific regulation of AMPA receptor function by PSD-95. Proc Natl Acad Sci USA 103:19535–19540
66. Huang YH, Sinha SR, Fedoryak OD, Ellis-Davies GCR, Bergles DE (2005) Synthesis and characterization of 4-methoxy-7-nitroindolinyl-D-aspartate, a caged compound for selective activation of glutamate transporters and *N*-methyl-D-aspartate receptors in brain tissue. Biochemistry 44:3316–3326
67. Fedoryak OD, Sul J-Y, Haydon PG, Ellis-Davies GCR (2005) Synthesis of a caged glutamate for efficient one- and two-photon photorelease on living cells. Chem Commun (Camb) 3664–3666
68. Ellis-Davies GCR, Matsuzaki M, Paukert M, Kasai H, Bergles DE (2007) 4-Carboxymethoxy-5,7-dinitroindolinyl-glu: an improved caged glutamate for expeditious ultraviolet and two-photon photolysis in brain slices. J Neurosci 27:6601–6604
69. Papageorgiou G, Lukeman M, Wan P, Corrie JET (2004) An antenna triplet sensitiser for 1-acyl-7-nitroindolines improves the efficiency of carboxylic acid photorelease. Photochem Photobiol Sci 3:366–373
70. Papageorgiou G, Ogden D, Corrie JET (2004) An antenna-sensitized nitroindoline precursor to enable photorelease of L-glutamate in high concentrations. J Org Chem 69:7228–7233
71. Papageorgiou G, Corrie JET (2005) Optimized synthesis and photochemistry of antenna-sensitized 1-acyl-7-nitroindolines. Tetrahedron 61:609–616
72. Furuta T (2005) Photoremovable protecting groups used for the caging of biomolecules. Coumarin-4-ylmethyl phototriggers. In: Goeldner M, Givens RS (eds) Dynamic studies in biology: phototriggers, photoswitches, and caged biomolecules. Wiley-VCH, Weinheim, Germany, pp 29–55
73. Schade B, Hagen V, Schmidt R, Herbich R, Krause E, Eckardt T, Bendig J (1999) Deactivation behavior and excited state properties of (coumarin-4-yl)methyl derivatives. 1. Photocleavage of (7-methoxycoumarin-4-yl) methyl-caged acids with fluorescence enhancement. J Org Chem 64:9109–9117
74. Eckardt T, Hagen V, Schade B, Schmidt R, Schweitzer C, Bendig J (2002) Deactivation behavior and excited-state properties of (coumarin-4-yl)methyl derivatives. 2. Photocleavage of selected (coumarin-4-yl) methyl-caged adenosine cyclic 3′,5′-

monophosphates with fluorescence enhancement. J Org Chem 67:703–710
75. Schmidt R, Geissler D, Hagen V, Bendig J (2005) Kinetics study of the photocleavage of (coumarin-4-yl)methyl esters. J Phys Chem A 109:5000–5004
76. Schmidt R, Geissler D, Hagen V, Bendig J (2007) Mechanism of photocleavage of (coumarin-4-yl)methyl esters. J Phys Chem A 111:5768–5774
77. Papageorgiou G, Corrie JET (1997) Synthesis and properties of carbamoyl derivatives of photolabile benzoins. Tetrahedron 53:3917–3932
78. Papageorgiou G, Barth A, Corrie JET (2005) Flash photolytic release of alcohols from photolabile carbamates or carbonates is rate-limited by decarboxylation of the photoproduct. Photochem Photobiol Sci 4:216–220
79. Furuta T, Noguchi K (2004) Controlling cellular systems with Bhc-caged compounds. Trends Anal Chem 23:511–519
80. Furuta T, Takeuchi H, Isozaki M, Takahashi Y, Kanehara M, Sugimoto M, Watanabe T, Noguchi K, Dore TM, Kurahashi T, Iwamura M, Tsien RY (2004) Bhc-cNMPs as either water-soluble or membrane-permeant photo-releasable cyclic nucleotides for both one and two-photon excitations. Chembiochem 5:1119–1128
81. Ando H, Furuta T, Okamoto H (2004) Photo-mediated gene activation by using caged mRNA in zebrafish embryos. Methods Cell Biol 77:159–171
82. Kawakami T, Cheng H, Hashiro S, Nomura Y, Tsukiji S, Furuta T, Nagamune T (2008) A caged phosphopeptide-based approach for photochemical activation of kinases in living cells. Chembiochem 9:1583–1586
83. Lin W, Lawrence DS (2002) A strategy for the construction of caged diols using a photolabile protecting group. J Org Chem 67:2723–2726
84. Suzuki AZ, Watanabe T, Kawamoto M, Nishiyama K, Yamashita H, Ishii M, Iwamura M, Furuta T (2003) Coumarin-4-ylmethoxycarbonyls as phototriggers for alcohols and phenols. Org Lett 5: 4867–4870
85. Quann EJ, Merino E, Furuta T, Huse M (2009) Localized diacylglycerol drives the polarization of the microtubule-organizing center in T cells. Nat Immunol 10: 627–635
86. Montgomery HJ, Perdicakis B, Fishlock D, Lajoie GA, Jervis E, Guillemette JG (2002) Photo-control of nitric oxide synthase activity using a caged isoform specific inhibitor. Bioorg Med Chem 10:1919–1927
87. Perdicakis B, Montgomery HJ, Abbott GL, Fishlock D, Lajoie GA, Guillemette JG, Jervis E (2005) Photocontrol of nitric oxide production in cell culture using a caged isoform selective inhibitor. Bioorg Med Chem 13:47–57
88. Goard M, Aakalu G, Fedoryak OD, Quinonez C, St Julien J, Schuman PSJ, EM DTM (2005) Light-mediated inhibition of protein synthesis. Chem Biol 12:685–693
89. Furuta T, Watanabe T, Tanabe S, Sakyo J, Matsuba C (2007) Phototriggers for nucleobases with improved photochemical properties. Org Lett 9:4717–4720
90. Katayama K, Tsukiji S, Furuta T, Nagamune T (2008) A bromocoumarin-based linker for synthesis of photocleavable peptidoconjugates with high photosensitivity. Chem Commun (Camb) 5399–5401
91. Wylie RG, Shoichet MS (2008) Two-photon micropatterning of amines within an agarose hydrogel. J Mater Chem 18:2716–2721
92. Lu M, Fedoryak OD, Moister BR, Dore TM (2003) Bhc-diol as a photolabile protecting group for aldehydes and ketones. Org Lett 5:2119–2122
93. Kilic F, Kashikar ND, Schmidt R, Alvarez L, Dai L, Weyand I, Wiesner B, Goodwin N, Hagen V, Kaupp UB (2009) Caged progesterone: a new tool for studying rapid nongenomic actions of progesterone. J Am Chem Soc 131:4027–4030
94. Geissler D, Antonenko YN, Schmidt R, Keller S, Krylova OO, Wiesner B, Bendig J, Pohl P, Hagen V (2005) (Coumarin-4-yl) methyl esters as highly efficient, ultrafast phototriggers for protons and their application to acidifying membrane surfaces. Angew Chem Int Ed 44:1195–1198
95. Geissler D, Kresse W, Wiesner B, Bendig J, Kettenmann H, Hagen V (2003) DMACM-caged adenosine nucleotides: ultrafast phototriggers for ATP, ADP, and AMP activated by long-wavelength irradiation. Chembiochem 4:162–170
96. Hagen V, Frings S, Wiesner B, Helm S, Kaupp UB, Bendig J (2003) [7-(Dialkylamino)coumarin-4-yl]methyl-caged compounds as ultrafast and effective long-wavelength phototriggers of 8-bromo-substituted cyclic nucleotides. Chembiochem 4:434–442
97. Hagen V, Bendig J, Frings S, Eckardt T, Helm S, Reuter D, Kaupp UB (2001) Highly efficient and ultrafast phototriggers for cAMP

and cGMP by using long-wavelength UV/Vis-activation. Angew Chem Int Ed Engl 40:1046–1048

98. Schönleber RO, Bendig J, Hagen V, Giese B (2002) Rapid photolytic release of cytidine 5′-diphosphate from a coumarin derivative: a new tool for the investigation of ribonucleotide reductases. Bioorg Med Chem Lett 10:97–101
99. Shembekar VR, Chen Y, Carpenter BK, Hess GP (2005) A protecting group for carboxylic acids that can be photolyzed by visible light. Biochemistry 44:7107–7114
100. Shembekar VR, Chen Y, Carpenter BK, Hess GP (2007) Coumarin-caged glycine that can be photolyzed within 3 micro s by visible light. Biochemistry 46:5479–5484
101. Hagen V, Dekowski B, Nache V, Schmidt R, Geissler D, Lorenz D, Eichhorst J, Keller S, Kaneko H, Benndorf K, Wiesner B (2005) Coumarinylmethyl esters for ultra-fast release of high concentrations of cyclic nucleotides upon one- and two-photon photolysis. Angew Chem Int Ed 44: 7887–7891
102. Hagen V, Dekowski B, Kotzur N, Lechler R, Wiesner B, Briand B, Beyermann M (2008) {7-[Bis(carboxymethyl)amino]coumarin-4-yl}methoxycarbonyl derivatives for photorelease of carboxylic acids, alcohols/phenols, thioalcohols/thiophenols, and amines. Chem Eur J 14:1621–1627
103. Senda N, Momotake A, Arai T (2007) Synthesis and photocleavage of 7-[{Bis(carboxymethyl)amino}coumarin-4-yl]methyl-caged neurotransmitters. Bull Chem Soc Jpn 80:2384–2388
104. Fernandes MJG, Goncalves MST, Costa SPG (2008) Comparative study of polyaromatic and polyheteroaromatic fluorescent photocleavable protecting groups. Tetrahedron 64:3032–3038
105. Fernandes MJG, Goncalves MST, Costa SPG (2008) Neurotransmitter amino acid-oxobenzo[f]benzopyran conjugates: synthesis and photorelease studies. Tetrahedron 64:11175–11179
106. Piloto AM, Rovira D, Costa SPG, Goncalves MST (2006) Oxobenzo[f]benzopyrans as new fluorescent photolabile protecting groups for the carboxylic function. Tetrahedron 62:11955–11962
107. Miller KA (2002) Synthesis and characterization of naphthocoumarins as caged compounds [Honors thesis]. University of Georgia, Athens, GA
108. Gagey N, Neveu P, Jullien L (2007) Two-photon uncaging with the efficient 3,5-dibromo-2,4-dihydroxycinnamic caging group. Angew Chem Int Ed 46:2467–2469
109. Gagey N, Neveu P, Benbrahim C, Goetz B, Aujard I, Baudin J-B, Jullien L (2007) Two-photon uncaging with fluorescence reporting: evaluation of the *o*-hydroxycinnamic platform. J Am Chem Soc 129:9986–9998
110. Gagey N, Emond M, Neveu P, Benbrahim C, Goetz B, Aujard I, Baudin J-B, Jullien L (2008) Alcohol uncaging with fluorescence reporting: evaluation of *o*-acetoxyphenyl methyloxazolone precursors. Org Lett 10:2341–2344
111. Fedoryak OD, Dore TM (2002) Brominated hydroxyquinoline as a photolabile protecting group with sensitivity to multiphoton excitation. Org Lett 4:3419–3422
112. Zhu Y, Pavlos CM, Toscano JP, Dore TM (2006) 8-Bromo-7-hydroxyquinoline as a photoremovable protecting group for physiological use: mechanism and scope. J Am Chem Soc 128:4267–4276
113. An H-Y, Ma C, Nganga JL, Zhu Y, Dore TM, Phillips DL (2009) Resonance Raman characterization of different forms of ground-state 8-bromo-7-hydroxyquinoline caged acetate in aqueous solutions. J Phys Chem A 113:2831–2837
114. Kim TG, Topp MR (2004) Ultrafast excited-state deprotonation and electron transfer in hydroxyquinoline derivatives. J Phys Chem A 108:10060–10065
115. Bardez E, Fedorov A, Berberan-Santos MN, Martinho JMG (1999) Photoinduced coupled proton and electron transfers. 2. 7-Hydroxyquinolinium ion. J Phys Chem A 103:4131–4136
116. Lee S-I, Jang D-J (1995) Proton transfers of aqueous 7-hydroxyquinoline in the first excited singlet, lowest triplet, and ground states. J Phys Chem 99:7537–7541
117. Davis MJ, Kragor CH, Reddie KG, Wilson HC, Zhu Y, Dore TM (2009) Substituent effects on the sensitivity of a quinoline photoremovable protecting group to one- and two-photon excitation. J Org Chem 74:1721–1729
118. Adams SR, Lev-Ram V, Tsien RY (1997) A new caged Ca^{2+}, azid-1, is far more photosensitive than nitrobenzyl-based chelators. Chem Biol 4:867–878
119. Sakata T, Jackson DK, Mao S, Marriott G (2008) Optically switchable chelates: optical control and sensing of metal ions. J Org Chem 73:227–233
120. Ford PC (2008) Polychromophoric metal complexes for generating the bioregulatory agent nitric oxide by single- and two-

photon excitation. Acc Chem Res 41: 190–200

121. Wecksler S, Mikhailovsky A, Ford PC (2004) Photochemical production of nitric oxide via two-photon excitation with NIR light. J Am Chem Soc 126:13566–13567
122. Wecksler SR, Mikhailovsky A, Korystov D, Ford PC (2006) A two-photon antenna for photochemical delivery of nitric oxide from a water-soluble, dye-derivatized iron nitrosyl complex using NIR light. J Am Chem Soc 128:3831–3837
123. Wecksler SR, Mikhailovsky A, Korystov D, Buller F, Kannan R, Tan L-S, Ford PC (2007) Single- and two-photon properties of a dye-derivatized Roussin's red salt ester (Fe2(micro -RS)2(NO)4) with a large TPA cross section. Inorg Chem 46:395–402
124. Zheng Q, Bonoiu A, Ohulchanskyy TY, He GS, Prasad PN (2008) Water-soluble two-photon absorbing nitrosyl complex for light-activated therapy through nitric oxide release. Mol Pharm 5:389–398
125. Hishikawa K, Nakagawa H, Furuta T, Fukuhara K, Tsumoto H, Suzuki T, Miyata N (2009) Photoinduced nitric oxide release from a hindered nitrobenzene derivative by two-photon excitation. J Am Chem Soc 131:7488–7489
126. Nikolenko V, Yuste R, Zayat L, Baraldo LM, Etchenique R (2005) Two-photon uncaging of neurochemicals using inorganic metal complexes. Chem Commun (Camb) 1752–1754
127. Nicolaou VKC, Dai WM, Wendeborn SV, Smith AL, Torisawa Y, Maligres P, Hwang CK (1991) Enediyne compounds with acid, base, and light sensitive trigger groups. Chemical simulation of dynemicin A reaction cascade. Angew Chem 103:1034–1038
128. Wender PA, Zercher CK, Beckham S, Haubold EM (1993) A photochemically triggered DNA cleaving agent: synthesis, mechanistic and DNA cleavage studies on a new analog of the anti-tumor antibiotic dynemicin. J Org Chem 58:5867–5869
129. Basak A, Bdour HM, Shain JC, Mandal S, Rudra KR, Nag S (2000) DNA-cleavage studies on N-substituted monocyclic enediynes: enhancement of potency by incorporation of intercalating or electron poor aromatic ring and subsequent design of a novel phototriggerable acyclic enediyne. Bioorg Med Chem Lett 10:1321–1325
130. Urdabayev NK, Poloukhtine A, Popik VV (2006) Two-photon induced photodecarbonylation reaction of cyclopropenones. Chem Commun (Camb) 454–456
131. Poloukhtine A, Popik VV (2006) Two-photon photochemical generation of reactive enediyne. J Org Chem 71:7417–7421
132. Urdabayev NK, Popik VV (2004) Wolff rearrangement of 2-diazo-1(2H)-naphthalenone induced by nonresonant two-photon absorption of NIR radiation. J Am Chem Soc 126:4058–4059
133. Dyer J, Jockusch S, Balsanek V, Sames D, Turro NJ (2005) Two-photon induced uncaging of a reactive intermediate. Multiphoton in situ detection of a potentially valuable label for biological applications. J Org Chem 70:2143–2147
134. Goodwin AP, Mynar JL, Ma Y, Fleming GR, Frechet JMJ (2005) Synthetic micelle sensitive to IR light via a two-photon process. J Am Chem Soc 127:9952–9953
135. Mynar JL, Goodwin AP, Cohen JA, Ma Y, Fleming GR, Frechet JMJ (2007) Two-photon degradable supramolecular assemblies of linear-dendritic copolymers. Chem Commun (Camb) 2081–2082
136. Babin J, Pelletier M, Lepage M, Allard J-F, Morris D, Zhao Y (2009) A new two-photon-sensitive block copolymer nanocarrier. Angew Chem Int Ed 48:3329–3332, S/1–S/5
137. Belfield KD, Bondar MV, Liu Y, Przhonska OV (2003) Photophysical and photochemical properties of 5,7-dimethoxycoumarin under one- and two-photon excitation. J Phys Org Chem 16:69–78
138. Kim HC, Kreiling S, Greiner A, Hampp N (2003) Two-photon-induced cycloreversion reaction of coumarin photodimers. Chem Phys Lett 372:899–903
139. Kim HC, Haertner S, Hampp N (2008) Single- and two-photon absorption induced photocleavage of dimeric coumarin linkers: therapeutic versus passive photocleavage in ophthalmologic applications. J Photochem Photobiol A 197:239–244
140. Echevarria W, Leite MF, Guerra MT, Zipfel WR, Nathanson MH (2003) Regulation of calcium signals in the nucleus by a nucleoplasmic reticulum. Nat Cell Biol 5:440–446
141. Milburn T, Matsubara N, Billington AP, Udgaonkar JB, Walker JW, Carpenter BK, Webb WW, Marque J, Denk W, McCray JA, Hess GP (1989) Synthesis, photochemistry, and biological activity of a caged photolabile acetylcholine receptor ligand. Biochemistry 28:49–55
142. Billington AP, Walstrom KM, Ramesh D, Guzikowski AP, Carpenter BK, Hess GP (1992) Synthesis and photochemistry of

photolabile *N*-glycine derivatives and effects of one on the glycine receptor. Biochemistry 31:5500–5507

143. Wieboldt R, Gee KR, Niu L, Ramesh D, Carpenter BK, Hess GP (1994) Photolabile precursors of glutamate: synthesis, photochemical properties, and activation of glutamate receptors on a microsecond time scale. Proc Natl Acad Sci USA 91: 8752–8756

Part II

Imparting Light Sensitivity on Cells Using Photosensitive Proteins

Chapter 5

Introduction to Part II: Natural Photosensitive Proteins

James J. Chambers and Richard H. Kramer

Abstract

Light sensing in nature has had the benefit of evolutionary time to optimize light-sensing and photoresponsive molecular systems. Molecular biologists and neurobiologists have recently discovered that these optimized systems can be adapted for very specific purposes and to fill a need in the neuroscience community. Myriad research groups have now employed natural photosensitive proteins in experiments where precise control of neuronal firing is desired without the use of traditional electrodes or perfusion devices. More recently, complementary light-sensing systems have been introduced that also allow researchers to apply light from afar that silences neuronal activity.

Key words: Light-sensing, Rhodopsin, Channelrhodopsin, Halorhodopsin, Melanopsin, Photosensitive proteins

Less than a decade ago, Nagel and colleagues discovered channelrhodopsin-2 (ChR2), a directly light-sensitive cation channel from the algae Chlamydomonas. After the announcement of their findings, at least a few astute neurobiologists recognized that ChR2 could potentially be employed as a very useful tool for direct photostimulation of neurons. If successful, the use of ChR2, coupled with a remote light source, could allow researchers to control neuronal excitability without the need for electrodes. However, there were two hurdles that first needed to be overcome: (1) the gene encoding ChR2 should be expressed and trafficked to the membrane of neurons in sufficiently high quantities and (2) sufficient amounts of the chromophore (all-*trans* retinal) should be available to convert the channelrhodopsin-2 apoprotein (usually called ChOP2) into a light-sensitive holoprotein (ChR2). Electroporation of the ChR2 gene into chick spinal cord neurons was sufficient to allow optical stimulation, but an important technical breakthrough came when Karl Deisseroth and colleagues "humanized" the codon usage of the ChR2 gene to optimize its expression in mammalian cells. Surprising to many, they also found that ordinary cell culture

James J. Chambers and Richard H. Kramer (eds.), *Photosensitive Molecules for Controlling Biological Function*, Neuromethods, vol. 55, DOI 10.1007/978-1-61779-031-7_5, © Springer Science+Business Media, LLC 2011

media had sufficient quantities of the required retinal to render cultured neurons highly sensitive to light, and remarkably, they later found that there is enough chromophore present in the mammalian brain, thus obviating the need for exogenously applied retinal.

ChR2 has several favorable properties that make it an effective tool for optically stimulating neuronal firing. The channels can be activated rapidly in response to visible light, and the channels close quickly at light offset. Thus, very brief pulses of light can trigger trains of neuronal spikes to the maximal firing frequency limit of a neuron. This allows precisely programmed patterns of spike activity, which may be useful for mimicking temporal coding in neural systems, and should be a benefit for studying synaptic events with strict temporal requirements, including spike timing-dependent plasticity.

A variety of gene delivery methods, including transfection, electroporation, viral transduction, and transgenesis, have been used to deliver the ChR2 gene into mammalian neurons successfully. Expression can be limited to a particular neuronal population by including a cell type-specific promoter element upstream of the ChR2 coding sequence. The combination of optical control and genetic targeting has given rise to a new term for this approach, called "optogenetics."

While ChR2 can effectively trigger neuronal firing in response to light, it would further benefit neurobiology to have a complementary tool for inhibiting activity with light. It did not take long for these two research groups to report the discovery and neuronal expression of another light-sensitive protein that suppresses action potentials in response to yellow light (λ_{max} = 570 nm). Halorhodopsin (NpHR), from halobacteria, is a light-driven chloride pump, and like ChR2, it employs all-*trans* retinal as its chromophore. Exogenous expression of NpHR enables light to suppress action potentials rapidly, such that brief flashes can "subtract out" individual action potentials from within a train of spikes. Since the excitation wavelength differs for ChR2 and NpHR, they can be co-expressed and selectively activated in the same neuron, giving the experimenter the ability to employ different wavelengths to increase or decrease the firing rate. So far, NpHR has been applied to far fewer systems than ChR2, largely because of difficulties with achieving high levels of exogenous expression needed for effective light-elicited inhibition. However, investigators are working to improve NpHR expression and reduce intracellular retention by adding an ER export signal and improving the signal peptide sequence.

Other modifications have been made to improve the performance of ChR2. The kinetics, wavelength sensitivity, and trafficking of ChR2 have all been modified through mutagenesis. ChR2 not only activates very rapidly with light, but it also partially inactivates over time, limiting its ability to evoke sustained neuronal activity during prolonged light exposure. To alleviate this problem, mutations have been introduced to convert ChR2 into a bistable switch that can generate sustained depolarization in response to short flashes of light. Chimeric combinations have also been made between ChR2 and ChR1, a related light-activated proton channel, to elucidate determinants of various functional properties. Some chimeras have faster inactivation kinetics, which might enable a faster spike repetition rate in response to brief light flashes, while another combination between ChR2 and a protein with a myosin-binding domain localizes the protein to somato-dendritic regions and excludes it from axons.

An alternate strategy that has also been adopted is to mine the genomes of microbes in the hope of finding light-regulated proteins with more desirable properties. This approach has resulted in the discovery of a light-activated channelrhodopsin homolog from Volvox (VChR1), whose peak activation wavelength is red-shifted by ~70 nm. This enables photostimulation of neurons with yellow light, which penetrates deeper into neural tissue and avoids spectral overlap with Ca^{2+} indicator dyes that are often used for monitoring neural activity. Investigators have also found in the flagellate *Euglena* a photosensitive adenylyl cylase (PAC), which uses as a blue-light sensitive chromophore, flavin adenine nucleotide, a common cofactor found in all cells. PAC can be exogenously expressed in neurons and other cells, and illumination leads to production of cyclic AMP and activation of downstream effects within just tens of milliseconds.

However, natural light-sensitive proteins are present in many species, and in fact, several are present in the retinal photoreceptors in our own eyes. Rhodopsin and related proteins regulate ion channel activity through a G-protein coupled biochemical cascade. These light-sensitive proteins are part of the large family of seven-transmembrane receptors that couple to several different G-proteins to exert myriad effects on cells. Recent work has shown that light sensitivity can be bestowed on two different G-protein signaling systems by generating chimeric combinations between rhodopsin and either the G_q-coupled α_{1a}-adrenergic receptor or the G_s-coupled β_2-adrenergic receptor. Exogenous neuronal expression of these chimeric receptors, called Opto-XRs, resulted in light regulation of the *appropriate signaling cascade*, leading to opposing effects on spike firing. This exciting study demonstrates that photochemical tools can be applied to metabotropic receptors as well, potentially regulating diverse and subtle aspects of electrophysiological function and intracellular signaling.

Chapter 6

Light-Activated Ion Pumps and Channels for Temporally Precise Optical Control of Activity in Genetically Targeted Neurons

Brian Y. Chow, Xue Han, Jacob G. Bernstein, Patrick E. Monahan, and Edward S. Boyden

Abstract

The ability to turn on and off specific cell types and neural pathways in the brain, in a temporally precise fashion, has begun to enable the ability to test the sufficiency and necessity of particular neural activity patterns, and particular neural circuits, in the generation of normal and abnormal neural computations and behaviors by the brain. Over the last 5 years, a number of naturally occurring light-activated ion pumps and light-gated ion channels have been shown, upon genetic expression in specific neuron classes, to enable the voltage (and internal ionic composition) of those neurons to be controlled by light in a temporally precise fashion, without the need for chemical co-factors. In this chapter, we review three major classes of such genetically encoded "optogenetic" microbial opsins – light-gated ion channels such as channelrhodopsins, light-driven chloride pumps such as halorhodopsins, and light-driven proton pumps such as archaerhodopsins – that are in widespread use for mediating optical activation and silencing of neurons in species from *C. elegans* to nonhuman primate. We discuss the properties of these molecules – including their membrane expression, conductances, photocycle properties, ion selectivity, and action spectra – as well as genetic strategies for delivering these genes to neurons in different species, and hardware for performing light delivery in a diversity of settings. In the future, these molecules will not only continue to enable cutting-edge science, but may also support a new generation of optical prosthetics for treating brain disorders.

Key words: Photosensitive proteins, Retinal, Halorhodopsin, Arch, Light-sensitive chloride pump, Photocontrol of behavior, Channelrhodopsin, Archaerhodopsin, Optogenetics

1. Introduction

The ability to turn on and off specific cell types and neural pathways in the brain, in a temporally precise fashion, has begun to enable the ability to test the sufficiency and necessity of particular

James J. Chambers and Richard H. Kramer (eds.), *Photosensitive Molecules for Controlling Biological Function*, Neuromethods, vol. 55, DOI 10.1007/978-1-61779-031-7_6, © Springer Science+Business Media, LLC 2011

neural activity patterns, and particular neural circuits, in the generation of normal and abnormal neural computations and behaviors by the brain. Most electrophysiological and imaging experiments in neuroscience are correlative – comparing a neural signal observed in the brain, to a behavior or pathology. In contrast, the power to manipulate specific cells and circuits is opening up the ability to understand their causal roles in brain functions. Over the last 5 years, a number of naturally occurring light-activated ion pumps and light-activated ion channels have been shown, upon genetic expression in specific neuron classes, to enable the voltage (and internal ionic composition) of those neurons to be controlled by light in a temporally precise fashion. These molecules are microbial (type I) opsins, seven-transmembrane proteins naturally found in archaea, algae, fungi, and other species, and which possess light-driven electrogenic activity or light-gated ion pores. These molecules, when heterologously expressed in neurons or other cells, translocate ions across cell membranes in response to pulses of light of irradiances that are easily achievable with common laboratory microscopes, LEDs, and lasers. These molecules have begun to find widespread use in neuroscience, due to their ease of handling and use (each is a single gene, less than 1-kb long, encoding for a monolithic protein), their lack of need for chemical supplementation in many species (they utilize the naturally occurring chromophore all-*trans* retinal, which appears to occur at sufficient quantities in the mammalian nervous system), and their high speed of operation (they can respond within tens of microseconds to milliseconds, upon delivery of light, and shut off rapidly upon cessation of light, as needed for neuroscience experiments).

Three major classes of such "optogenetic" microbial opsins have been described to date. The first class, channelrhodopsins, is exemplified by the light-gated inwardly rectifying nonspecific cation channel channelrhodopsin-2 (ChR2) from the green algae *Chlamydomonas reinhardtii* (1), which, when expressed in neurons, can mediate sizeable currents up to 1,000 pA in response to millisecond-timescale pulses of blue light (2–5), thus enabling reliable spike trains to be elicited in ChR2-expressing neurons by blue light pulse trains (Fig. 1b). Several additional channelrhodopsins useful to biologists and bioengineers have been discovered or engineered, with faster or slower kinetics, red-shifted activation, and cell region-specific targeting, explored in detail below (6–10). The channelrhodopsins have been used to activate neurons in neural circuits in animals from worms to monkeys, and have proven to be powerful and easy to use. The second class of microbial opsins utilized for biological control to date, halorhodopsins, is exemplified by the light-driven inwardly directed chloride pump halorhodopsin, from the archaeal species *Natronomas pharaonis* (Halo/NpHR/pHR; (11)), which, when

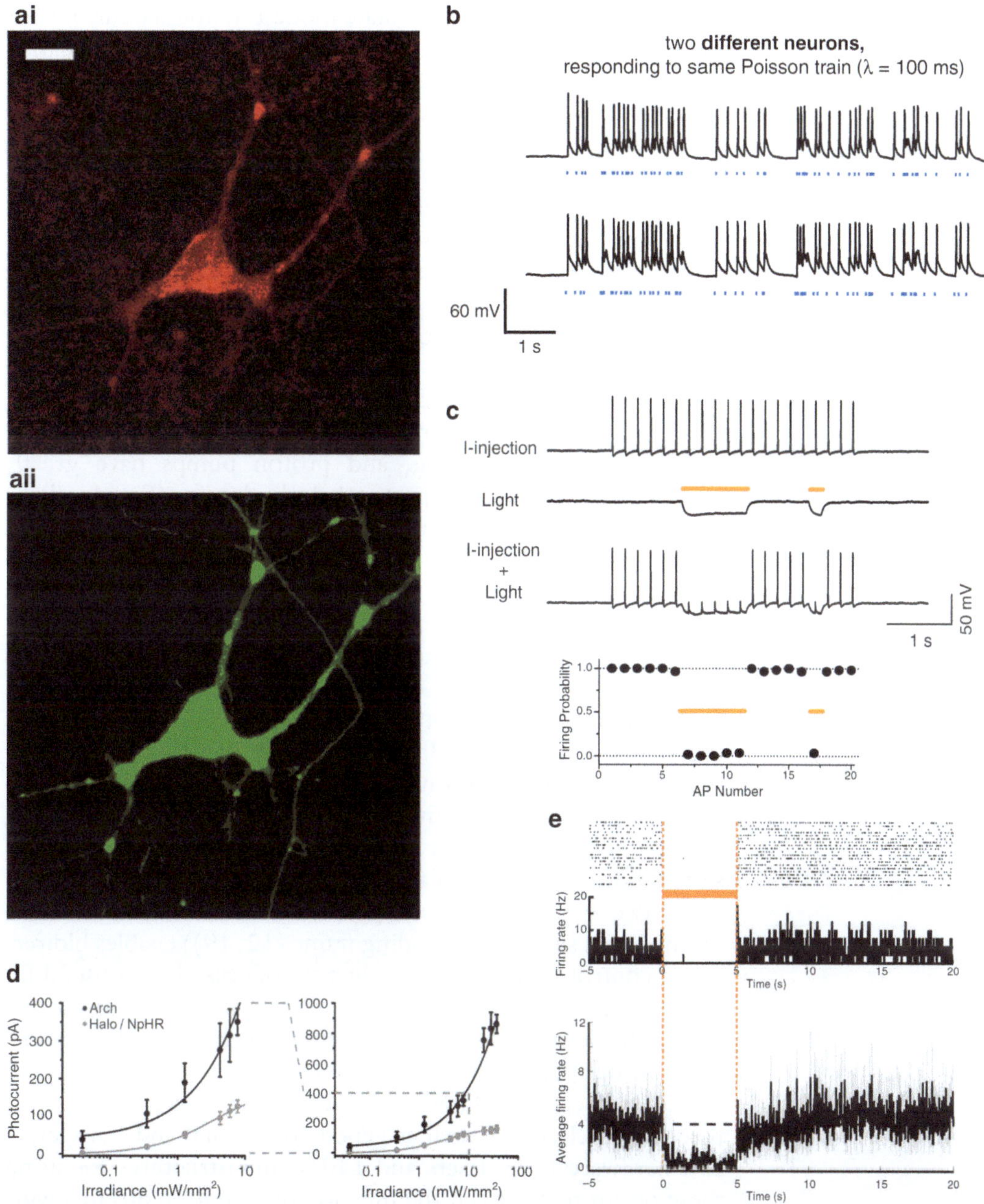

Fig. 1. Three classes of microbial opsins that enable optical neural activation and silencing tools. (**a**) Neuron expressing hChr2-mCherry (**ai**; bar, 20 μm) and Halo-GFP (**aii**). (**b**) Poisson trains of spikes elicited by pulses of blue light (*blue dashes*), in two different neurons. (**c**) Light-driven spike blockade, demonstrated (*top*) for a representative hippocampal neuron, and (*bottom*) for a population of neurons ($n=7$). *I-injection*, neuronal firing induced by pulsed somatic current injection (300 pA, 4 ms). *Light*, hyperpolarization induced by periods of yellow light (*yellow dashes*). *I-injection + light*, yellow light drives Halo to block neuron spiking, leaving spikes elicited during periods of darkness intact. (**a–c**) Adapted from (3) and (12). (**d**) Photocurrents of Arch vs. Halo measured as a function of 575 ± 25 nm light irradiance (or effective light irradiance), in patch-clamped cultured neurons ($n=4$–16 neurons for each point), for low (**i**) and high (**ii**) light powers. The line is a single Hill fit to the data. (**e**) *Top*, Neural activity in a representative neuron before, during, and after 5 s of yellow-light illumination, shown as a spike raster plot (*top*) and as a histogram of instantaneous firing rate averaged across trials (*bottom*; bin size, 20 ms). *Bottom*, population average of instantaneous firing rate before, during, and after yellow-light illumination (*black line*, mean; *gray lines*, mean ± SE; $n=13$ units). (**d–e**) Adapted from (18).

expressed in neurons, can mediate modest inhibitory currents on the order of 40–100 pA in response to yellow-light illumination (12, 13), enabling moderate silencing of neural activity (Fig. 1c). Halorhodopsins have same intrinsic kinetic limitations, with some photocycles taking tens of minutes to complete (e.g., Fig. 4a, b; (12, 14)), and halorhodopsins also require improved trafficking for expression at high levels (15–17). A third class of microbial opsin, the bacteriorhodopsins, is exemplified by the light-driven outward proton pump archaerhodopsin-3 (Arch/aR-3), from the archaeal species *Halorubrum sodomense*. Arch can mediate strong inhibitory currents of up to 900 pA (Fig. 1d), capable of mediating near-100% silencing of neurons in the awake-behaving mouse or monkey brain in response to yellow-green light (Fig. 1e, (15, 18)). Protons are extremely effective as a charge carrier for mediating neural silencing, and proton pumps have greatly improved kinetics with respect to halorhodopsins (Fig. 4c, d), as well as a fast photocycle and efficient trafficking to membranes. Furthermore, outward proton pumps, perhaps surprisingly, do not alter pH to a greater extent than do other opsins (such as ChR2) or than does normal neural activity. The broad and ecologically diverse class of outward proton pumps, which includes blue-green light-drivable outward proton pumps such as the *Leptosphaeria maculans* opsin Mac (Fig. 5), enables, alongside the yellow-red-drivable Halo, multi-color silencing of independent populations of neurons (15). Also, because the neural silencers Halo and Arch are activated by yellow or yellow-green light, and the neural depolarizer ChR2 is driven by blue light, expression of both a silencer and a depolarizer in the same cell (either by using two viruses, or by using the 2A linker to combine two opsins into a single open reading frame (12, 19)) enables bidirectional control of neural activity in a set of cells. This is useful for testing necessity and sufficiency of a given set of neurons in the same animal, or for disruption of neural synchrony and coordination through "informational lesioning" (19).

In the sections below, we describe the properties of these three opsin classes, as well as genetic (e.g., viral and transgenic) and hardware (e.g., lasers and LEDs) infrastructures for using these opsins to parse out the function of neural circuits in a wide variety of animal nervous systems. A theme of this field is that extremely rapid progress and adoption of these technologies have been driven by technology development curves in other fields such as gene therapy and optical imaging. Our hope is to convey a snapshot of this rapidly moving field as of 2009–2010, summarizing the first half-decade of its existence, to teach both neuroengineers hoping to innovate by inventing new tools, and neuroscientists hoping to utilize these tools to answer new scientific questions. We will first survey the general properties of these opsins (Sect. 2), then go into the channelrhodopsins (Sect. 3), followed by the neural silencing pumps (halorhodopsins and bacteriorhodopsins,

Sect. 4), the molecular strategies for delivering these genes to cells for appropriate expression in vitro and in vivo (Sect. 5), and the hardware for illuminating these opsin-expressing neurons in vitro and in vivo (Sect. 6).

2. Properties of "Optogenetic" Microbial (Type I) Opsins

The three classes of molecules described to date are from organisms such as unicellular algae, fungi, and archaea, whose native environments and membrane lipid composition are very different from those of mammalian neurons. Thus, the performance of these molecular tools in neurons can be difficult to predict based solely upon their properties in other species, and must be assessed empirically for assurance of efficacy and safety. Nevertheless, there are several molecular properties that contribute to efficacious, temporally precise optical control of neurons, which can be explored in a unified and logical fashion:

- *Initial protein expression levels*: The efficiency of ribosomal translation of a molecule is largely affected by codon optimization. It is recommended that genes codon-optimized for the target species be used.
- *Membrane insertion properties, protein folding, and interactions with local environment*: Increased membrane localization will result in more functional molecules and thus increased photocurrents. This property may also be inversely associated with the potential property of *toxicity*, since poorly trafficked or folded molecules may aggregate in the cytosol and endoplasmic reticulum. On the contrary, if a molecule has adverse intrinsic side effects, enhanced trafficking may exaggerate them. Furthermore, any given channel or pump will best operate under defined conditions (e.g., chloride conductance, pH, and lipid environment), which may not exist in a given target cell type.
- *Innate conductance and permeability*: Channels translocate more ions per photocycle than pumps, because channels open up a pore in the membrane, unlike pumps. On the other hand, pumps can move ions against concentration gradients. Each opsin furthermore passes a precise set of ions in a specific cellular context, and not others.
- *Photocycle kinetics*: Both light-driven channels and pumps are described by a photocycle, the list of states that a molecule goes through after light exposure, including ion-translocating or ion pore-forming steps. The faster the photocycle, the more temporally precise the molecular function might be, and for a pump, the more ions will be translocated. (For a channel, a faster photocycle may result in the channel entering the ion

pore-forming state more often, but may also reduce the time spent in the open state). If a molecule enters an inactive photocycle state for an enduring period of time, it may be effectively nonfunctional.

- *Photosensitivity*: Molecules may require different amounts of light to begin moving through their photocycle, based on the chromophore absorption efficiency. Furthermore, from an end-user standpoint, effective photosensitivity will appear to be a function of the overall photocycle; for example, a pump that has a slow photocycle may appear to be light insensitive (because incident photons may have no effect on the molecule during the photocycle), whereas a channel that inactivates extremely slowly may appear to be light sensitive (because each photon will result in large charge transfers).
- *Action spectrum*: Different molecules are driven by different colors of light. Multiple cell types can be orthogonally addressed with different colors of light if they express opsins whose action spectra minimally overlap.
- *Ion selectivity*: Unlike traditional electrodes, microbial opsins can generate ion-specific currents, since they will pass specific ions such as chloride (Cl^-) or calcium (Ca^{+2}). This opens up novel kinds of experimental capability, such as the ability to test the sufficiency of a given ion, in a given location, for a given biological function.

We will, in the following sections, frame current knowledge about cell type-specific optical control of neurons, in the context of decades of research in structure–function relationships of microbial (type I, or archaeal) opsins. In many ways, these molecules are similar in tertiary structure to mammalian (type II) rhodopsins (20), the pigments that confer photosensitivity to the rods and cones of the human retina. Both types are composed of seven-transmembrane (7-TM) α-helices, linked by six loop segments, and their photosensitivity is enabled by a retinal bound to a specific lysine residue near the C-terminus, forming a Schiff base that undergoes a *trans-cis* or *cis–trans* isomerization upon illumination, which then induces conformational changes in the protein. However, they are evolutionarily unrelated, and their differences have important implications for their use in perturbing neuronal activity. Mammalian rhodopsins (21) are very sensitive photon detectors, optimized for sensitivity rather than speed. They utilize 11-*cis* retinal as the primary chromophore, which isomerizes to all-*trans* retinal upon absorbing a photon. The resultant structural change activates an associated G-protein, transducin, which then initiates a cascade of secondary messengers. The all-*trans* retinal dissociates from the opsin, is converted back to its 11-*cis* form, and then re-associates with the apoprotein to reconstitute a functional molecule – a process that typically takes hundreds of milliseconds, too slow to enable fast control of neurons in the central nervous system. On the contrary, a microbial opsin utilizes all-*trans* retinal as its chromophore, which isomerizes to 13-*cis* retinal upon absorbing a photon. The chromophore does not undergo a quasi-irreversible dissociation event, but rather thermally relaxes to its active all-*trans* form in the dark (although this process can be facilitated by light). The *trans–cis* isomerization sets off several coupled structural rearrangements within the molecule that accommodate the passive conduction or active pumping of ions (22–24). Ultimately, this means that at the expense of light sensitivity, archaeal opsins can deflect the membrane potential of a cell on the scale of hundreds of microseconds to a few milliseconds. However, it should be noted that genetically targetable, optical neural silencing has also been demonstrated using mammalian G-protein-coupled receptors, which can couple to potassium channels (4), and genetically targetable optical neural activation has been demonstrated using melanopsins and invertebrate-style rhodopsins, at the price of temporal precision (25, 26).

As an exemplar of a well-characterized microbial opsin reagent, with crystal structure and photocycle both well-described, Fig. 2 shows the crystal structure of the light-driven chloride-pump

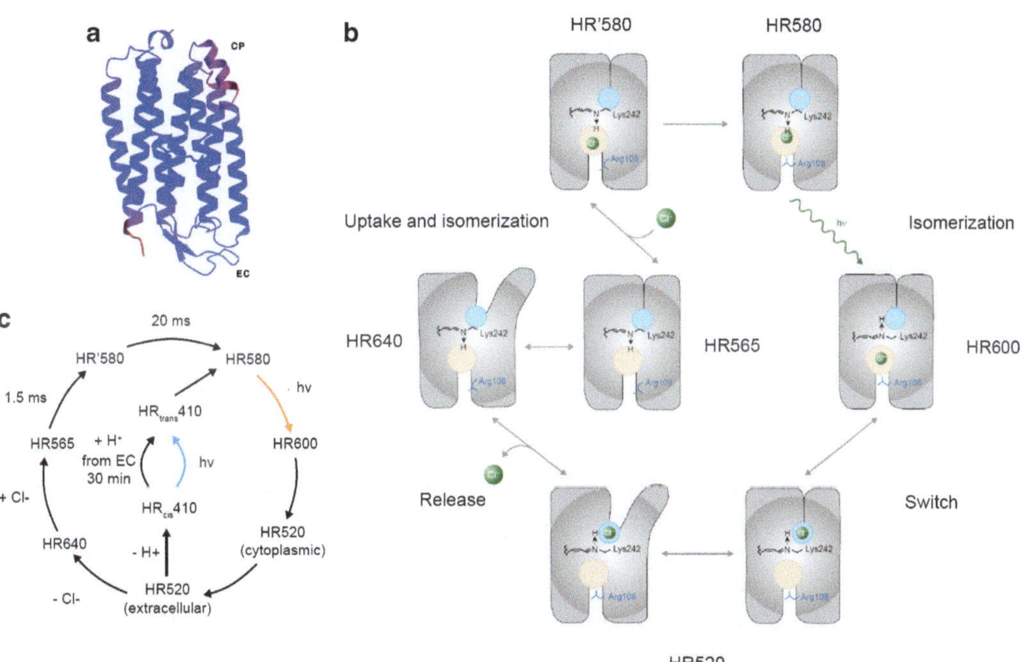

Fig. 2. Structure and function of halorhodopsin. (**a**) Crystal structure of halorhodopsin, which is composed of seven-transmembrane α-helices (7-TM) and a retinal chromophore that forms a Schiff base to a lysine near the C-terminus (from (27)). (**b**) Schematic of the halorhodopsin structural rearrangements and their relation to pumping activity at various points in the photocycle. Image modified from (24). (**c**) The halorhodopsin photocycle at high-power continuous illumination on the timescale of a typical photocycle (i.e., conditions used for neural silencing, >few ms). The HR410 intermediate is the origin of the long-lived inactivation in neural silencing (e.g., (12, 14)).

halorhodopsin (Fig. 2a; (27)), as well as schematized structural rearrangements that are hypothesized to occur as the molecule pumps a chloride ion across the membrane and into the cytoplasm (Fig. 2b; (24)). While it is convenient to consider the molecular tools discussed here as toggle switches for turning neurons on and off, it is critical for use of these opsins to realize that the translocation of ions by microbial opsins is not as simple as that in a two-state toggle switch. The structural rearrangements constitute an active advancement through a complex photocycle with various intermediate states beyond the initial phototransition (Fig. 2c). There exists a very rich literature on type I microbial opsins from an evolutionary and protein structure–function perspective. The canonical molecules include the proton-pumping bacteriorhodopsin (BR), the chloride-pumping halorhodopsin (HR/sHR/HsHR) from *Halobacterium salinarum* (*halobium*), and the halorhodopsin from *N. pharaonis* (Halo/NpHR/pHR). These were some of the first membrane proteins crystallized, and a myriad of structure–function studies have been performed on these molecules (11, 14, 20, 22–24, 27–45). These studies will not be reviewed here, but it is important to point out their existence because much of what we know with respect to the photocycle and structure of opsins comes from these studies and from sequence homology of novel opsins to these canonical molecules. As will be shown later, for example, a deep understanding of the literature has enabled some researchers to derive powerful new variants of channelrhodopsin (5, 6, 8, 10), even though no crystal structure exists for this molecule.

3. Optical Neural Stimulation: Channelrhodopsins

Channelrhodopsins are the primary photoreceptors in the eyespot of the unicellular algae that are responsible for phototactic and photophobic responses (46–48). Their name is derived from the fact that despite the sensory function, the 7-TM segment is in itself a light-activated ion channel. While the channel pore and properties remain poorly understood, it has been realized recently that channelrhodopsins are likely proton pumps like many other microbial opsins, but with a leaky step in the photocycle during which the opsin lets positive charge into cells (49). In *C. reinhardtii*, two separate channelrhodopsins were originally identified (48), one with fast kinetics but poor light sensitivity, channelrhodopsin-1 (ChR1) (50), and another with slower kinetics but improved sensitivity, channelrhodopsin-2 (ChR2) (1). Two more channelrhodopsins from *V. carteri* (VChR1 and VChR2) have also been identified (7, 51), and as will later be discussed, many more ChRs are expected to exist. ChR1-style channelrhodopsins have red-shifted action spectra (peak $\lambda_{ChR1}=500$ nm, $\lambda_{VChR1}=535$ nm) relative to ChR2 (peak $\lambda_{ChR2}=470$ nm), and thus in principle, ChR1-style and ChR2-style opsins could be used together to drive separate sets of neurons with two different colors of light, if suitably spectrally separated opsin pairs could be found.

3.1. Conductance, Permeability, and Context

Channelrhodopsins (abbreviated as ChRs, chops) are light-activated, inwardly rectifying cation channels that are, at neutral pH, permeable to physiologically relevant cations such as H^+, Na^+, K^+, and Ca^{2+}, with permeabilities (relative to sodium) of 1×10^6, 1, 0.5, and 0.1, respectively (1, 6, 46, 50). It is of particular note that the proton conductance (G_{H+}) is 10^6-fold larger than the sodium conductance (G_{Na+}), and thus it is expected that at near physiological pH, perhaps half the photocurrent is carried by protons (46); therefore, ChRs may rapidly equilibrate the intracellular pH with its environment (10). Kinetic selectivity analysis has shown that the mechanism of ion selectivity is likely to be due to differential binding affinity of channelrhodopsin channel residues for different ions, not differential ion transport rates (46).

ChR1 was originally believed to be a selective proton channel (50); however, it was later discovered that the poor photocurrents at mammalian pH were likely attributable to poor membrane localization (6), and the apparent lack of sodium currents in the original report was due to the low pH used to perform experiments in that study; the sodium conductance of ChR1 lessens at low pH (6, 46), unlike that of ChR2 (6). This highlights what will be a recurrent theme throughout this chapter: effective conductance in a heterologous system is determined not only by the innate kinetic and transport properties of the molecule, but also by its trafficking and performance in the environment of the heterologous system.

The single ion channel conductance of ChR2 has been estimated at 50 fS (1), which corresponds to approximately 3×10^4 ions per second, or 300 ions per photocycle event, assuming a 10-ms turnover. This is considerably less than that of a typical voltage-dependent sodium channel that may have a conductance on the order of ~10 pS. It has been estimated from electrophysiological data that 10^5–10^6 membrane-embedded ChR2 molecules are required to cause reliable spiking in cultured rodent hippocampal neurons (52), with saturation of blue-light densities of several mW/mm^2 both in vitro (10) and in vivo (53).

3.2. Basic View of Kinetics and Wavelength Selectivity

Figure 3a shows a typical photocurrent trace from a voltage-clamped neuron expressing ChR2 (top), and the spiking pattern that would result in current-clamp mode (bottom). There is a large transient peak with an opening time constant near 1 ms (1, 6, 10), although photocurrent onset can be measured at <200 μs (1, 3, 54); this transient peak quickly decays to a stationary component

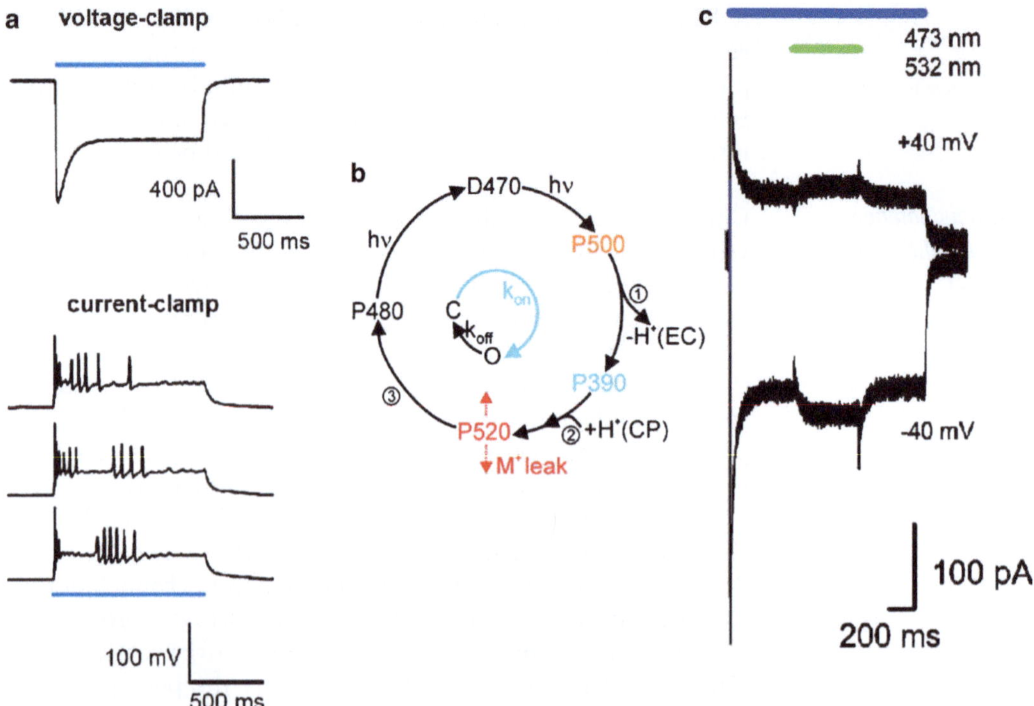

Fig. 3. Channelrhodopsin kinetic and photocycle properties, and impact on neural activity. (**a**) ChR2 currents elicited in a voltage-clamped hippocampal neuron expressing ChR2 and illuminated by blue light (*top*), and ChR2-driven spikes elicited in a current-clamped hippocampal neuron (three repetitions of the same blue-light pulse in the same neuron) (*bottom*), under 1 s of blue-light illumination. Adapted from (3). (**b**) The photocycle of ChR2 determined by a combination of spectroscopy, site-directed mutagenesis, and electrophysiology, adapted from (49). The *inner circle* summarizes the effective appearance of the photocycle, an approximation to the outer photocycle. (**c**) The interplay between photocycle, wavelength, and electrophysiological activity. ChR2 is excited with blue light for a brief period, and then a green light is turned on. The photocurrent initially diminishes because the channel is forced to close, but then increases because the green light also pumps the molecule into its most highly efficient state. Image modified from (54).

that is typically <20–50% of the initial peak photocurrent (1, 3, 6, 10). Upon removing the light, ChR2 closes with a time constant of 10–20 ms (1, 6, 10). The transient photocurrent peak is highly dependent on the illumination intensity (51, 55) and history (1, 3, 10); the history dependence results from a desensitization of the transient component that takes ~5 s to recover from in the dark (3). The stationary component on the contrary is less photosensitive and effectively history independent (55). ChR2 absorbs maximally at 460 nm (1, 10), and the action spectra of both temporal components are nearly identical in ChR2.

The large and fast-onset peak enables ChR2-expressing neurons to spike with exquisite temporal precision on the millisecond timescale (Fig. 1b), the timescale of an action potential. However, the large inactivation (or alternatively, the small stationary component) and its slow recovery in the dark, as well as the slow closing rate of ~10–15 ms, ultimately limit the ability to drive reliable spike rates >25 Hz (3, 10) because (1) the stationary photocurrent may be too small to depolarize a neuron to spike threshold sufficiently and (2) the channel cannot physically close quickly enough to enable de-inactivation of sodium channels. It should be noted, though, that many neurons, such as pyramidal cells, seldom fire action potentials at this rate on the individual neuron level (vs. population synchrony or rhythmogenesis).

ChR1-style channelrhodopsins (VChR1 and ChR1) (7, 50) on the contrary demonstrate dramatically faster kinetics than ChR2-style channelrhodopsins (VChR2 and ChR2). The stationary photocurrents of ChR1s are >70% of the peak photocurrents, and the channels open and close approximately two- to threefold faster than does ChR2. Therefore, one would expect that given comparable expression, protein folding, membrane localization, and photosensitivity (i.e., factors contributing to effective conductance), ChR1s would be capable of driving spike rates with greater fidelity than ChR2s. However, poor membrane expression limits the performance of natural ChR1-style channelrhodopsins (7, 50). Chimeras composed of the first five helices of ChR1 and last two helices of ChR2 have been constructed (6, 10, 56), and these new variants exhibit the small inactivation and action spectrum of ChR1, but the overall effective conductance of ChR2. These structure–function studies will be discussed in detail later in this chapter. A point mutant of this chimera dubbed "ChIEF" (based on its composition as a [Ch]annelrhodopsin chimera with an [I]190V substitution with domains swapped between ChR1 helix-[E] and ChR2 helix-[F]), developed by Tsien and coworkers (10), appears to be a highly improved tool for stimulating neurons. Its large stationary photocurrent and very fast channel closing, the latter conferred by the I190V mutation, contribute to far more reliable spiking (up to 100 Hz) than that of ChR2.

Based on the available characterization of the channelrhodopsins from *V. carteri* (7, 51), the general characteristics are similar to those of the analogous molecules in *C. reinhardtii*. VChR2 and ChR2 have nearly identical photocycles and action spectrum (51). VChR1 and ChR1 exhibit similar reduced inactivation, and are both red-shifted from their respective VChR2/ChR2 counterparts. It has been proposed that VChR1 could be used for multicolor optical stimulation in conjunction with ChR2, which is blue-shifted by ~70 nm, but further improvements are likely required for reliable spiking because the VChR1 photocurrents are unfortunately approximately four- to fivefold smaller (7), and also there is significant spectral overlap between VChR1 and ChR2.

3.3. Detailed Models of Kinetics and Wavelength Selectivity

As previously mentioned, it is critical to realize that the translocation of ions by opsins is not as simple as the operation of an on/off switch, but rather these opsins traverse a complex photocycle with various intermediate states beyond the initial opening of the channel.

Figure 3b shows the photocycle of ChRs based on photophysical studies performed primarily by laser flash spectroscopy, physiology, and site-directed mutagenesis (8, 49, 51, 54, 57). Importantly, the intermediates of the photocycles themselves can also undergo photoreactions, and thus they may be optically driven or "short circuited" (54) between photointermediates at much faster rates (Fig. 3c). The ChR2 photocycle begins in its closed dark-adapted state D470 (where the number in the state name corresponds to the peak light absorption, in nanometer, of the molecule in that state). The channel opens when D470 absorbs a photon, after which the molecule will become a green absorbing photoproduct or P-intermediate, P520, via thermal relaxation from short-lived photoproducts. This initial cascade of events takes 0.2–1.5 ms, depending on the transmembrane potential. The open ChR2 can be closed by either optically pumping P520→D470 with green light, or by decaying to P480 (via a yet-to-be-determined intermediate), a process that takes ~6 s. The inactivation toward the stationary photocurrent may be due to molecules making the P520→P480 transition, rather than the optically induced P520→D470 transition that would allow the molecule to open again quickly. Assuming that ChR1 and ChR2 photocycles are topologically similar, e.g., the ChR2 D470 and P480 equate to the ChR1 peaks at 464 and 505 nm, this interpretation of the transient and stationary photocurrents is consistent with the finding that for ChR1, the stationary photocurrent is red-shifted from the transient photocurrent (6).

The various wavelengths of absorption of rhodopsins and their intermediates throughout the photocycle arise from the different conformations of the chromophore and its local environment, which influences the chromophore charge distribution and the protonation of the retinylidene Schiff base. Figure 3c demonstrates this complex interplay in an experiment by Bamberg and coworkers (54). After blue-light-excited ChR2 has reached its steady state, a green-light costimulation is introduced. The photocurrent briefly diminishes because the open channel is forced to close, but the stationary photocurrent quickly improves because many molecules have been pumped back into their highly efficient, peak-producing state. Thus, it is possible that slightly red-shifted or broadband illumination of ChR2 may strike a balance between optimally exciting the dark state (transient component) and repriming the dark state (driving the red-shifted intermediate photoproduct). As will be discussed, optimal silencing with *N. pharanois* halorhodopsin is analogously achieved by using both yellow light to hyperpolarize the neuron and blue light to drive the molecule out of its inactive state (12, 14).

3.4. Mutants and Variants

As previously mentioned, even though no ChR crystal structure exists at the time of this writing, useful structure–function studies have been performed based largely on sequence homology to *H. salinarum* bacteriorhodopsin. The E90Q mutation (57) has increased sodium selectivity (with respect to G_{H+}) vs. wild-type ChR2, and the H134R mutant (5) demonstrates increased conductance by approximately twofold. Various mutations to C128 (8) corresponding to bacteriorhodopsin T90 drastically slow down the rate of ChR2 closure from the open state, thus effectively creating a bistable open P520 state until illuminated with green light. By lengthening the time that ChR2 spends open on a per-photon basis, this mutation effectively decreases the amount of light needed to activate the channel, at the expense of temporal precision.

Chimeras of ChR1 and ChR2 have been constructed by several researchers (6, 10, 56), one of which was that composed of ChR1 helices A–E and ChR2 helices F–G (called abcdeFG, ChEF, or ChR1/$2_{5/2}$ by various investigators). These chimeras displayed the small inactivation of ChR1, but the large photocurrents of ChR2 on account of improved membrane localization and light sensitivity (based on quantitative confocal fluorescence microscopy, (6)). An I190V substitution to ChEF led to the molecule "ChIEF" that is capable of driving more reliable fast spiking due to the much larger stationary current and faster channel closing kinetics after light offset (10). During these studies, it was also discovered that a single point mutation to wild-type ChR1, E87Q, eradicates its pH-dependent spectral shifts and increases inactivation during illumination (56).

The fact that the poor effective conductance of ChR1 can be largely attributed to membrane localization rather than its photophysical properties highlights the importance of considering and improving the trafficking of heterologously expressed molecules. In particular, ChRs are not localized to the outer membrane but rather the eyespot in *C. reinhardtii*, and the membrane composition of the organism is less than 20% phosphoglyceride (58), a primary lipid type in mammalian neurons. As will be discussed later, in the context of halorhodopsins, the use of signaling peptides can improve outer membrane localization and reduce aggregation in the cytosol, endoplasmic reticulum, and Golgi apparatus.

Along similar lines of using signal peptides to alter trafficking, the myosin-binding domain (MBD) peptide promotes subcellular localization of opsins to neuronal dendrites (9). This subcellular localization strategy may prove to be helpful for enabling driving of electrical activity in specific neural compartments, or for high-resolution connectomic mapping in vivo. Two-photon excitation is a powerful laser excitation technique that enables submicron resolution in 3D (59) relatively deep in the brain (~750 μm, or a significant fraction of the thickness of the mouse cortex), but its ability to induce action potentials in a neuron expressing ChR2 is limited by the interplay between molecule density and the extent of optical depolarization with respect to time (60, 61). The probability of inducing an action potential, at low powers that are not destructive to tissue, is relatively low using a traditional raster scan because the fraction of molecules excited at any point in time is small, and most photons that do hit the membrane are wasted (since the open time of ChR2 is long relative to a femtosecond laser photon delivery rate). Thus, with most conventional two-photon laser scanning methods, the aggregate contributions of the serially excited molecules never sufficiently depolarize the whole neuron to spike threshold. However, Rickgauer and Tank have demonstrated that neurons expressing ChR2 can be reliably excited by two-photon microscopy by optimizing the scan pattern to deliver light optimally to the cell membrane, in a fashion that reaches the maximum surface area while minimizing wastage of photons on already-light-driven channelrhodopsin molecules (61).

3.5. Diversity

Unlike microbial rhodopsins from archaea, significantly less is known about the photo-electrogenic molecules of unicellular algae, the only organisms known to date to have naturally occurring light-activated channels. Photoelectric responses have been measured in several green flagellates, as well as in phylogenetically distant cryptophytes (47, 62–64). Interestingly, the two-component phototaxis strategy employed by *C. reinhardtii*, in which the response is mediated by a fast (ChR1) and slow (ChR2) rhodopsin, appears to be general (47), which begs the question whether chimeras of their respective rhodopsins will also result in kinetic improvements and variants with interesting properties. Thus, as more phototaxis-mediating rhodopsins are isolated and sequenced, or as perhaps new depolarizing rhodopsin types are discovered, new molecular tools for controlling neurons will surely emerge.

4. Optical Neural Silencing: Halorhodopsins and Bacteriorhodopsins

Whereas traditional electrodes can stimulate neurons with temporal precision (albeit without cell type specificity), they are incapable of silencing neurons in order to assess their necessity for given neural computations, behaviors, and pathologies. Therefore, there is a large need for spatio-temporally precise methods for optical inhibition of neurons. Inwardly rectifying chloride pumps and outwardly rectifying proton pumps, halorhodopsins (HRs, hops) and bacteriorhodopsins (BRs, bops), respectively, are electrogenic pumps that when heterologously expressed are capable of sufficiently hyperpolarizing a neuron to silence its activity (Fig. 1c–e; (12, 15, 65)). They are thus far known to exist in every kingdom except in animals: archaea (22, 23, 66–68), bacteria (69–75), fungi (76, 77), and algae (78). In addition to their opposite electrophysiological effect, HRs and BRs differ primarily from channelrhodopsins in that their physiological functions are chiefly due to their role as pumps as opposed to operating as passive channels, and thus can translocate ions against concentration gradients (but typically only one ion per photocycle). Much is known about the photocycles and structure–function relationships of HRs and BRs because they have been crystallized (24, 27, 28, 79, 80) and heavily characterized via spectroscopy, mutagenesis, and physiology for decades.

This section will focus on two molecules in particular: *N. pharaonis* halorhodopsin (Halo/NpHR) and *H. sodomense* bacteriorhodopsin (Arch/AR-3), also known as an archaerhodopsin (i.e., a bacteriorhodopsin from the *Halorubrum* genus). Halorhodopsins were shown in 2007 to be capable of mediating modest optical neural hyperpolarizations, and since then have been improved in trafficking to boost their currents (12, 13, 16); bacteriorhodopsins were shown in 2009 to be able to mediate very powerful and kinetically versatile silencing of multiple neural populations with different colors of light (18). We will discuss in the following section "Conductance, permeability, and context" and "Kinetics and wavelength selectivity" of halorhodopsins and bacteriorhodopsins for these two classes separately, followed by a joint discussion of the "Mutants and variants" and genomic "Diversity" in a unified section.

4.1. Halorhodopsins: Conductance, Permeability, and Context

N. pharanois halorhodopsin (NpHR, Halo) is a highly selective, inwardly rectifying chloride pump, which can also conduct larger monovalent anions (81). It has a reversal potential of approximately −400 mV (81), and its chloride dependence of pumping activity (full- and half-saturating chloride concentrations: $[Cl^-]_{saturation}$ = 20 mM, $[Cl^-]_{1/2}$ = 2.5 mM) (82) is appropriate for operation in mammalian cells, whereas *H. salinarum* halorhodopsin is not capable of effective operation in mammalian neurons (15), presumably because of its large chloride dependency: $[Cl^-]_{saturation}$ = 5 M and $[Cl^-]_{1/2}$ = 200 mM (82, 83). In the absence of any signal peptide sequences to improve trafficking and membrane localization, Halo has been reported to generate 40–100 pA of hyperpolarizing current (12, 18, 65), with the differences in measured photocurrents between studies largely attributable to the power and wavelength of excitation used. This photocurrent is approximately 10- to 25-fold less than typical peak depolarizing currents generated by ChR2, highlighting one potential disadvantage inherent to a pump that translocates one ion per photocycle (e.g., Halo) vs. a channel that conducts 300 ions per photocycle (e.g., ChR2). To mediate these currents, there are an estimated ten million membrane-embedded Halo molecules per neuron (as assessed in hippocampal neuron culture). Because the expression levels are so high, Halo is known to form puncta or intracellular blebs, aggregating in the endoplasmic reticulum (ER) and Golgi apparatus (16, 17). These issues are somewhat addressed by attaching trafficking enhancement sequences to the molecule, e.g., a C-terminal ER-export sequence from the KiR2.1 protein (eNpHR), which increases the effective conductance by about 70% by increasing membrane expression (16).

Recently, we discovered that the crux-halorhodopsin (HR from the *Haloarcula* genus) from *Haloarcula marismortui*, canonically known as cHR-5 (68, 84), produces photocurrents similar to Halo with more uniform expression; even when highly overexpressed under high copy number transfection conditions, no puncta or intracellular blebbing is observed (15). This molecule may better express than Halo in vivo, but it remains unknown at this moment whether it will ultimately be more efficacious at altering mammalian behavior, given their statistically insignificant difference in photocurrent. However, the prolactin (Prl) ER-location sequence in conjunction with a signal sequence from an MHC class I antigen triples the Halo photocurrent (15, 18); we are now trying out multiple trafficking sequences in combination to see if they boost current further. However, it is important to note that if halorhodopsins have other side effects that are due to the protein's intrinsic properties – for example, one paper quantitates the significantly altered neuronal capacitance that results from expressing halorhodopsin in neurons in vivo (85) – then boosting expression may only make such side effects worse.

4.2. Halorhodopsins: Kinetics and Wavelength Selectivity

N. pharaonis halorhodopsin is capable of silencing weakly firing neurons on the millisecond timescale with its ~100 pA scale currents, with rapid onset and offset (12), but during long periods of illumination, all halorhodopsins that we have tested so far and that have current (from *Natronomas pharaonis*, *Halorubrum sodomense*, *Haloarcula vallismortis*, *Haloarcula marismortui*, and *Salinibacter ruber*) inactivate by approximately 30% every 15 s of illumination at 1–10 mW/mm^2 of yellow (593 nm) light (Fig. 4a, b; (12, 18)). This slow inactivation stands in contrast to that in ChR2, which responds to light with a large transient peak that decays within seconds, followed by a stable stationary photocurrent. For all of the halorhodopsins named above, recovery in the dark from light-induced inactivation is slow, with a time constant of tens of minutes, as has been described for some halorhodopsins earlier (12, 14, 39) (Fig. 4a, b). This long-lasting inactivation property may hinder the use of halorhodopsins for silencing for prolonged periods, e.g., during repeated behavioral trials. Importantly, for all halorhodopsins investigated, the inactive photoproduct can be driven back to its active pumping state with a short (e.g., subsecond duration) pulse of blue or UV light (12, 14); thus, optimal use of Halo for neural silencing requires both yellow and blue light to be delivered to the same set of neurons, which is possible (86) but can complicate optics setups.

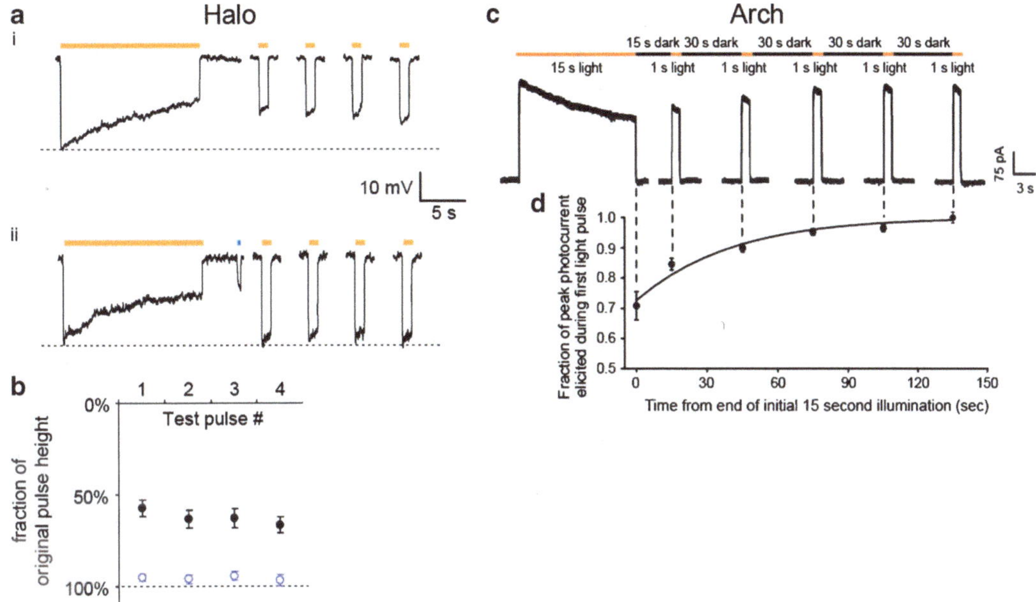

Fig. 4. Kinetic Comparisons between halorhodopsins and archaerhodopsins. (**a**) (**i**) Timecourse of Halo-mediated hyperpolarizations in a representative current-clamped hippocampal neuron during 15 s of continuous yellow light, followed by four 1-s test pulses of yellow light (one every 30 s, starting 10 s after the end of the first 15-s period of yellow light). (**ii**) Timecourse of Halo-mediated hyperpolarization for the same cell exhibited in (**i**), but when Halo function is facilitated by a 400-ms pulse of blue light in between the 15-s period of yellow light and the first 1-s test pulse. (**b**) Population data for blue-light facilitation of Halo recovery (*n*=8 neurons). Plotted are the hyperpolarizations elicited by the four 1-s test pulses of yellow light, normalized to the peak hyperpolarization induced by the original 15-s yellow-light pulse. *Dots* represent mean ± SEM. *Black dots* represent experiments when no blue-light pulse was delivered (as in Fig. 6.5ai). *Open blue dots* represent experiments when 400 ms of blue light was delivered to facilitate recovery (as in Fig. 6.5aii). (**c**) Raw current trace of a neuron lentivirally infected with Arch, illuminated by a 15-s light pulse (575 ± 25 nm, irradiance 7.8 mW/mm^2), followed by 1-s test pulses delivered starting 15, 45, 75, 105, and 135 s after the end of the 15-s light pulse. (**d**) Population data of averaged Arch photocurrents (*n*=11 neurons) sampled at the times indicated by the *vertical dotted lines* that extend into Fig. 4c.

The Halo photocycle is shown in schematic form in Fig. 2c. The time constants listed are the limiting ones, with all other transitions <100 μs. It should be noted that the names for canonical spectroscopic states of *H. salinarum* halorhodopsin – the K, L, N, and O photointermediates – have not been used here because the order of the N and O states in *N. pharaonis* halorhodopsin is still somewhat debated (42, 87, 88). In the dominant photocycle, Halo absorbs a photon and then within tens of microseconds, quickly releases a chloride ion into the cytoplasm during the HR520→H640 transition, via short-lived intermediates that "switch" the chloride location within the molecule from the extracellular loading domain to the cytoplasmic release domain. The molecule then re-isomerizes from the HR640 state and takes up a chloride ion from the extracellular side, a process that takes

~1.5 ms; it then forms the HR' state, which finally relaxes to the active ground state with a time constant of ~20 ms. However, halorhodopsins can enter an alternate photocycle (middle trajectory within Fig. 2c), most notably under prolonged or bright illumination, as might occur during in vivo neural silencing. The 13-*cis* retinylidene Schiff base becomes deprotonated (releasing a proton into the cytoplasm and thus introducing a small depolarizing proton current) and forms a long-lived intermediate HR410 (14, 39). In the dark, the halorhodopsin will remain in this inactive state for a duration on the order of 30 min (39). This formation of HR410 is the origin of the long inactivation observed in neurons expressing halorhodopsin and the ability to recover the active state using the short blue light pulse (12, 14). In contrast, as will be discussed in detail in the next section, archaerhodopsins (bacteriorhodopsins from the *Halorubrum* genus) spontaneously recover in the dark under physiological conditions.

4.3. Bacteriorhodopsins: Conductance, Permeability, and Context

Arch, canonically known as archaerhodopsin-3 (AR-3) from *H. sodomense*, is a yellow-green light-sensitive, outwardly rectifying proton pump with nearly an order of magnitude increase in hyperpolarizing current over any characterized natural halorhodopsin (15, 18), attaining neuronal currents up to 900 pA in response to light powers easily achievable in vitro or in vivo. The efficacy of these proton pumps is surprising, given that protons occur, in mammalian tissue, at a millionfold lower concentration than the ions carried by the optical control molecules described above. This high efficacy may not only be due to the fast photocycle of Arch (see also (89, 90)), but may also be due to the ability of high-pK_a residues in proton pumps to mediate proton uptake (89, 91).

Arch is a highly efficacious tool in vivo, with cortical neurons in the awake-behaving mouse undergoing a median of 97.1% reductions in firing rate for periods of seconds to minutes (Fig. 1e) (18), and safely expresses for months in both mice and monkeys when virally delivered in vivo. Due to the larger currents, Arch enables very large (e.g., order of magnitude scale) increases in addressable volume of silenceable tissue, over earlier reagents. We thoroughly investigated the safety of Arch function. To date, blebbing issues that have affected the usage of halorhodopsins have not been observed in vitro or in vivo for Arch, but membrane trafficking sequences may still prove effective at boosting expression beyond the natural state (the Prl sequence, which greatly magnifies Halo current, slightly increases Arch current in neurons). Furthermore, we have not observed changes in cell membrane capacitance or other passive neural properties, as has been reported with halorhodopsin expression (see above). From an end-user standpoint, illumination of Arch neurons was safe: spike rates measured in vivo were not significantly different before vs. after periods of optical neural silencing. Biophysically, pH changes in neurons expressing Arch and undergoing illumination were minimal, plateauing rapidly at alkalinizations of 0.1–0.15 pH units; the fast stabilization of pH_i may reflect a self-limiting influence that rapidly limits proton concentration swings, and may contribute to the safe operation of Arch in neurons, as observed in mice and monkeys. Indeed, the changes in pH observed in cells expressing Arch and being illuminated are comparable in magnitude to those observed during illumination of ChR2-expressing cells (10) (due to the proton currents carried by ChR2 (1, 46)) and are also within the magnitudes of changes observed during normal neural activity (92–95). We have observed that other archaerhodopsins from other *Halorubrum* strains are also particularly powerful molecular reagents (work in progress). In contrast, the canonical *H. salinarum* bacteriorhodopsin, well known to function poorly in *E. coli*, successfully produced modest photocurrents in mammalian neurons, which highlights the importance of not assuming that all molecules will express and traffic in the same manner in different organisms. (In contrast, *E. coli* does not support any detectable expression of ChR2, which expresses well in neurons; target species' influences in modulation of opsin function should not be underestimated).

4.4. Bacteriorhodopsins: Kinetics and Wavelength Selectivity

Unlike all of the halorhodopsins we have screened to date (including not only the natural halorhodopsins described above, but also products of halorhodopsin site-directed mutagenesis aimed at improving kinetics), which after illumination remained inactivated for tens of minutes, Arch spontaneously recovers its function in seconds in the dark (Fig. 4c, d), more like the light-gated cation channel channelrhodopsin-2 (ChR2) than like halorhodopsins. This feature is particularly useful for in vivo behavior work because it dramatically simplifies the necessary optical hardware; the need to use only one wavelength of light also increases the available bandwidth for multicolor silencing in multiple cell types. We have also observed this spontaneous recovery with other archaerhodopsins, and thus it may be a general feature of archaerhodopsins as a whole (work in progress).

Arch is maximally excited with green-yellow light ($\lambda = 561$ nm), a fairly common peak wavelength for proton pumps. Thus it is backwards compatible with halorhodopsin-driving equipment. Proton pumps naturally exist that are activated by many colors of light, in contrast to chloride pumps, which are primarily driven by yellow-orange light (even with significant mutagenesis of retinal-flanking residues, (15)). The light-driven proton pump from *L. maculans*, abbreviated Mac, has a strongly blue-shifted action spectrum relative to that of the light-driven chloride pump Halo (Fig. 5a). We found that Mac-expressing neurons could undergo 4.1-fold larger hyperpolarizations with blue light than with red light,

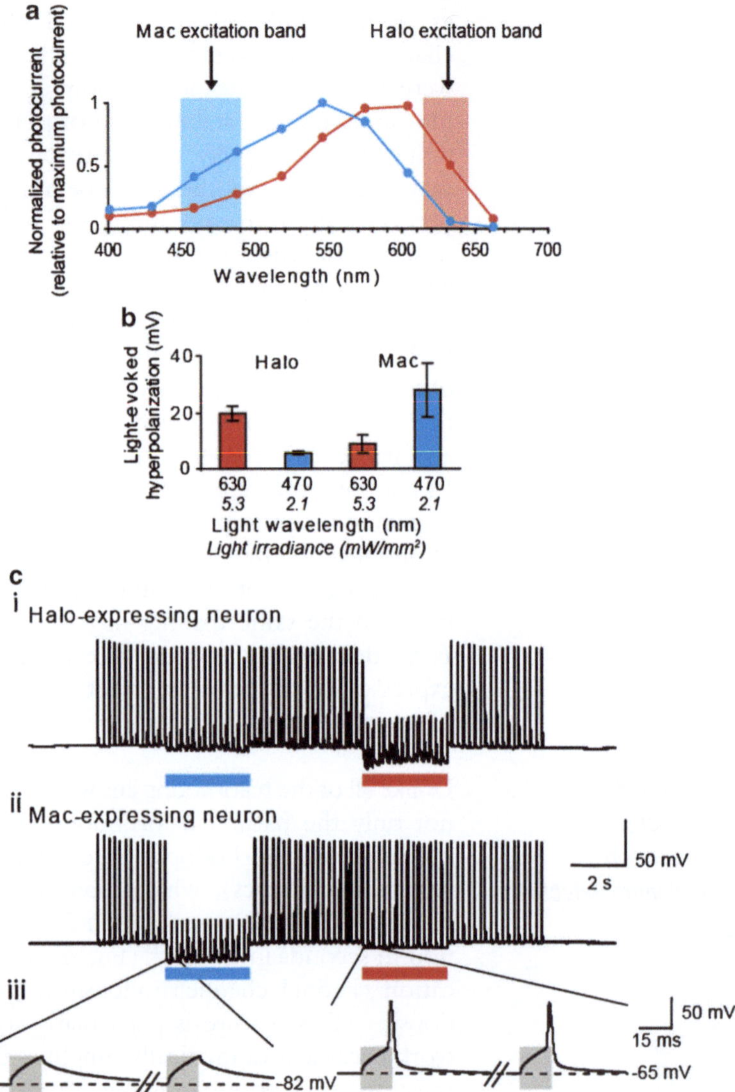

Fig. 5. Multicolor silencing of two neural populations, enabled by blue- and red-light drivable ion pumps of different classes. (**a**) Action spectra of Mac vs. Halo; *rectangles* indicate filter bandwidths used for multi-color silencing in vitro. Blue-light power is via a 470 ± 20 nm filter at 5.3 mW/mm^2, and red-light power is via a 630 ± 15 nm filter at 2.1 mW/mm^2. (**b**) Membrane hyperpolarizations elicited by blue vs. red light, in cells expressing Halo or Mac (n = 5 Mac-expressing neurons, n = 6 Halo-expressing neurons). (**c**) Action potentials evoked by current injection into patch-clamped cultured neurons transfected with Halo (**i**) were selectively silenced by the red light but not by the blue light, and vice versa in neurons expressing Mac (**ii**). *Gray boxes* in *inset* (**iii**) indicate periods of patch clamp current injection.

and Halo-expressing neurons could undergo 3.3-fold larger hyperpolarizations with red light than with blue light, when illuminated with appropriate powers and filters (Fig. 5b). Accordingly, we could demonstrate selective silencing of spike firing in Mac-expressing neurons in response to blue light, and selective silencing of spike firing in Halo-expressing neurons in response to red light (Fig. 5c). Thus, the spectral diversity of proton pumps points the way towards independent multicolor silencing of separate neural populations. This result opens up novel kinds of experiment, in which, for example, two neuron classes, or two sets of neural projections from a single site, can be independently silenced during a behavioral task.

Figure 6 shows the photocycle of the canonical *H. salinarum* bacteriorhodopsin, as representative of the photocycle of the class of proton pumps (the photocycles of archaerhodopsins are not as well characterized and may well be different) (22, 91). The data that have led to the synthesis of the modern model of the bacteriorhodopsin photocycle provide many of the insights that have led to solid investigations into this field as a whole. As in Figs. 2c and 3b, it has been simplified to represent the dominant photocycle expected at large and continuous illumination on the timescale of a typical photocycle (e.g., many milliseconds, as

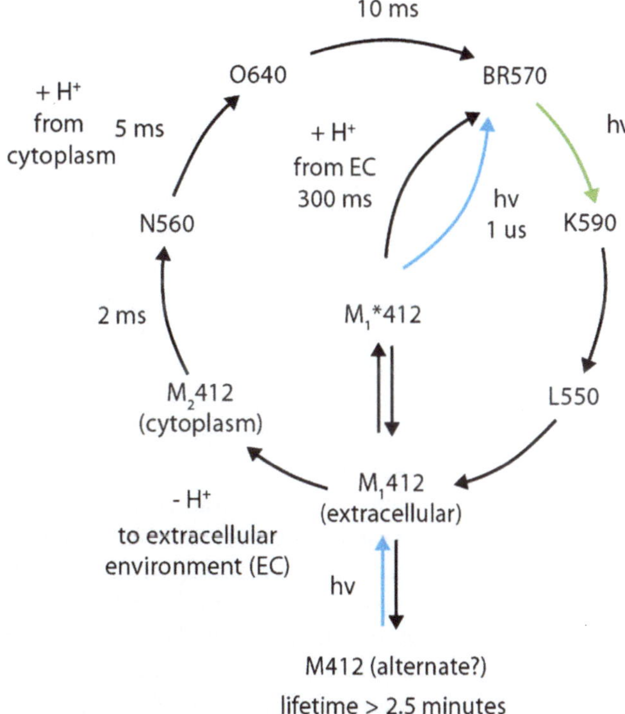

Fig. 6. The photocycle of the *H. salinarum* bacteriorhodopsin. As in Figs. 2c and 3b, the photocycle has been simplified to reflect the dominant photocycle at large continuous illumination on the timescale of a typical photocycle (i.e., conditions used for neural silencing, >few ms). The M412 alternate intermediate is the origin of long-lived inactivation with bacteriorhodopsin. In contrast, Arch spontaneously quickly recovers from this state in the dark.

would be used for neural silencing). Upon absorbing a photon, bacteriorhodopsin forms its L550 intermediate (L = lumi) within microseconds, after quickly transferring a proton from the retinylidene Schiff base (at the Lys-216 position) to the Asp-85 proton acceptor. This transfer triggers the proton releasing group (PRG), containing Glu-204 and Glu-194, to release its own proton (96, 97), during the L→M transition. After a "switching" step during which the Schiff base reorients itself on the cytoplasmic side, it is then reprotonated during the M→N transition via the Asp-96 donor residue, which in turn picks up a proton from the cytoplasm in the next transition (N→O). The chromophore re-isomerizes to the all-*trans* state during this transition as well. Finally, the bacteriorhodopsin relaxes back to its ground state as the proton release group is reloaded from the Asp-85 residue. Like halorhodopsin, bacteriorhodopsin can become trapped in a long-lasting light-unresponsive state (Fig. 6, "M412 alternate") that requires blue light to re-enter the normal photocycle; this could partly explain the lower currents observed with BR when compared to that in Arch.

4.5. Halorhodopsins and Bacteriorhodopsins: Mutants and Variants

For decades, researchers have been making mutants of bacteriorhodopsins and halorhodopsins for structure–function studies reviewed in many earlier publications (e.g., (22, 24, 91)). However, point mutations that improve these molecules as neural silencing tools have yet to be reported. Given that the conductance of a pump is limited by the fact that they move only one ion per photocycle, it would be highly desirable to mutate halorhodopsin into a channel, or channelrhodopsin into an anion channel. Tuning the action spectrum of both bacteriorhodopsins and halorhodopsins may be achieved by mutating the retinal-flanking residues (74, 75, 98, 99). Mutations that alter ion selectivity, such as the bacteriorhodopsin D85T mutation that converts bacteriorhodopsin into a chloride pump (100), could allow ion-specific currents to be mimicked. A Ca^{2+} selective pump, for example, would have powerful impact on enabling powerful studies of plasticity, synaptic transmission, and cellular signaling. Given that crystal structures for many bacteriorhodopsins and halorhodopsins exist, we anticipate that function-oriented and applied site-directed mutagenesis will be a highly active field of research in neuroengineering. The fact that archaerhodopsins on the whole represent a class of molecules that express particularly well in mammalian cells possibly indicates properties of their lipid-interacting residues, or perhaps "signal sequence"-like activity by their loop regions. The known crystal structures of these molecules (79, 80) may provide key insights into the trafficking of heterologously expressed molecules and their membrane insertion.

4.6. Diversity

Light-activated proton and chloride pumps are known to exist in far more organisms than do light-gated cation channels, and proton pumps are particularly prevalent, as all opsins described to date likely have at least some proton pumping capability (22, 23, 66–78, 101, 102). Even though many proton pumps maximally absorb blue-green to green wavelengths (which opens up, as shown in Fig. 5, the possibility, alongside yellow light-driven chloride pumps, for multicolor silencing of distinct neural populations), light-activated hyperpolarizing currents carried by protons have been observed across the whole visible spectrum, from deep blue (via a sensitizer) (103) to far red (104) (>650 nm), although the red light-sensitive current likely originates from a receptor that triggers an H^+-ATPase, as opposed to direct light-mediated ion translocation.

The discovery or creation of a purely genetically encoded light-activated inhibitory channel would be highly desirable. In addition to seeking natural molecules and creating site-directed mutants, linking a non-light-gated ion channel to a type I archaeal rhodopsin may be another promising approach to creating such a molecular tool (105). In this way, a light-activated shunt could be created that would more closely mimic natural mechanisms of neural inhibition in the brain.

5. Molecular Targeting of Microbial Opsins to Different Cell Types

The number of papers on optical neural control in species ranging from *C. elegans* to mouse to nonhuman primate is increasing exponentially each year, and so we will not attempt to review the literature comprehensively. In each species, opsins have been used to test the necessity and sufficiency of neurons, cell types, muscles, neural pathways, brain regions, and other entities in behaviors, pathologies, and neural computations. Opsins have proven valuable in exploring neural dynamics in multiple mammalian brain structures as well. We will focus on highlighting principles that govern how to best use these opsins in various settings, from a molecular biology standpoint (this section, Sect. 5) and from a physical–optical standpoint (next section, Sect. 6). From a molecular biology standpoint, these opsins can be delivered to neurons in almost any conventional way that genes are delivered into cells or into organisms. Transgenic mice have been made with ChR2, for example (106); mice and monkeys have been injected with lentiviruses, adeno-associated viruses (AAV), and other viruses encoding for light-gated proteins (2, 52, 107–109); and rodents and chicks have been electroporated in utero with plasmids encoding for light-gated proteins (4, 53, 110). In each case, different parameters of the technique can be selected so as to enable specific cell types, pathways, or regions to be labeled selectively or to express the opsin selectively. Transgenic *C. elegans* expressing ChR2 have been made using conventional methods (5), as have transgenic Drosophila (111) and zebrafish (112). For these latter species, supplementation with all-*trans* retinal may be beneficial, whereas mammalian brains seem to operate microbial opsins without the need for supplementation. (It is possible that in the future, genetically engineering retinal-lacking organisms, such as invertebrates, to produce retinal within their nervous systems may be of use, for example, by expressing within them enzymes that can produce retinal from vitamin precursors (113)).

For mammalian nervous systems, a large variety of possible strategies exists for conveying opsin genes into specific cell types. For example, transgenic mice can be made through BAC transgenic, knock-in, or other methods, but such strategies are not common for other species yet. For viral delivery, cell type-specific promoters can be inserted upstream of the opsin to target various excitatory, inhibitory, and modulatory neurons (e.g., (19, 114–117)); the size of the promoter is limited by the virus type (AAV viruses hold typically 4–4.5 kb total, whereas lentiviruses hold typically 8–10 kb maximum). The surface or coat proteins that a virus bears can also modulate which cell types will take up the virus; for example, lentiviruses may favor excitatory neurons of the cortex, whereas certain AAV serotypes may favor inhibitory cells (118). Lentiviruses can be pseudotyped – fabricated with a coat protein of desired targeting capacity, e.g., with rabies glycoprotein that leads to lentivirus that travels retrogradely (119) – whereas AAVs can be engineered with biotinylation sites that enable, upon streptavidin conjugation, targeting of potentially arbitrary substrates (120). Retroviruses, which preferentially label dividing cells, have been used to deliver ChR2 to newborn neurons of the dentate gyrus of the hippocampus (121). Other viruses such as rabies virus and pseudorabies virus can be used, with unique tracing capabilities including the ability to go retrogradely across multiple synapses (122, 123). Viral particles typically have to be injected directly into the brain, often through stereotactic targeting to a specific brain area, since the potent blood–brain barrier typically precludes systemic delivery of large viral particles (although see (124)). We have recently described a parallel injector array that can deliver viruses into complex three-dimensional configurations (125). Finally, in utero electroporation of ChR2-GFP into embryonic mice at embryonic day ~15.5 has been found to label pyramidal cells selectively in layers 2/3 (53, 110).

Of potential interest is the possibility of a new generation of neural prosthetics, which can accomplish the synthetic neurobiology mission of repairing the nervous system by enabling optical input of information to sculpt neural dynamics and overcome pathology.

To deliver these genes in a safe, efficacious, and enduring way, viruses such as AAV may be valuable; AAV has been used in over 600 human patients in gene therapy clinical trials without a single serious adverse event (126), and has been successfully used with opsin delivery. The ability to control specific targets optically within the brain may enable more potent and side-effect free therapies than possible with existing electrical and magnetic neuromodulation therapies, or with drugs that are often nonspecific and have side effects. Several groups have already prototyped blindness therapies that may enable new approaches to a currently intractable set of disorders, those in which photoreceptors degenerate within the retina (127, 128). To that end, the recent assessment of brain and immune function in nonhuman primates after ChR2 expression and activation, which showed a lack of harmful effects in a preliminary study (107), may pave the way towards new ideas for neural prosthetics for humans.

One strategy that has become widely popular is to inject the brain of a mouse expressing Cre recombinase in a specific cell type, with either a virus that bears an opsin preceded by a lox-flanked stop cassette, or a virus that bears an opsin reversed and flanked by pairs of lox sites in a specific configuration (Fig. 7; (108, 109)). Given the very large number of Cre transgenic mice in existence, and that are being generated, this strategy is likely to be very useful, at least for mice. Transgenic mice that express Cre in extremely

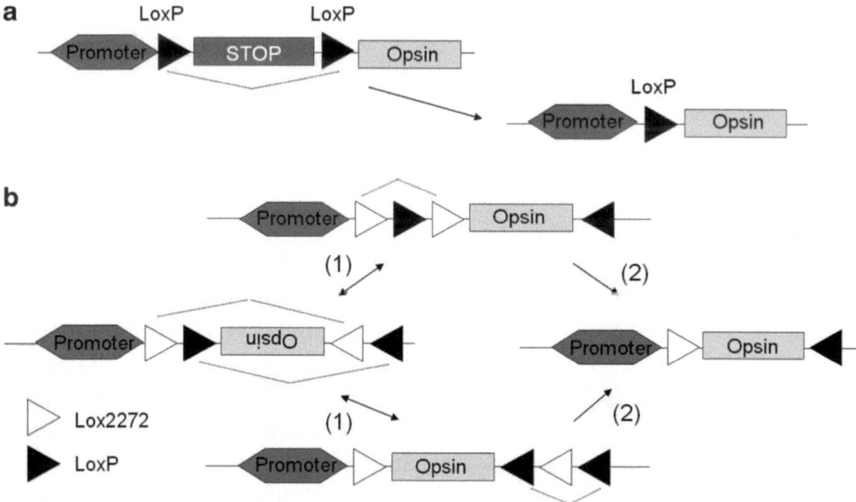

Fig. 7. Transgenic mouse expressing Cre within specific cells, coupled to lox-containing AAV viruses, enables cell type-specific opsin expression. (**a**) A pair of loxP recombination sequences meditate the removal of the transcriptional and translational stop cassette containing multiple poly-adenylation signals (STOP), in the presence of Cre (provided in transgenic mice within specific cell types), to initiate gene expression. (**b**) Two pairs of loxP-type recombination sequences (FLEX) for stable inversion proceed in two steps: (1) inversion followed by (2) excision. loxP and lox2272 are orthogonal recombination sites. (Adapted from (109) and (108)).

cell-specific ways (e.g., through 3′ UTR knockins at the ends of cell type-specific genes) can be made, and then viruses can be rapidly made as new opsins are created, thus enabling cell type-specific expression without requiring the difficult process of trimming cell-specific promoters to fit within the small viral payload, or the difficult process of making transgenic mice for each new opsin tool that is developed (given the rapid pace of innovation, as shown in Sects. 1–4).

6. Hardware for Optical Neural Control

For in vitro use, xenon lamps (e.g., Sutter DG-4, Till Photonics Polychrome) equipped with fast-moving mirrors or monochromators can be used for flexible delivery of fast (millisecond timescale), bright light pulses to biological samples on microscopes. Fluorescence filters can be used to deliver light of the appropriate wavelength (e.g., GFP excitation filter for ChR2, rhodamine or Texas Red excitation filter for Arch, Texas Red excitation filter for Halo, and GFP or YFP excitation filter for Mac). Recently, many companies such as Thorlabs have begun to sell LEDs or LED arrays compatible with microscope fluorescence illuminators, which sell for a small fraction of the price of a full lamp setup. Or fiber-coupled LEDs can simply be placed nearby to the sample (129). Confocal and two-photon microscopes, or more generally scanning laser methodologies, can be used to drive opsins, as have been described in a variety of papers (110, 130, 131). Recently, digital micromirror displays (DMDs) have come forth as potentially useful for photostimulating in complex patterns, comprising millions of individual pixels that can be toggled either on or off (132, 133).

In vivo, the brain scatters light starting within a few hundred microns of an optical source, and absorbs light starting within a few millimeters of distance. Thus, most efforts for in vivo neuromodulation focus on delivering light to a volume of tissue, ranging from a very small volume containing a few hundred cells to a large volume (say, a few cubic millimeters) containing many thousands of cells. One widespread method is to use a laser coupled to an optical fiber (Fig. 8a shows a versatile setup that couples multiple colors of laser light into a single fiber; simpler commercially available single-color laser-coupled fibers with TTL control are also available), and to insert the fiber into a cannula implanted in the brain (Fig. 8b shows a hand-built one for mice; commercial versions from companies such as Plastics One can also be built) or directly into the brain (Fig. 8c shows a setup for monkey), or to couple the fiber to an implanted fiber via a ferrule. An optical commutator (e.g., from Doric Lenses) can be placed between the

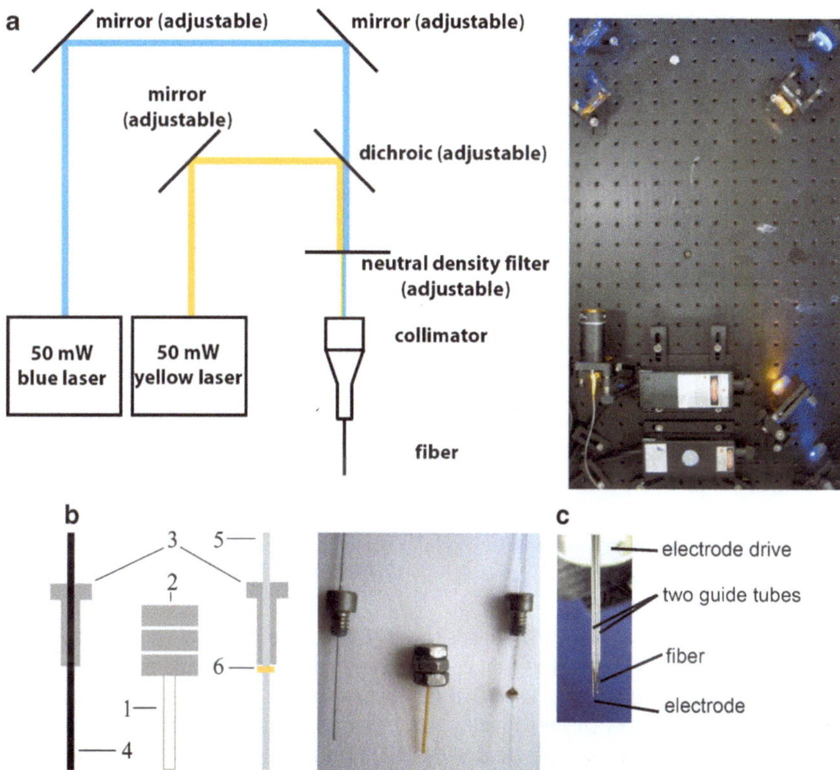

Fig. 8. (**a**) Schematic design (*left*) and picture (*right*) of an optics assembly used to couple blue and yellow laser light into a single optical fiber for in vivo neural modulation. A pictured assembly, lacking a neutral density filter, shows the hardware laid out on a standard optical breadboard. (**b**) Schematic design (*left*) and picture (*right*) of a system for targeting and securing optical fibers within the brain. A polyimide cannula (**1**, 250 μm ID), designed to terminate at the locus of optical modulation, is epoxied to a stack of hex nuts (**2**, sized 2–56) which will be secured to the skull with dental cement. Vented screws (**3**, sized 2–56), which have holes in their centers, screw into the nuts while leaving a path open to the cannula. A dummy wire (**4**, 230-μm stainless steel wire) may be epoxied to the screw to seal the craniotomy when the optics are not in use. An optical fiber (**5**, 230-μm OD silica fiber) is allowed to rotate freely without vertical displacement by a plastic washer (**6**, homemade), which is epoxied to the fiber and sandwiched in between the vented screw above and the cannula below. (**c**) Apparatus for optical activation and electrical recording. Photograph, showing optical fiber (200-μm diameter) and electrode (200-μm shank diameter) in guide tubes. Adapted from (19).

fiber that inserts into the brain and the fiber that connects to the laser, to allow free rotation (65). Arrays of custom-targetable optical fibers, each coupled to a miniaturized light source (e.g., a raw die LED) and targeted to a unique target, will open up the ability to drive activity in complex 3D patterns, enabling the perturbation of complex-shaped structures as well as the ability to perturb targets in a patterned fashion (134, 135).

Aside from the key advantage of being able to manipulate a specific cell type, another key advantage of optical stimulation is the lack of electrical artifact compared to conventional electrical stimulation methods. However, despite the lack of electrical artifact, light does produce a voltage deflection when the electrode tip is illuminated (19, 136). For example, Fig. 9 shows traces recorded

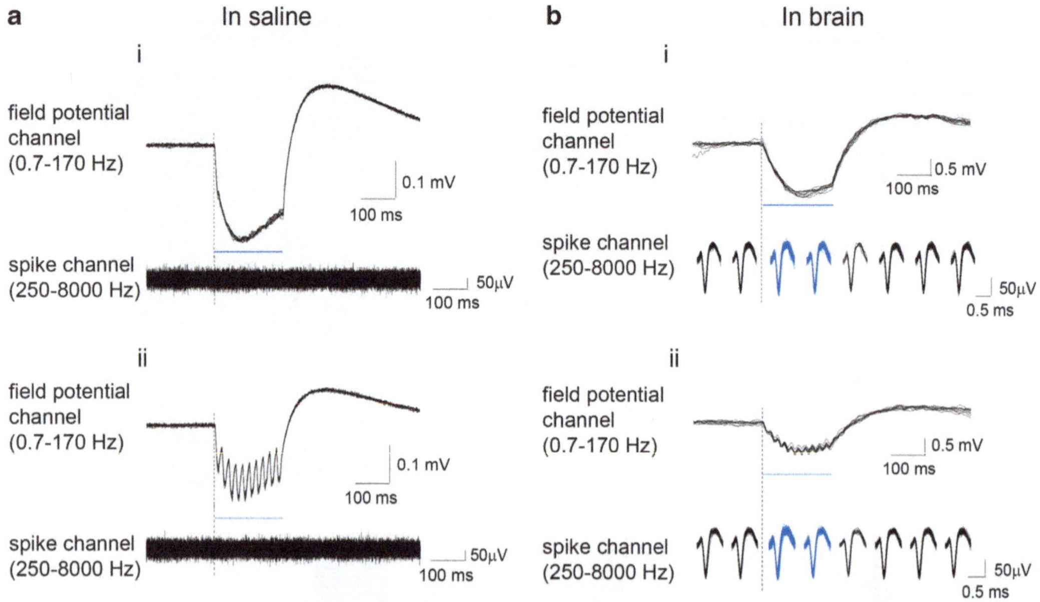

Fig. 9. Voltage deflections observed on tungsten electrodes immersed in saline (**a**) or brain (**b**), upon tip exposure to 200-ms blue-light pulses (**i**) or trains of 10-ms blue-light pulses delivered at 50 Hz (**ii**). Light pulses are indicated by *blue dashes*. Electrode data were hardware filtered using two data acquisition channels operating in parallel, yielding a low-frequency component ("field potential channel") and a high-frequency component ("spike channel"). For the "spike channel" traces taken in brain (**b**), spikes were grouped into 100-ms bins, and then the binned spikes were displayed beneath corresponding parts of the simultaneously acquired "field potential channel" signal (59 and 53 repeated light exposures for **bi** and **bii**, respectively). (Shown are the spikes in eight such bins – the two bins before light onset, the two bins during the light delivery period, and the four bins after light cessation.) For all other signals shown, ten overlaid traces are plotted. Adapted from (19).

on a tungsten electrode in saline illuminated by a pulsed laser beam, as adapted from (19). This voltage deflection slowly evolved over many tens of milliseconds, and accordingly was only recorded when the electrode voltage filtered at 0.7–170 Hz to examine local field potentials (Fig. 9, top traces of each panel). This voltage deflection was not recorded when the electrode voltage was filtered at 250–8,000 Hz to detect spike signals (Fig. 9, bottom traces of each panel). When light illuminated parts of the electrode other than the tip, no artifact was recorded; only illumination of the tip–saline interface resulted in the voltage transient. This phenomenon is consistent with a classical photoelectrochemical finding, the Becquerel effect, in which illumination of an electrode placed in saline can produce a voltage on the electrode (137, 138). Consistent with the generality of the Becquerel effect as a property of electrode–electrolyte interfaces, this artifact is observed on various electrode materials, such as stainless steel, platinum–iridium, silver/silver chloride, gold, nichrome, or copper. Similar slowly evolving voltage deflections were observed when tungsten electrodes were used to record neural activity in the brain during optical stimulation. Because the optical artifact was slowly evolving over many tens of milliseconds, spike waveforms were detected without corruption by the artifact. However, local field potentials and field oscillations, which reflect coherent neural dynamics in the range of Hertz to tens of Hertz, may be difficult to isolate from this Becquerel artifact using the electrodes tested here. Notably, we have not seen the artifact with pulled glass micropipettes (such as previously used in (3) and (12), or in the mouse recordings with pulled glass pipettes in (107)). Thus, for recordings of local field potentials and other slow signals of importance for neuroscience, hollow glass electrodes may prove useful.

References

1. Nagel G et al (2003) Channelrhodopsin-2, a directly light-gated cation-selective membrane channel. Proc Natl Acad Sci U S A 100:13940–13945
2. Ishizuka T, Kakuda M, Araki R, Yawo H (2006) Kinetic evaluation of photosensitivity in genetically engineered neurons expressing green algae light-gated channels. Neurosci Res 54:85–94
3. Boyden ES, Zhang F, Bamberg E, Nagel G, Deisseroth K (2005) Millisecond-timescale, genetically targeted optical control of neural activity. Nat Neurosci 8:1263–1268
4. Li X et al (2005) Fast noninvasive activation and inhibition of neural and network activity by vertebrate rhodopsin and green algae channelrhodopsin. Proc Natl Acad Sci U S A 102:17816–17821
5. Nagel G et al (2005) Light activation of channelrhodopsin-2 in excitable cells of *Caenorhabditis elegans* triggers rapid behavioral responses. Curr Biol 15:2279–2284
6. Wang H et al (2009) Molecular determinants differentiating photocurrent properties of two channelrhodopsins from *Chlamydomonas*. J Biol Chem 284:5685–5696
7. Zhang F et al (2008) Red-shifted optogenetic excitation: a tool for fast neural control derived from *Volvox carteri*. Nat Neurosci 11:631–633
8. Berndt A, Yizhar O, Gunaydin LA, Hegemann P, Deisseroth K (2009) Bi-stable neural state switches. Nat Neurosci 12:229–234
9. Lewis TL Jr, Mao T, Svoboda K, Arnold DB (2009) Myosin-dependent targeting of transmembrane proteins to neuronal dendrites. Nat Neurosci 12:568–576
10. Lin JY, Lin MZ, Steinbach P, Tsien RY (2009) Characterization of engineered channelrhodopsin variants with improved properties and kinetics. Biophys J 96:1803–1814
11. Lanyi JK, Duschl A, Hatfield GW, May K, Oesterhelt D (1990) The primary structure of a halorhodopsin from *Natronobacterium pharaonis*. Structural, functional and evolutionary implications for bacterial rhodopsins and halorhodopsins. J Biol Chem 265:1253–1260
12. Han X, Boyden ES (2007) Multiple-color optical activation, silencing, and desynchronization of neural activity, with single-spike temporal resolution. PLoS ONE 2:e299
13. Zhang F et al (2007) Multimodal fast optical interrogation of neural circuitry. Nature 446:633–639
14. Bamberg E, Tittor J, Oesterhelt D (1993) Light-driven proton or chloride pumping by halorhodopsin. Proc Natl Acad Sci U S A 90:639–643
15. Chow BY, Han X, Dobry AS, Qian X, Chuong AS, Li M, Henninger MA, Belfort GM, Lin Y, Monahan PE, Boyden ES (2010) High-performance genetically targetable optical neural silencing by light-driven proton pumps. Nature 463:98–102
16. Gradinaru V, Thompson KR, Deisseroth K (2008) eNpHR: a *Natronomonas* halorhodopsin enhanced for optogenetic applications. Brain Cell Biol 36:129–139
17. Zhao S et al (2008) Improved expression of halorhodopsin for light-induced silencing of neuronal activity. Brain Cell Biol 36:141–154
18. Chow B, Han X, Qian X, Boyden ES (2009) High-performance halorhodopsin variants for improved genetically-targetable optical neural silencing. Front Syst Neurosci. Conference abstract: computational and systems neuroscience. doi:10.3389/conf.neuro.10.2009.03.347
19. Han X, Qian X, Stern P, Chuong AS, Boyden ES (2009) Informational lesions: optical perturbation of spike timing and neural synchrony via microbial opsin gene fusions. Front Mol Neurosci 2:12. doi:10.3389/neuro.02.012.2009

20. Henderson R, Schertler GF (1990) The structure of bacteriorhodopsin and its relevance to the visual opsins and other seven-helix G-protein coupled receptors. Philos Trans R Soc Lond B Biol Sci 326:379–389
21. Palczewski KG (2006) Protein-coupled receptor rhodopsin. Annu Rev Biochem 75:743–767. doi:10.1146/annurev.biochem.75.103004.142743
22. Lanyi JK (2004) Bacteriorhodopsin. Annu Rev Physiol 66:665–688. doi:10.1146/annurev.physiol.66.032102.150049
23. Lanyi JK (1986) Halorhodopsin: a light-driven chloride ion pump. Annu Rev Biophys Biophys Chem 15:11–28. doi:10.1146/annurev.bb.15.060186.000303
24. Essen LO (2002) Halorhodopsin: light-driven ion pumping made simple? Curr Opin Struct Biol 12:516–522
25. Zemelman BV, Lee GA, Ng M, Miesenbock G (2002) Selective photostimulation of genetically chARGed neurons. Neuron 33:15–22
26. Lin B, Koizumi A, Tanaka N, Panda S, Masland RH (2008) Restoration of visual function in retinal degeneration mice by ectopic expression of melanopsin. Proc Natl Acad Sci U S A 105:16009–16014
27. Kolbe M, Besir H, Essen LO, Oesterhelt D (2000) Structure of the light-driven chloride pump halorhodopsin at 1.8 Å resolution. Science 288:1390–1396
28. Luecke H, Schobert B, Richter HT, Cartailler JP, Lanyi JK (1999) Structure of bacteriorhodopsin at 1.55 Å resolution. J Mol Biol 291:899–911
29. Braiman MS, Stern LJ, Chao BH, Khorana HG (1987) Structure–function studies on bacteriorhodopsin. IV. Purification and renaturation of bacterio-opsin polypeptide expressed in *Escherichia coli*. J Biol Chem 262:9271–9276
30. Gilles-Gonzalez MA, Engelman DM, Khorana HG (1991) Structure–function studies of bacteriorhodopsin XV. Effects of deletions in loops B-C and E-F on bacteriorhodopsin chromophore and structure. J Biol Chem 266:8545–8550
31. Mogi T, Stern LJ, Chao BH, Khorana HG (1989) Structure–function studies on bacteriorhodopsin. VIII. Substitutions of the membrane-embedded prolines 50, 91, and 186: the effects are determined by the substituting amino acids. J Biol Chem 264:14192–14196
32. Mogi T, Marti T, Khorana HG (1989) Structure–function studies on bacteriorhodopsin. IX. Substitutions of tryptophan residues affect protein-retinal interactions in bacteriorhodopsin. J Biol Chem 264:14197–14201
33. Mogi T, Stern LJ, Hackett NR, Khorana HG (1987) Bacteriorhodopsin mutants containing single tyrosine to phenylalanine substitutions are all active in proton translocation. Proc Natl Acad Sci U S A 84:5595–5599
34. Marti T et al (1991) Bacteriorhodopsin mutants containing single substitutions of serine or threonine residues are all active in proton translocation. J Biol Chem 266:6919–6927
35. Marinetti T, Subramaniam S, Mogi T, Marti T, Khorana HG (1989) Replacement of aspartic residues 85, 96, 115, or 212 affects the quantum yield and kinetics of proton release and uptake by bacteriorhodopsin. Proc Natl Acad Sci U S A 86:529–533
36. Mogi T, Stern LJ, Marti T, Chao BH, Khorana HG (1988) Aspartic acid substitutions affect proton translocation by bacteriorhodopsin. Proc Natl Acad Sci U S A 85:4148–4152
37. Subramaniam S, Greenhalgh DA, Khorana HG (1992) Aspartic acid 85 in bacteriorhodopsin functions both as proton acceptor and negative counterion to the Schiff base. J Biol Chem 267:25730–25733
38. Brown LS, Needleman R, Lanyi JK (1996) Interaction of proton and chloride transfer pathways in recombinant bacteriorhodopsin with chloride transport activity: implications for the chloride translocation mechanism. Biochemistry 35:16048–16054
39. Hegemann P, Oesterhelt D, Steiner M (1985) The photocycle of the chloride pump halorhodopsin. I: Azide-catalyzed deprotonation of the chromophore is a side reaction of photocycle intermediates inactivating the pump. EMBO J 4:2347–2350
40. Blanck A, Oesterhelt D (1987) The halo-opsin gene. II. Sequence, primary structure of halorhodopsin and comparison with bacteriorhodopsin. EMBO J 6:265–273
41. Rudiger M, Oesterhelt D (1997) Specific arginine and threonine residues control anion binding and transport in the light-driven chloride pump halorhodopsin. EMBO J 16:3813–3821
42. Varo G et al (1995) Light-driven chloride ion transport by halorhodopsin from *Natronobacterium pharaonis*. 1. The photochemical cycle. Biochemistry 34:14490–14499
43. Tittor J et al (1997) Chloride and proton transport in bacteriorhodopsin mutant D85T: different modes of ion translocation in a retinal protein. J Mol Biol 271:405–416
44. Tittor J, Oesterhelt D, Bamberg E (1995) Bacteriorhodopsin mutants D85N, D85T and D85,96N as proton pumps. Biophys Chem 56:153–157

45. Varo G, Needleman R, Lanyi JK (1995) Light-driven chloride ion transport by halorhodopsin from *Natronobacterium pharaonis*. 2. Chloride release and uptake, protein conformation change, and thermodynamics. Biochemistry 34:14500–14507
46. Berthold P et al (2008) Channelrhodopsin-1 initiates phototaxis and photophobic responses in *Chlamydomonas* by immediate light-induced depolarization. Plant Cell 20:1665–1677. doi:10.1105/tpc.108.057919
47. Sineshchekov OA, Govorunova EG, Spudich JL (2009) Photosensory functions of channelrhodopsins in native algal cells. Photochem Photobiol 85:556–563
48. Sineshchekov OA, Jung KH, Spudich JL (2002) Two rhodopsins mediate phototaxis to low- and high-intensity light in *Chlamydomonas reinhardtii*. Proc Natl Acad Sci U S A 99:8689–8694
49. Feldbauer K et al (2009) Channelrhodopsin-2 is a leaky proton pump. Proc Natl Acad Sci U S A 106:12317–12322
50. Nagel G et al (2002) Channelrhodopsin-1: a light-gated proton channel in green algae. Science 296:2395–2398
51. Ernst OP et al (2008) Photoactivation of channelrhodopsin. J Biol Chem 283:1637–1643
52. Zhang F, Wang LP, Boyden ES, Deisseroth K (2006) Channelrhodopsin-2 and optical control of excitable cells. Nat Methods 3:785–792
53. Huber D et al (2008) Sparse optical microstimulation in barrel cortex drives learned behaviour in freely moving mice. Nature 451:61–64
54. Bamann C, Kirsch T, Nagel G, Bamberg E (2008) Spectral characteristics of the photocycle of channelrhodopsin-2 and its implication for channel function. J Mol Biol 375:686–694
55. Nikolic K, Degenaar P, Toumazou C (2006) Modeling and engineering aspects of channelrhodopsin-2 system for neural photostimulation. Conf Proc IEEE Eng Med Biol Soc 1:1626–1629
56. Tsunoda SP, Hegemann P (2009) Glu 87 of channelrhodopsin-1 causes pH-dependent color tuning and fast photocurrent inactivation. Photochem Photobiol 85:564–569
57. Ritter E, Stehfest K, Berndt A, Hegemann P, Bartl FJ (2008) Monitoring light-induced structural changes of channelrhodopsin-2 by UV-visible and Fourier transform infrared spectroscopy. J Biol Chem 283:35033–35041. doi:10.1074/jbc.M806353200
58. Harwood JL, Guschina IA (2009) The versatility of algae and their lipid metabolism. Biochimie 91:679–684
59. Zipfel WR, Williams RM, Webb WW (2003) Nonlinear magic: multiphoton microscopy in the biosciences. Nat Biotechnol 21:1369–1377
60. Mohanty SK et al (2008) In-depth activation of channelrhodopsin 2-sensitized excitable cells with high spatial resolution using two-photon excitation with a near-infrared laser microbeam. Biophys J 95:3916–3926
61. Rickgauer JP, Tank DW (2009) Two-photon excitation of channelrhodopsin-2 at saturation. Proc Natl Acad Sci U S A 106:15025–15030
62. Sineshchekov OA et al (2005) Rhodopsin-mediated photoreception in cryptophyte flagellates. Biophys J 89:4310–4319
63. Sineshchekov OA, Litvin FF, Keszthelyi L (1990) Two components of photoreceptor potential in phototaxis of the flagellated green alga *Haematococcus pluvialis*. Biophys J 57:33–39
64. Litvin FF, Sineshchekov OA, Sineshchekov VA (1978) Photoreceptor electric potential in the phototaxis of the alga *Haematococcus pluvialis*. Nature 271:476–478
65. Gradinaru V et al (2007) Targeting and readout strategies for fast optical neural control in vitro and in vivo. J Neurosci 27:14231–14238
66. Ihara K et al (1999) Evolution of the archaeal rhodopsins: evolution rate changes by gene duplication and functional differentiation. J Mol Biol 285:163–174
67. Mukohata Y, Ihara K, Tamura T, Sugiyama Y (1999) Halobacterial rhodopsins. J Biochem 125:649–657
68. Klare JP, Chizhov I, Engelhard M (2008) Microbial rhodopsins: scaffolds for ion pumps, channels, and sensors. Results Probl Cell Differ 45:73–122
69. Antón J et al (2005) Salinibacter ruber: genomics and biogeography. In: Gunde-Cimerman N, Plemenitas A, Oren A (eds) Adaptation to life in high salt concentrations in archaea, bacteria and eukarya. Springer pp 257–266
70. Balashov SP et al (2005) Xanthorhodopsin: a proton pump with a light-harvesting carotenoid antenna. Science 309:2061–2064
71. Beja O, Spudich EN, Spudich JL, Leclerc M, DeLong EF (2001) Proteorhodopsin phototrophy in the ocean. Nature 411:786–789
72. Beja O et al (2000) Bacterial rhodopsin: evidence for a new type of phototrophy in the sea. Science 289:1902–1906
73. Friedrich T et al (2002) Proteorhodopsin is a light-driven proton pump with variable vectoriality. J Mol Biol 321:821–838

74. Kelemen BR, Du M, Jensen RB (2003) Proteorhodopsin in living color: diversity of spectral properties within living bacterial cells. Biochim Biophys Acta 1618:25–32
75. Kim SY, Waschuk SA, Brown LS, Jung KH (2008) Screening and characterization of proteorhodopsin color-tuning mutations in *Escherichia coli* with endogenous retinal synthesis. Biochim Biophys Acta 1777: 504–513
76. Brown LS (2004) Fungal rhodopsins and opsin-related proteins: eukaryotic homologues of bacteriorhodopsin with unknown functions. Photochem Photobiol Sci 3:555–565
77. Waschuk SA, Bezerra AG, Shi L, Brown LS (2005) Leptosphaeria rhodopsin: bacteriorhodopsin-like proton pump from a eukaryote. Proc Natl Acad Sci U S A 102:6879–6883. doi:10.1073/pnas.0409659102
78. Tsunoda SP et al (2006) H^+-Pumping rhodopsin from the marine alga *Acetabularia*. Biophys J 91:1471–1479
79. Yoshimura K, Kouyama T (2008) Structural role of bacterioruberin in the trimeric structure of archaerhodopsin-2. J Mol Biol 375:1267–1281
80. Enami N et al (2006) Crystal structures of archaerhodopsin-1 and -2: common structural motif in archaeal light-driven proton pumps. J Mol Biol 358:675–685
81. Seki A et al (2007) Heterologous expression of *Pharaonis halorhodopsin* in *Xenopus laevis* oocytes and electrophysiological characterization of its light-driven Cl^- pump activity. Biophys J 92:2559–2569
82. Okuno D, Asaumi M, Muneyuki E (1999) Chloride concentration dependency of the electrogenic activity of halorhodopsin. Biochemistry 38:5422–5429
83. Muneyuki E, Shibazaki C, Wada Y, Yakushizin M, Ohtani H (2002) Cl(−) concentration dependence of photovoltage generation by halorhodopsin from *Halobacterium salinarum*. Biophys J 83:1749–1759
84. Baliga NS et al (2004) Genome sequence of *Haloarcula marismortui*: a halophilic archaeon from the Dead Sea. Genome Res 14:2221–2234
85. Tonnesen J, Sorensen AT, Deisseroth K, Lundberg C, Kokaia M (2009) Optogenetic control of epileptiform activity. Proc Natl Acad Sci U S A 106:12162–12167
86. Bernstein JG et al (2008) Prosthetic systems for therapeutic optical activation and silencing of genetically-targeted neurons. Proc Soc Photo Opt Instrum Eng 6854:68540H
87. Ludmann K, Ibron G, Lanyi JK, Varo G (2000) Charge motions during the photocycle of pharaonis halorhodopsin. Biophys J 78:959–966
88. Chizhov I, Engelhard M (2001) Temperature and halide dependence of the photocycle of halorhodopsin from *Natronobacterium pharaonis*. Biophys J 81:1600–1612
89. Ming M et al (2006) pH dependence of light-driven proton pumping by an archaerhodopsin from Tibet: comparison with bacteriorhodopsin. Biophys J 90:3322–3332
90. Lukashev EP et al (1994) pH dependence of the absorption spectra and photochemical transformations of the archaerhodopsins. Photochem Photobiol 60:69–75
91. Lanyi JK (2006) Proton transfers in the bacteriorhodopsin photocycle. Biochim Biophys Acta – Bioenergetics 1757:1012–1018
92. Bevensee MO, Cummins TR, Haddad GG, Boron WF, Boyarsky G (1996) pH regulation in single CA1 neurons acutely isolated from the hippocampi of immature and mature rats. J Physiol 494(Pt 2):315–328
93. Chesler M (2003) Regulation and modulation of pH in the brain. Physiol Rev 83:1183–1221
94. Meyer TM, Munsch T, Pape HC (2000) Activity-related changes in intracellular pH in rat thalamic relay neurons. NeuroReport 11:33–37
95. Trapp S, Luckermann M, Brooks PA, Ballanyi K (1996) Acidosis of rat dorsal vagal neurons in situ during spontaneous and evoked activity. J Physiol 496(Pt 3):695–710
96. Brown LS et al (1995) Glutamic acid 204 is the terminal proton release group at the extracellular surface of bacteriorhodopsin. J Biol Chem 270:27122–27126
97. Phatak P, Ghosh N, Yu H, Cui Q, Elstner M (2008) Amino acids with an intermolecular proton bond as proton storage site in bacteriorhodopsin. Proc Natl Acad Sci USA 105:19672–19677
98. Henderson R et al (1990) Model for the structure of bacteriorhodopsin based on high-resolution electron cryo-microscopy. J Mol Biol 213:899–929
99. Man-Aharonovich D et al (2004) Characterization of RS29, a blue-green proteorhodopsin variant from the Red Sea. Photochem Photobiol Sci 3:459–462
100. Sasaki J et al (1995) Conversion of bacteriorhodopsin into a chloride ion pump. Science 269:73–75
101. Iwamoto M et al (2004) Proton release and uptake of pharaonis phoborhodopsin (sensory

rhodopsin II) reconstituted into phospholipids. Biochemistry 43:3195–3203
102. Sudo Y, Iwamoto M, Shimono K, Sumi M, Kamo N (2001) Photo-induced proton transport of pharaonis phoborhodopsin (sensory rhodopsin II) is ceased by association with the transducer. Biophys J 80:916–922
103. Boichenko VA, Wang JM, Antón J, Lanyi JK, Balashov SP (2006) Functions of carotenoids in xanthorhodopsin and archaerhodopsin, from action spectra of photoinhibition of cell respiration. Biochim Biophys Acta – Bioenergetics 1757:1649–1656
104. Serrano EE, Zeiger E, Hagiwara S (1988) Red light stimulates an electrogenic proton pump in Vicia guard cell protoplasts. Proc Natl Acad Sci U S A 85:436–440
105. Moreau CJ, Dupuis JP, Revilloud J, Arumugam K, Vivaudou M (2008) Coupling ion channels to receptors for biomolecule sensing. Nat Nanotechnol 3:620–625
106. Wang H et al (2007) High-speed mapping of synaptic connectivity using photostimulation in channelrhodopsin-2 transgenic mice. Proc Natl Acad Sci U S A 104:8143–8148
107. Han X et al (2009) Millisecond-timescale optical control of neural dynamics in the nonhuman primate brain. Neuron 62:191–198
108. Atasoy D, Aponte Y, Su HH, Sternson SM (2008) A FLEX switch targets channelrhodopsin-2 to multiple cell types for imaging and long-range circuit mapping. J Neurosci 28:7025–7030
109. Kuhlman SJ, Huang ZJ (2008) High-resolution labeling and functional manipulation of specific neuron types in mouse brain by Cre-activated viral gene expression. PLoS ONE 3:e2005
110. Petreanu L, Huber D, Sobczyk A, Svoboda K (2007) Channelrhodopsin-2-assisted circuit mapping of long-range callosal projections. Nat Neurosci 10:663–668
111. Schroll C et al (2006) Light-induced activation of distinct modulatory neurons triggers appetitive or aversive learning in *Drosophila larvae*. Curr Biol 16:1741–1747
112. Douglass AD, Kraves S, Deisseroth K, Schier AF, Engert F (2008) Escape behavior elicited by single, channelrhodopsin-2-evoked spikes in zebrafish somatosensory neurons. Curr Biol 18:1133–1137
113. Yan W et al (2001) Cloning and characterization of a human beta, beta-carotene-15,15¢-dioxygenase that is highly expressed in the retinal pigment epithelium. Genomics 72:193–202
114. Dittgen T et al (2004) Lentivirus-based genetic manipulations of cortical neurons and their optical and electrophysiological monitoring in vivo. Proc Natl Acad Sci U S A 101:18206–18211
115. Chhatwal JP, Hammack SE, Jasnow AM, Rainnie DG, Ressler KJ (2007) Identification of cell-type-specific promoters within the brain using lentiviral vectors. Gene Ther 14:575–583
116. Adamantidis AR, Zhang F, Aravanis AM, Deisseroth K, de Lecea L (2007) Neural substrates of awakening probed with optogenetic control of hypocretin neurons. Nature 450:420–424
117. Tan W et al (2008) Silencing preBotzinger complex somatostatin-expressing neurons induces persistent apnea in awake rat. Nat Neurosci 11:538–540
118. Nathanson JL, Yanagawa Y, Obata K, Callaway EM (2009) Preferential labeling of inhibitory and excitatory cortical neurons by endogenous tropism of adeno-associated virus and lentivirus vectors. Neuroscience 161:441–450
119. Wickersham IR, Finke S, Conzelmann KK, Callaway EM (2007) Retrograde neuronal tracing with a deletion-mutant rabies virus. Nat Methods 4:47–49
120. Stachler MD, Chen I, Ting AY, Bartlett JS (2008) Site-specific modification of AAV vector particles with biophysical probes and targeting ligands using biotin ligase. Mol Ther 16:1467–1473
121. Toni N et al (2008) Neurons born in the adult dentate gyrus form functional synapses with target cells. Nat Neurosci 11:901–907
122. Wickersham IR et al (2007) Monosynaptic restriction of transsynaptic tracing from single, genetically targeted neurons. Neuron 53:639–647
123. Banfield BW, Kaufman JD, Randall JA, Pickard GE (2003) Development of pseudorabies virus strains expressing red fluorescent proteins: new tools for multisynaptic labeling applications. J Virol 77:10106–10112
124. Foust KD et al (2009) Intravascular AAV9 preferentially targets neonatal neurons and adult astrocytes. Nat Biotechnol 27: 59–65
125. Chan SY, Bernstein JG, Boyden ES (2010) Scalable fluidic injector arrays for viral targeting of intact 3-D brain circuits. J Vis Exp 35:1489. doi:10.3791/1489, http://www.jove.com/index/details.stp?id=1489
126. (2007) Retracing events. Nat Biotechnol 25:949

127. Bi A et al (2006) Ectopic expression of a microbial-type rhodopsin restores visual responses in mice with photoreceptor degeneration. Neuron 50:23–33
128. Lagali PS et al (2008) Light-activated channels targeted to ON bipolar cells restore visual function in retinal degeneration. Nat Neurosci 11:667–675
129. Campagnola L, Wang H, Zylka MJ (2008) Fiber-coupled light-emitting diode for localized photostimulation of neurons expressing channelrhodopsin-2. J Neurosci Methods 169:27–33
130. Rickgauer JP, Tank DW (2009) Two-photon excitation of channelrhodopsin-2 at saturation, PNAS 106(35):15025–30
131. Petreanu L, Mao T, Sternson SM, Svoboda K (2009) The subcellular organization of neocortical excitatory connections. Nature 457:1142–1145
132. Farah N, Reutsky I, Shoham S (2007) Patterned optical activation of retinal ganglion cells. Conf Proc IEEE Eng Med Biol Soc 2007:6369–6371
133. Guo ZV, Hart AC, Ramanathan S (2009) Optical interrogation of neural circuits in *Caenorhabditis elegans.* Nat Methods 6:891–896
134. Bernstein J et al (2008) A scalable toolbox for systematic, cell-specific optical control of entire 3-D neural circuits in the intact mammalian brain. Society for Neuroscience, online
135. Bernstein JG et al (2009) Modulation of fear behavior via optical fiber arrays targeted to bilateral prefrontal cortex. Society for Neuroscience, online
136. Ayling OG, Harrison TC, Boyd JD, Goroshkov A, Murphy TH (2009) Automated light-based mapping of motor cortex by photoactivation of channelrhodopsin-2 transgenic mice. Nat Methods 6:219–224
137. Honda K (2004) Dawn of the evolution of photoelectrochemistry. J Photochem Photobiol A Chem 166:63–68
138. Gratzel M (2001) Photoelectrochemical cells. Nature 414:338–344

Chapter 7

Vertebrate and Invertebrate Rhodopsins: Light Control of G-Protein Signaling

Davina V. Gutierrez, Eugene Oh, and Stefan Herlitze

Abstract

Vertebrate and invertebrate rhodopsins are G-protein-coupled receptors that are involved in sensing light. Light activation of these receptors leads to intracellular responses via activation of G-proteins. Both receptor types belong to the Class A (rhodopsin-like) family of seven transmembrane domain receptors. Differences between vertebrate and invertebrate rhodopsins exist according to their light cycles and the G-proteins utilized to sense and transduce light into a cellular response. While vertebrate rhodopsin uses 11-*cis*-retinal as the light-sensing compound coupled to an extrinsic photocycle to activate a G-protein of the Gi family, invertebrate rhodopsin uses all-*trans*-retinal coupled to an intrinsic photocycle to activate a G-protein of the Gq family. We will now discuss the difference between the two receptor types, their applications and expression outside of the vertebrate and invertebrate eye in regard to the control of cellular signaling.

Key words: Opsins, Rhodopsin, Melanopsin, Photoreception, Visual cycle

1. Introduction

From an evolutionary standpoint, photoreception is a process that is ironically both divergent and conserved. While the proteins responsible for detecting light are essentially similar in structure and are found in vertebrates, fungi, bacteria, and green algae, the amino acid sequences of each are rather diverse (1). An additional point of interest involves the idea that the molecules, which distinguish and respond to illumination, appeared in biological systems before the development of eyes (2). Despite the divergence of photoreceptive cellular structures between vertebrates and invertebrates, the basic idea of the production and recycling of the visual chromophore, photon absorption, conversion to an electrical signal, and the activation of signaling cascades is present in both systems and is critical for vision.

James J. Chambers and Richard H. Kramer (eds.), *Photosensitive Molecules for Controlling Biological Function*, Neuromethods, vol. 55, DOI 10.1007/978-1-61779-031-7_7, © Springer Science+Business Media, LLC 2011

Briefly, a transmembrane protein moiety comprised of an opsin and a light-sensitive chromophore are linked together to form the photosensitive molecule rhodopsin; a prototypical G-protein-coupled receptor (GPCR). While both the vertebrate and invertebrate photoresponses utilize a GPCR, a G-protein and an effector protein in response to light stimulation, there are noteworthy distinctions between the groups that include the structure–function relationship of rhodopsin, downstream effects of G-protein activation, and modifications in membrane properties after photon absorption (3–5). Specifically, the two kinds of phototransduction systems that have been reported are vertebrate hyperpolarizing photoreceptor cells that are mediated by G_t coupled proteins and the G_q-mediated scheme of invertebrate depolarization (6).

The aims of the following sections are to provide the reader with information about vertebrate and invertebrate rhodopsin, the mechanisms of photoactivation, how the systems differ in regard to structure and function, and the possibility of using light-activated GPCRs to control neuronal G-protein signaling.

2. Opsins, Subfamilies, and Function

Visual pigments have a molecular mass ranging from 30 to 50 kDa and are distinguished as being membrane proteins consisting of a polypeptide opsin and a covalently linked (via Schiff base) retinal chromophore (6, 7). The opsin follows the characteristic GPCR description in that it includes seven transmembrane α-helices. In animal models the opsin, or more specifically its related protein moiety rhodopsin, functions as a light sensor and receptor (8). Conversely, a non-animal opsin will act as light-driven ion pump or sensor and is found in a variety of biological organisms (8).

Overall, about 1,000 opsins have been distinguished via molecular genetic studies and are found in both vertebrate and invertebrate families (4). The opsins are divided into three clusters in the phylogenetic tree and are further organized into seven subfamilies. The subfamilies are categorized based on the type of G-proteins utilized and its respective GPCR. Within a given subfamily, amino acid similarity is less than 25% but increases to more than 40% within a single opsin family (4, 8). The seven subfamilies consist of vertebrate visual (transducin-coupled) and nonvisual opsins, encephalopsin/tmt-opsins, G_q-coupled opsins/melanopsins, G_o-coupled opsins, neuropsins, peropsins, and the retinal photoisomerase subfamily (4, 8). There is evidence for an evolutionary diversification of the subfamilies before the deuterostome-protostome split and is supported by the presence of certain opsin subfamilies belonging to both vertebrate and invertebrate species (4, 8).

Each opsin interacts and binds a chromophore. Specifically, it is known that vertebrate visual and nonvisual opsins, G_q- and G_o-coupled

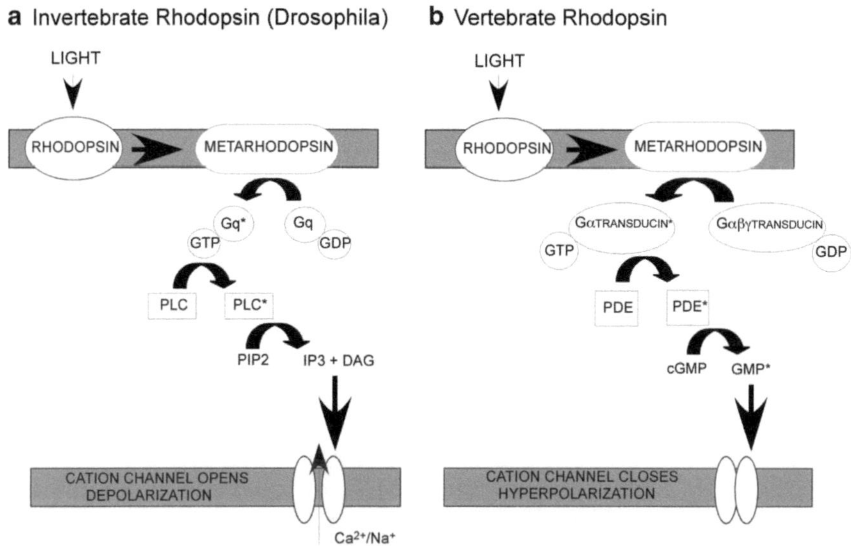

Fig. 1. Comparison of the G-protein signaling cascade during phototransduction between invertebrate and vertebrate rhodopsin. (**a**) Activation of the invertebrate rhodopsin by light leads to the activation of the G-protein (Gq-type), which activates the enzyme phosholipase C (PLC). PLC breaks down PIP2 into IP3 and DAG. These intracellular second messengers are involved in the opening of nonselective cation channels. The opening of these channels causes membrane depolarization. (**b**) Activation of the vertebrate rhodopsin by light leads to the activation of the G-protein transducin (Gi-type), which activates the enzyme phosphodiesterase (PDE). PDE breaks down cGMP, resulting in cGMP-gated cation channel closure and membrane hyperpolarization.

opsins bind 11-*cis*-retinal, while photoisomerases and peropsins bind all-*trans*-retinal; the chromophores linked to other opsin subfamilies are unknown (8). The principal task of visual pigments is to recognize, acquire and relay an extracellular, photic signal to a retinal G-protein; subsequently initiates an intracellular G-protein activation (6) (Fig. 1). Absorbing visible light functionally drives most opsins; retinal, however, has a maximal absorption range that extends into the UV region (8). Any potential problems that may arise by having an optimal activation range not in the visible spectrum are resolved by structural arrangements within the visual pigment.

2.1. Rhodopsins and Phototransduction

Phototransduction is initiated by the absorption of a single photon by a retinal aldehyde of vitamin A that is bound to a critical lysine residue in the rhodopsin apoprotein. Light causes photoisomerization of retinal from 11-*cis* to the all-*trans* conformation, which in turn leads to dissociation of the G-protein from the GPCR and subsequent activation of downstream signaling pathways. Phototransduction in vertebrates and invertebrates differ in two fundamental ways. The transduction cascades employ divergent G-protein pathways; hence, signaling involves different enzymes and different second messengers. Furthermore, photoreceptors respond with opposite electric polarity to stimulation. A notable difference between vertebrates and invertebrates is also the cellular structure of photoreceptors. Photoreceptors of vertebrates are ciliated, whereas photoreceptors of invertebrates are organized in a rhabdomeric fashion.

2.2. Invertebrate Photoreceptor

2.2.1. Structure

Invertebrate photoreceptors possess a rhabdomere, a specialized microvilli configuration whose exclusive purpose is to enclose the apparatus required for phototransduction signaling (5, 9). As previously stated, the invertebrate visual pigment contains the light-absorbing chromophore 11-*cis*-retinal. This retinal is covalently linked to the apoprotein by a protonated Schiff base by the means of a lysine residue in helix VII and is stabilized by a tyrosine (Y) residue (position 113) that functions as a counterion (6, 8, 10). The presence of Y113 is conserved throughout invertebrate visual pigments but is replaced by phenylalanine in some ultraviolet absorbing receptors (6, 10). Invertebrate rhodopsin has a spectral sensitivity that ranges from λ_{max} 350 to 550 nm (4).

2.2.2. Visual Cycle

Invertebrate rhodopsins in contrast to vertebrate rhodopsin (see next paragraph) possess the ability to regenerate 11-*cis*-retinal through a photochemical reaction. In insects, the retinal chromophore remains bound to the rhodopsin apoprotein after light activation and the visual cycle does not require intercellular transport of retinoids. Photoconversion of 11-*cis*-retinal to the all-*trans* conformation generates a thermally stable *meta*-rhodopsin and a second photon (of higher wavelength) catalyzes reisomerization to 11-*cis* (11). The relative stability of invertebrate *meta*-rhodopsin is not an intrinsic property but imparted by its interaction with arrestin (12, 13). Hence, the need for accessory retinoid-binding proteins to stabilize and traffic retinal is circumvented. The visual systems of cephalopods (which include marine invertebrates such as squid, octopus, and cuttlefish) are similar to insects in that all-*trans*-retinal is reisomerized to 11-*cis*-retinal via photochemical reaction but differ from insects in that the Schiff base bond between all-*trans*-retinal and rhodopsin is hydrolyzed after activation. Unlike in insects, *meta*-rhodopsins of cephalopods are thermally unstable and the spent chromophore binds soluble retinaldehyde-binding protein (RALBP) after *meta*-rhodopsin decay. RALBP facilitates delivery of all-*trans*-retinaldehyde to the retinochrome complex, present in the myeloid bodies of the photoreceptor inner segments (14, 15). Absorption of a second photon allows retinochrome to convert all-*trans*-retinaldehyde back to the 11-*cis* form, which then reassociates with RALBP and is recycled to the outer segments (15). Thus, in cephalopods, a rhodopsin–retinochrome conjugate system maintains photoreceptive function in visual cells.

2.2.3. Signaling

Once the invertebrate rhodopsin is activated, the *meta*-rhodopsin allows for an interaction with the G-protein (heterotrimeric protein with α-, β-, and γ-subunits) to occur, prompting an exchange of GTP for GDP on the transducin α-subunit (5). Activated G-protein dissociates into a βγ-subunit and an α-GTP unit that interacts with effector enzymes, second messengers and ion channels. Invertebrate rhodopsins use G-protein α-subunits of the Gq family as their G-protein of choice. Selective activation of G-protein subtypes by invertebrate (Gq) and vertebrate (Gi/o) rhodopsin is dictated by binding specificity to α-subunits and not by relative availability of different G-protein subtypes (16). Crystal structure of squid rhodopsin reveals a peculiar feature that is not seen in vertebrate rhodopsin that may be crucial for Gq coupling. The fifth and sixth transmembrane domains extend ~25 Å into the intracellular space (17). This differs from the structure of bovine rhodopsin (18, 19).

Invertebrates utilize a multitude of second messenger signaling systems with a phosphoinositide (PI) pathway being the main visual excitation cascade (20). After dissociation of transducin occurs, the α-subunit activates a phospholipase C (PLC), resulting in the rapid generation of inositol-1,4,5-triphosphate (IP_3) and diacylglycerol (DAG) (4). Intracellular Ca^{2+} release has also been noted to contribute to the photoexcitation response (5). While the exact mechanism that underlies the coupling of IP_3 to Ca^{2+} is not entirely clear, the end result of Ca^{2+} influx across the plasma membrane is an amplification in membrane conductance; ultimately producing a depolarizing receptor potential (4, 5). The activation of the signaling cascade is terminated via the phosphorylation of the rhodopsin catalyzed by a rhodopsin kinase, followed by the binding of an arrestin to the GPCR (5). Essentially, arrestin acts by blocking *meta*-rhodopsin dephosphorylation to inhibit G-protein activation.

2.3. Vertebrate Rhodopsin

2.3.1. Structure

In regard to vertebrate vision, the two ciliary photoreceptor cell types (rods and cones) participate in vision and are discriminated by their shapes and opsins. Within the plasma membrane of a rod outer segment are stacks of flattened membrane vesicles; cones have outer segments that are comprised of extremely invaginated plasma membrane (5). Rods house rhodopsin and initiate twilight vision, while cones contain opsins that manage color (daylight) vision (6). Spatially, vertebrate rhodopsin is organized in the following manner: the chromophore localized to the center, a cytosolic C-terminus with phosphorylation sites, and an N-terminus that associates with extracellular interhelical loops (4, 7). Structural analysis of bovine rhodopsin has revealed the composition of this photoreceptor to consist of a transmembrane apoprotein, an opsin, and 11-*cis*-retinal that is bound to a lysine side-chain (K296) in helix VII through a Schiff base (7). Crystallography further revealed that the nitrogen atom of K296 forms a double bond with the carbon atom at one end of retinal (8).

In vertebrate visual pigments, the Schiff base is protonated and positively charged. A negatively charged lysine residue (E113) stabilizes the Schiff base and acts as a counterion (6). Bovine rhodopsin has a molecular mass of 39,007 and is composed of 348 amino acids (7). Posttranslational modifications not only significantly alter rhodopsin but also augment the total molecular mass. It is suspected that this protein is relatively hydrophobic based upon the prevalence of the amino acids Phe, Val, Ala, and Leu (7). Because the retinal binding site is relatively hydrophobic, several charged resides are localized around the chromophore to ensure accurate rhodopsin action (7). As is the case for a majority of opsins, the primary functions of rhodopsin are light absorption and G-protein activation. The spectral sensitivity of vertebrate rhodopsin ranges from l_{max} 450 to 530 nm (4).

2.3.2. Visual Cycle

To maintain the function of rhodopsin, the visual system must supply the light-sensitive chromophore, 11-*cis*-retinal. The series of transport and enzymatic reactions that govern retinoid homeostasis within the visual system is referred to as the visual cycle or retinoid cycle. In vertebrates, the retinal pigment epithelium (RPE), the tissue adjacent to photoreceptor cells, contains isomerases that convert vitamin A (all-*trans*-retinol) to 11-*cis*-retinal and supplies rhodopsin with photosensitive chromophore. Exposure to light then converts 11-*cis*-retinal bound to lysine 296 to an all-*trans* conformation. All-*trans*-retinal must then be reconverted to 11-*cis*-retinal for rhodopsin to regain activity. Retinal regeneration after light stimulus starts when the Schiff base bond between all-*trans*-retinal and rhodopsin is hydrolyzed. The all-*trans*-retinaldehyde is reduced to all-*trans* retinol by retinol dehydrogenase (21). All-*trans* retinol then leaves photoreceptor cells, crosses the interphotoreceptor matrix (apical RPE microvilli and outer segment plasma membranes of photoreceptors) before entering the RPE cells. The intercellular transport of retinal is facilitated by binding to interphotoreceptor or interstitial retinoid-binding protein (IRBP), which is thought to solubilize retinoids in the extracellular space and target the delivery of all-*trans* retinol to the RPE (22, 23). Vitamin A (all-*trans* retinol) from the serum also enters the RPE but enters from the basal surface (24). Once inside the RPE, all-*trans* retinol binds to the cellular retinol-binding protein (CRBP) and is processed by an enzyme localized to the endoplasmic reticulum called lecithin retinol acyl transferase (LRAT). LRAT adds a fatty acid chain to all-*trans* retinol forming all-*trans* retinyl palmitate. This in turn becomes the substrate for a retinol isomerohydrolase, which catalyzes the formation of 11-*cis* retinol and palmitate (25, 26). 11-*cis* retinol dehydrogenase oxidizes 11-*cis* retinol to 11-*cis*-retinaldehyde, which is subsequently released from the RPE. IRPB again facilitates extracellular transport by binding and targeting 11-*cis*-retinaldehyde to photoreceptor cells where it reassociates with rhodopsin.

2.3.3. Signaling

As indicated above the absorption of a single photon of light by rhodopsin induces the transformation of 11-*cis*-retinal to an all-*trans* form and is followed by the quick formation of *photo*-, *batho*-, *lumi*- and *meta*-rhodopsin I, II and III intermediates (4, 8). Once a stable transition state is established within milliseconds, *meta*-rhodopsin II is able to activate the G-protein (transducin, G_t) signaling cascade (7). Before *meta*-rhodopsin II activates transducin, it interacts with the G-protein αβγ trimer of transducin that is GDP-bound (3). Once GDP is exchanged for GTP on the Gα subunit, transducin becomes activated and dissociates from *meta*-rhodopsin II; active Gα-GTP also detaches from Gβγ (3). Interaction between downstream effectors, Gα-GTP and Gβγ subunits is the next step in phototransduction. Gα-GTP couples *meta*-rhodopsin II to cyclic guanosine monophosphate (cGMP) phosphodiesterase (PDE) and removes inhibitory molecules, leading to cGMP hydrolysis (3–5). It should be noted that during dark periods, vertebrate rods experience sodium (Na^+) and Ca^{2+} influx into the outer segments via cGMP-activated channels and a outflow of potassium (K^+) through inner segment channels (3). Ionic gradients are maintained through the use of ion exchangers in both outer and inner segments. PDE activation induces a reduction of cytoplasmic cGMP that directly results in the closure of light-dependent cGMP-gated ion channels (4). Once the cGMP channels are closed, a hyperpolarizing receptor potential results due to the lack of Na^+ and Ca^{2+} influx (4). Subsequently, the synaptic transmitter glutamate experiences diminished release at the photoreceptor terminal (3). To terminate the G-protein signaling cascade photoactivated vertebrate rhodopsin undergoes G-protein-coupled receptor kinase phosphorylation at several sites near the C-terminus of rhodopsin (5). Phosphorylated *meta*-rhodopsins are further subdued by arrestins that compete with transducin (5).

An astonishing event that transpires during vertebrate phototransduction is the finding that one light photon can activate one rhodopsin, which in turn catalyzes the activation of several hundred G-proteins and ultimately an equivalent concentration of PDE; the end result being closure of about 1,000 cGMP channels (4).

3. Light-Activated GPCRs for Control of Neuronal Signaling

Rhodopsins and other GPCRs embody a substantial, multifaceted collection of protein sensors that contribute to and influence a multitude of physiological and biochemical processes (Fig. 2).

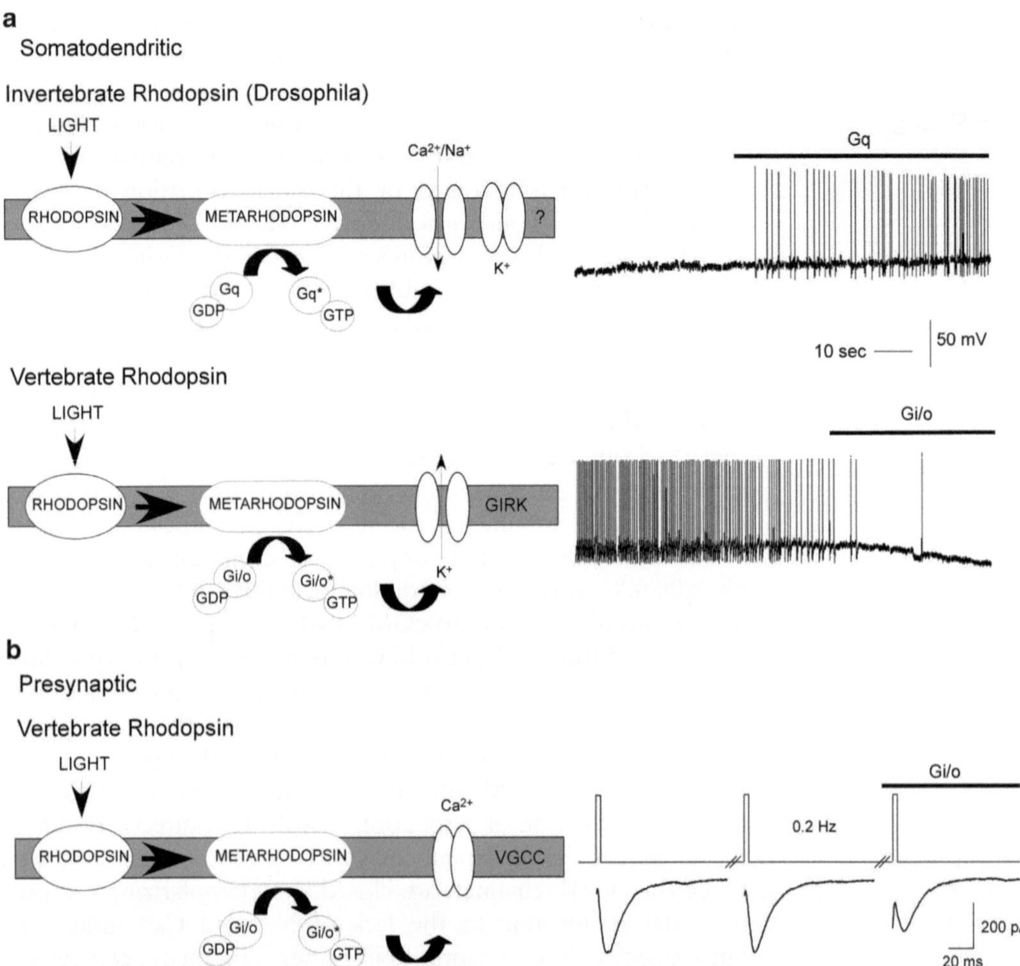

Fig. 2. Somatodendritic and presynaptic modulation of neuronal activity and signaling by invertebrate and vertebrate rhodopsin. (**a**) Somatodendritic activation of invertebrate and vertebrate rhodopsin increases or decreases neuronal firing. (*Upper*) Activation of the invertebrate rhodopsin by light in neurons leads to the activation of the G-protein (Gq-type). Activation of the Gq pathway may cause the activation of nonselective cation channels or the reduction of K^+ conductances. Both mechanisms may contribute to the depolarization of the cell membrane and the induction of neuronal firing (*right*). (*Lower*) Activation of the vertebrate rhodopsin by light in neurons leads to the activation of the G-protein (Gi/o-type). Activation of the Gi/o pathway causes the activation of G-protein-coupled inward rectifying K^+ channels (GIRK) and subsequent hyperpolarization of the cell membrane and the reduction of neuronal firing (*right*). (**b**) Presynaptic activation of vertebrate rhodopsin decreases transmitter release. Activation of the vertebrate rhodopsin by light at the presynaptic terminal leads to the activation of the Gi/o pathway. This causes the inhibition of the presynaptic voltage-gated Ca^{2+} channels (VGCC), a reduction in the Ca^{2+} influx and a reduction in the transmitter release. Therefore, the excitatory postsynaptic current (EPSC) is reduced once the Gi/o pathway is activated (*right*).

GPCRs are divided into three major families based on receptor homology and include: Family 1-rhodopsin homologous receptors, Family 2-secretin homologous receptors, and Family 3-GABA homologous receptors (7). As described previously, GPCRs function by recognizing and transmitting an extracellular signal to a cytosolic G-protein, thereby coupling the two together.

Once activated by a GDP/GTP exchange, G-proteins proceed to initiate the activation of second messenger systems and downstream effectors like enzymes and ion channels. If one were to consider the basic idea behind the phototransduction mechanism utilized in vertebrate and invertebrate rhodopsin, it should become quite evident that the use of these GPCR is ideal for controlling G-protein signaling in neurons. Specifically, the application of light to specific systems that exclusively express rhodopsins can be employed to noninvasively influence neuronal activity and signaling.

Establishing the correlation between one specific neuronal population to cellular function and animal behavior seems to be a rather arduous and overwhelming task. Equally daunting is determining the extent of neuronal projections and how one circuit may impact a particular behavior and/or function of another cellular group. The complexity that is presented with these concepts is based on several factors that include an intricate neuroanatomical arrangement, difficulty in controlling neuronal activity and the idea that processing centers can receive an array of diverse inputs that are anatomically dispersed and regulated on multiple levels. While it is clear that focusing on an single group of functionally related neurons is critical for gaining a complete appreciation and sense of how the development, maintenance, and plasticity of the nervous system occurs, the approaches taken and methods utilized in the past have concurrently been extremely beneficial and limiting. Traditional procedures have attempted to associate neural circuits to behavior or processing by invasively and irreversibly inactivating functionally and physically entangled neurons. Furthermore, pharmacological and genetic approaches have presented kinetic and expression restraints. The feasibility of neuronal manipulations that overcome some of these challenges has recently been demonstrated.

The basic idea behind the noninvasive control of neuronal, cell-type specific signaling in a vertebrate animal model is dependent upon the use of nonintrinsic channels or receptors that can be specifically activated by a stimulus that does not activate endogenous receptors. Example candidates that could control the neuronal signaling include GPCRs and ion channels activated by species-specific transmitters (e.g., allatostatin (27)), GPCRs and ion channels activated by synthesized agonists (e.g., clozapine-*N*-oxide (28)) and GPCRs and ion channels activated by organ/receptor-specific stimuli (e.g., vision (29, 30), smell, and taste (31)).

3.1. Control of Gq Signaling to Increase Neuronal Activity

As described above, activation of the invertebrate phototransduction pathway (via rhodopsin) results in stimulation of the Gq protein, PLC activation, IP3, and DAG production; ultimately leading to membrane depolarization. Phototransmission is deactivated on the binding of arrestin and the regeneration of 11-*cis*-retinal.

By building on these ideas, Miesenbock and associates were able to design a construct that allowed for the direct examination of neural circuits (30). Specifically, they generated a technique for stimulating neurons via illumination by co-expressing *Drosophila* photoreceptor genes that encode for arrestin-2, rhodopsin (NinaE), and the G-protein α_q-subunit (called chARGe). chARGe functioned as a depolarizing signaling complex, which stimulated electrical activity and sensitized vertebrate neurons to illumination. The experimental design called for the expression of chARGe in cultured hippocampal neurons. Under whole cell current clamp conditions, the application of light-induced action potential firing which was reduced once light was switched off. By identifying the potential for cellular depolarization that invertebrate rhodopsin encompasses, this group was able to create a noninvasive method (light-controlled) for regulating firing patterns via the transgenic expression of invertebrate transduction components. Additionally, the framework for creating a light-activated system to be used in studying neuronal development, circuitry and behavior was established. Despite the advances this construct produced, there also were numerous limitations. For example, the utilization of three separate genes resulted in a complex construct for transgenic studies. Second, the intrinsic electrophysiological properties were problematic in that the activation and deactivation kinetics were rather slow, in comparison, for example, to channelrhodopsin (ChR2) (29) – making the exact manipulation of neuronal firing problematic. In spite of these disadvantages, chARGe establish the use of light to control neuronal activity and allows for precise control of the Gq pathway in neurons.

3.2. Control of Gi/o Signaling to Decrease Neuronal Activity

Continuing with the recognition that understanding and defining the correlation between neuronal circuitry, perception and behavior is significant and necessary, the option of selectively inactivating a specific cellular type within an intricate neural organization was explored. Previous disadvantages with invasive design and approach, physical limitations and kinetic restraints were all taken into consideration. The result was the exogenous application of AlstR to slices of the neonatal ferret visual cortex (27). Briefly, AlstR is a *Drosophila* allatostatin GPCR that regulates juvenile hormone synthesis in insects (27). When expressed in *Xenopus* oocytes, it was found to activate mammalian G-protein-coupled inward rectifier K^+ channels (GIRK) channel subunits 1 and 2 (27), thus coupling to the Gi/o pathway. In cortical neurons expressing AlstR, the application of allatostatin (peptide ligand) induced rapid and reversible hyperpolarization. The use of AlstR furthered the search for a simple construct that explicitly regulated neuronal firing properties in an expeditious and reversible manner. The advantage of using a ligand/chemical-gated GPCR is based on the application of the ligand to experimental animals and the control of a large neuronal population. The downside of this approach lies in the kinetics and specificity of the signaling pathway, which is often slow and difficult to control and has a poor spatial resolution.

In regard to establishing a noninvasive and reversible method that is able to control neuronal signaling from a millisecond to a second, we investigated the employment of vertebrate rhodopsin in both in vitro and in vivo studies. The success of this project lies in the development and utilization of constructs that function as on/off switches; thus is the case with the employment of vertebrate rhodopsin as an inhibitory/modulating light switch.

To review, since photoactivation of vertebrate rhodopsin results in the activation of the Gi/o protein pathway, we postulated that vertebrate rhodopsin activation can activate GIRK channels and inhibit presynaptic voltage-gated Ca^{2+} channels. Presynaptically, this would lead to the inhibition of transmitter release, while postynaptically the membrane would be hyperpolarized and action potential firing reduced. To examine this idea, we expressed vertebrate rhodopsin with GIRK or Ca^{2+} channels in HEK293 cells and found that K^+ currents were activated and Ca^{2+} currents were inhibited with photostimulation (29). The application of light to cultured hippocampal neurons expressing vertebrate rhodopsin-induced membrane hyperpolarization (within 1 s) and reduced cellular firing. Presynaptically, vertebrate rhodopsin-induced paired-puse facilitation, suggesting that the Ca^{2+} influx into the presynaptic terminal is reduced once light is switched on. The results suggested that vertebrate rhodopsin is transported into subcellular structures, such as the presynaptic terminal and that it interacts with the downstream components of the Gi/o pathway. In order to examine this method of cellular control at the next level, vertebrate rhodopsin was electroporated into the spinal cord of embryonic chicken. It has been observed that the motor units fire asynchronous, spontaneous bursts of action potentials every several minutes (29). When light pulses were applied to the spinal cord, we discovered that the firing activity was suppressed and hyperpolarization ensued. The light-induced hyperpolarization allowed the synchronization of the motor neuron firing in the chicken spinal cord.

4. Retinoid Processing in Heterologous Expression Systems, Neurons and Neuronal Circuits: The Big Surprise

In order to reconstitute function of invertebrate and vertebrate rhodopsins in nonvisual cell types, retinoid processing enzymes and binding proteins must be present to sufficiently supply rhodopsin with a photoactive substrate. Surprisingly, unlike the photoreceptors of the vertebrate eye, it appears that many nonvisual cell types possess the intrinsic capability to regenerate 11-*cis*-retinal from all-*trans*-retinal. In human embryonic kidney cells (HEK293), activation of heterologously expressed rhodopsin (in HEK293 cells) can be achieved for over 4 h once the cells are loaded with 11-*cis*-retinal (29). Furthermore, application of all-*trans*-retinal, which presumably has to be reconverted to 11-*cis*-retinal seems to be sufficient to achieve light responses in mammalian cells (29). Vertebrate rhodopsin expressed in HEK293S cells respond with early receptor currents (ERC), a conformation-associated charge shift, when exposed to light in the presence of 11-*cis*-retinal, 9-*cis*-retinal, 13-*cis*-retinal, and even all-*trans*-retinal or vitamin A (all-*trans* retinol) (32). However, ERC responses could only be elicited after overnight incubations with all-*trans*-retinal or vitamin A. With respect to chromophore delivery, retinoids not only bind IRBP but also bind extracellular albumin in interphotoreceptor matrix, possibly facilitating retinal trafficking within the visual context (33). In a heterologous setting, albumin may be able to take on the traditional role of binding proteins by solubilizing and successfully delivering retinal. In fact, fatty acid free bovine serum albumin has been used to successfully deliver retinal compounds to photoreceptors and HEK293 cells (33, 34). For invertebrate rhodopsins expressed in non-photo-sensitive cells, a retinal compound must be initially supplied to confer functionality, but the photopigment should be able to auto-regenerate active receptor since the chromophore remains bound to rhodopsin. Indeed, only a single 15 min application of all-*trans*-retinal is sufficient to generate a receptor that is functional for several hours (30).

In the chicken spinal cord, sufficient retinal compounds are available to drive light-activated currents even without an exogenous supply of chromophore (29). This is consistent with the observation that in hippocampal neurons expressing the green algae Channelrhododopsin-2 (ChR2) activation was not dependent on retinal application, even though ChR2 requires all-*trans*-retinal as chromophore (35). This suggests that endogenous retinoids (and/or what is provided in the culture medium) are sufficient for rhodopsin loading, though it is possible that providing exogenous retinal may enhance light responses in nonvisual cell types. Therefore, a single application of 11-*cis*-retinal, its analogs 9-*cis* or 13-*cis*, or even all-*trans*-retinal is sufficient to regenerate the active compounds necessary for repetitive light activation of both vertebrate and invertebrate rhodopsins.

5. The Therapeutic Potential of Light-Activated GPCRs

Several studies have demonstrated the practicality of exploiting light-activated GPCRs in both in vitro and in vivo investigations to regulate cellular activity via control of the Gq and the Gi/o pathway. Importantly and surprisingly, expression of these GPCRs does not require application of light-sensitive compounds in neurons and intact neuronal circuits giving the opportunity to control the Gq and Gi/o pathways in vivo over a long time periods. Therefore, light-activated GPCRs coupling to a specific pathway in a receptor-specific manner have great potential for treating disease in the CNS but in particular in the periphery, which is easily accessible to light. Various diseases such as Parkinson's disease, schizophrenia, anxiety and depression have been associated with transmitters activating GPCRs, such as dopamine, glutamate, and serotonin receptors. Therefore, expression of light-activated receptors in neurons and neuronal circuits involved in these diseases gives the opportunity to switch on and switch of the GPCR pathway, whenever needed.

In summary, the control of G-protein-coupled receptor signaling by light will be useful for the basic characterization of Gi/o and Gq pathways for neuronal/cell function and behavior and will provide tools for developing externally, light-controlled molecular machines with the potential to treat diseases involving a multitude GPCR malfunctions such as asthma, heart arrhythmia, schizophrenia, anxiety, and depression.

References

1. Sineshchekov OA, Jung KH, Spudich JL (2002) Two rhodopsins mediate phototaxis to low- and high-intensity light in Chlamydomonas reinhardtii. Proc Natl Acad Sci USA 99:8689–8694
2. Fernald RD (2004) Evolving eyes. Int J Dev Biol 48:701–705
3. Arshavsky VY, Lamb TD, Pugh EN (2002) G proteins and phototransduction. Annu Rev Physiol 64:153–187
4. Santillo S, Orlando P, De Petrocellis L, Cristino L, Guglielmotti V, Musio C (2006) Evolving visual pigments: hints from the opsin-based proteins in a phylogenetically old "eyeless" invertebrate. Biosystems 86:3–17
5. Yarfitz S, Hurley JB (1994) Transduction mechanisms of vertebrate and invertebrate photoreceptors. J Biol Chem 269: 14329–14332
6. Shichida Y, Imai H (1998) Visual pigment: G-protein-coupled receptor for light signals. Cell Mol Life Sci 54:1299–1315
7. Filipek S, Stenkamp RE, Teller DC, Palczewski K (2003) G protein-coupled receptor rhodopsin: a prospectus. Annu Rev Physiol 65:851–879
8. Terakita A (2005) The opsins. Genome Biol 6:213
9. Frechter S, Minke B (2006) Light-regulated translocation of signaling proteins in Drosophila photoreceptors. J Physiol Paris 99:133–139
10. Nakagawa M, Iwasa T, Kikkawa S, Tsuda M, Ebrey TG (1999) How vertebrate and invertebrate visual pigments differ in their mechanism of photoactivation. Proc Natl Acad Sci USA 96:6189–6192
11. Byk T, Bar-Yaacov M, Doza YN, Minke B, Selinger Z (1993) Regulatory arrestin cycle secures the fidelity and maintenance of the fly photoreceptor cell. Proc Natl Acad Sci USA 90:1907–1911
12. Kiselev A, Subramaniam S (1994) Activation and regeneration of rhodopsin in the insect visual cycle. Science 266:1369–1373
13. Kiselev A, Subramaniam S (1996) Modulation of arrestin release in the light-driven regeneration of Rh1 Drosophila rhodopsin. Biochemistry 35:1848–1855
14. Molina TM, Torres SC, Flores A, Hara T, Hara R, Robles LJ (1992) Immunocytochemical localization of retinal binding protein in the octopus retina: a shuttle protein for 11-*cis* retinal. Exp Eye Res 54:83–90
15. Terakita A, Hara R, Hara T (1989) Retinal-binding protein as a shuttle for retinal in the

rhodopsin-retinochrome system of the squid visual cells. Vis Res 29:639–652

16. Terakita A, Yamashita T, Tachibanaki S, Shichida Y (1998) Selective activation of G-protein subtypes by vertebrate and invertebrate rhodopsins. FEBS Lett 439:110–114
17. Murakami M, Kouyama T (2008) Crystal structure of squid rhodopsin. Nature 453:363–367
18. Palczewski K, Kumasaka T, Hori T, Behnke CA, Motoshima H, Fox BA, Le Trong I, Teller DC, Okada T, Stenkamp RE, Yamamoto M, Miyano M (2000) Crystal structure of rhodopsin: a G protein-coupled receptor. Science 289:739–745
19. Teller DC, Okada T, Behnke CA, Palczewski K, Stenkamp RE (2001) Advances in determination of a high-resolution three-dimensional structure of rhodopsin, a model of G-protein-coupled receptors (GPCRs). Biochemistry 40:7761–7771
20. Oday PM, Bacigalupo J, Vergara C, Haab JE (1997) Current issues in invertebrate phototransduction – second messengers and ion conductances. Mol Neurobiol 15:41–63
21. Saari JC, Garwin GG, Van Hooser JP, Palczewski K (1998) Reduction of all-*trans*-retinal limits regeneration of visual pigment in mice. Vis Res 38:1325–1333
22. Okajima TI, Pepperberg DR, Ripps H, Wiggert B, Chader GJ (1990) Interphotoreceptor retinoid-binding protein promotes rhodopsin regeneration in toad photoreceptors. Proc Natl Acad Sci USA 87:6907–6911
23. Pepperberg DR, Okajima TL, Wiggert B, Ripps H, Crouch RK, Chader GJ (1993) Interphotoreceptor retinoid-binding protein (IRBP). Molecular biology and physiological role in the visual cycle of rhodopsin. Mol Neurobiol 7:61–85
24. Gonzalez-Fernandez F (2002) Evolution of the visual cycle: the role of retinoid-binding proteins. J Endocrinol 175:75–88
25. McBee JK, Palczewski K, Baehr W, Pepperberg DR (2001) Confronting complexity: the interlink of phototransduction and retinoid metabolism in the vertebrate retina. Prog Retin Eye Res 20:469–529
26. Rando RR (1996) Polyenes and vision. Chem Biol 3:255–262
27. Lechner HAE, Lein ES, Callaway EM (2002) A genetic method for selective and quickly reversible silencing of mammalian neurons. J Neurosci 22:5287–5290
28. Armbruster BN, Li X, Pausch MH, Herlitze S, Roth BL (2007) Evolving the lock to fit the key to create a family of G protein-coupled receptors potently activated by an inert ligand. Proc Natl Acad Sci USA 104:5163–5168
29. Li X, Gutierrez DV, Hanson MG, Han J, Mark MD, Chiel H, Hegemann P, Landmesser LT, Herlitze S (2005) Fast noninvasive activation and inhibition of neural and network activity by vertebrate rhodopsin and green algae channelrhodopsin. Proc Natl Acad Sci USA 102:17816–17821
30. Zemelman BV, Lee GA, Ng M, Miesenbock G (2002) Selective photostimulation of genetically chARGed neurons. Neuron 33:15–22
31. Wicher D, Schafer R, Bauernfeind R, Stensmyr MC, Heller R, Heinemann SH, Hansson BS (2008) Drosophila odorant receptors are both ligand-gated and cyclic-nucleotide-activated cation channels. Nature 452:1007–1011
32. Brueggemann LI, Sullivan JM (2002) HEK293S cells have functional retinoid processing machinery. J Gen Physiol 119:593–612
33. Adler AJ, Edwards RB (2000) Human interphotoreceptor matrix contains serum albumin and retinol-binding protein. Exp Eye Res 70:227–234
34. Li Z, Zhuang J, Corson DW (1999) Delivery of 9-*Cis* retinal to photoreceptors from bovine serum albumin. Photochem Photobiol 69:500–504
35. Boyden ES, Zhang F, Bamberg E, Nagel G, Deisseroth K (2005) Millisecond-timescale, genetically targeted optical control of neural activity. Nat Neurosci 8:1263–1268

Chapter 8

Restoring Visual Function After Photoreceptor Degeneration: Ectopic Expression of Photosensitive Proteins in Retinal Neurons

Bin Lin and Richard H. Masland

Abstract

A leading cause of blindness worldwide is degeneration of the retinal photoreceptor cells. The two large classes of such disorders are retinitis pigmentosa, which affects ≃100,000 individuals in the USA, and macular degeneration, which affects ≃3,000,000. The causes of both disorders are diverse, but the initial lesion in both cases is to the rod and cone photoreceptor cells, leaving a retina in which many neurons appear functionally intact, but the retina – either the entire tissue or specific regions of it – can no longer detect light. A strategy for restoring at least a minimal level of vision is to engineer the expression of a photosensitive molecule in the surviving, nonphotoreceptor, neurons. This has been achieved at the level of proof of principle in the rd strain of mice, which undergoes photoreceptor degeneration similar to retinitis pigmentosa. In separate experiments, Channelrhodopsin-2 or melanopsin were introduced into retinal neurons and restoration of electrophysiological responsiveness and simple visually guided behaviors was demonstrated. There is reason for cautious optimism that vision aided in this way may eventually be of use for humans suffering from photoreceptor degenerations.

Key words: Photoreceptor degeneration, Gene therapy, AAV, Ganglion cells, Bipolar cells, Melanopsin, Channelrhodopsin

1. Introduction

The rod and cone photoreceptor cells of mammalian retinas are relatively fragile neurons, exquisitely specialized for the detection of light. Two of these specializations are particularly costly. First, the amplification of small photic signals is accomplished at the cost of a large resting ionic current, which in the dark must continuously be pumped out of the rod or cone cell. Second, both cell types but most notably the rods contain huge amounts of photosensitive membrane proteins, opsins, which are the

James J. Chambers and Richard H. Kramer (eds.), *Photosensitive Molecules for Controlling Biological Function*, Neuromethods, vol. 55, DOI 10.1007/978-1-61779-031-7_8, light-detecting molecules, and must be continually replaced. Both of these processes are energetically costly and the photoreceptor cells consume >70% of the retina's total energy metabolism (1). Furthermore, the large amounts of opsin protein that must be packaged and transported render these small cells susceptible to mutations that interfere with protein synthesis, folding, or transport. Many mutations of the gene coding for rhodopsin or other components of the phototransduction machinery lead to a degeneration of the rod or cone photoreceptors.

If the rods degenerate, the cones follow: for reasons that are poorly understood, loss of the normal microenvironment after rod degeneration leads to cone degeneration. Since acute vision in humans depends primarily on the cones, it is the loss of the cones, which often are not the site of the primary mutation, that is the most handicapping. These conditions, lumped together as *retinitis pigmentosa* (for the retinal pigment that becomes visible opthalmoscopically in the absence of the photoreceptor cells) often manifest first in the retinal periphery and lead initially to tunnel vision. In later stages, the process can cover the entire retina and the afflicted individuals become densely blind.

Macular degenerations follow the opposite spatial pattern, with the initial degeneration occurring in the central retina. They are also a heterogeneous collection of disorders. Many of them occur late in life and are collectively termed age-related maculopathies. Except for a few rare inherited forms, the cause of macular degeneration is not known; the commonest forms may be considered as diseases of aging. (Interestingly, the major risk factors are very similar to those for cardiovascular disease). It is not clear whether the initial pathology lies in the retina or in its supporting tissues, but the initial visual impairment results from damage to the rods and cones of the central retina. Although much of the retina remains normal, loss of central vision is severely handicapping; the central retina is where acute vision occurs, and individuals with macular degeneration often lose the ability to recognize faces, read text, or watch television. As is the case for retinitis pigmentosa, the nonphotoreceptor neurons of the retina are substantially preserved. After photoreceptor degeneration, the cells which receive direct input from the photoreceptors, the horizontal and bipolar cells, undergo remodeling, especially of their now-deafferented dendritic structures. The inner retina is more robust and many retinal ganglion cells survive unchanged (2, 3).

The initial spur to the experiments to be reviewed here was the quite simple notion that rendering the remaining neurons sensitive to light could restore some amount of vision to individuals whose photoreceptor cells had degenerated. The possibility had existed for some time. By the early 2000s the advent of reliable gene therapy vectors (4–6) allowed one to imagine using various light-sensitive proteins – bacterial rhodopsin, for example, whose existence has been known since the 1970s (7) – to render nonphotoreceptor cells light sensitive. At least three groups have pursued this sort of "photoreplacement" as a practical possibility. They were spurred by two developments: the advent of robust and nontoxic viral vectors for ocular gene therapy (reviewed by (8, 9)) and the discovery of photosensitive proteins that can be expressed and retain their functions in a wide spectrum of different neurons (10–15). Here, we will summarize three recent experiments, from different laboratories. In each, channelrrhodopsin-2 or melanopsin were used to restore light sensitivity to retinas in which the photoreceptor cells had degenerated. The advantages and disadvantages of different proteins will be discussed, as will some of the issues for developing a photoreplacement therapy that could be of use to human patients.

2. Channelrhodopsin Expressed in Retinal Ganglion Cells of *rd/rd* Mice

All of the studies to be discussed here used a model of human retinitis pigmentosa, the rd1 mouse. In this well-known strain, a mutation of the beta subunit of the rod-specific phosphodiesterase gene leads to the massive degeneration of the rod photoreceptors (Fig. 1). The degeneration of rods begins at about postnatal day 8–10 and is virtually complete by day 30. This leads eventually to degeneration of the cones and to remodeling of the horizontal and bipolar cells (16–18). A few cones survive in the far periphery for many weeks after rod degeneration, but the final outcome is their almost complete disappearance (19–27) (Fig. 2).

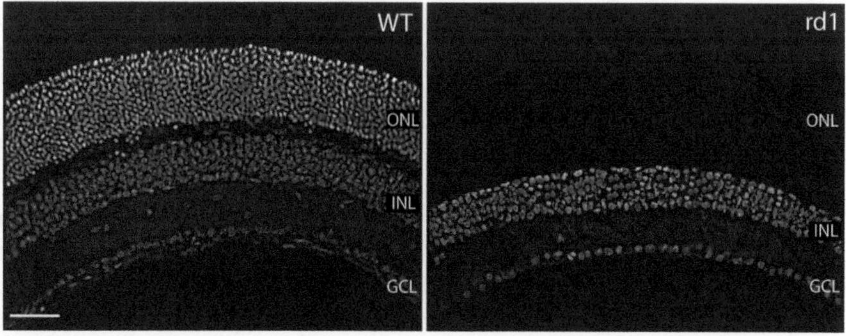

Fig. 1. Degeneration of photoreceptor cells in the *rd/rd* mouse. *WT* normal adult mouse retina, *Rd1* retina from an *rd/rd* mouse. The rod and cone photoreceptors are entirely lost in this section from the central retina of a mouse at 50 days postnatal. A few terminally degenerating cones may be present in the far periphery of the retina (26, 51), but most of the retina is photoreceptor-free. Not visible in this nuclear stain (DAPI) is the remodeling of the outer plexiform layer that follows rod and cone degeneration (see Fig. 2). Scale, 50 μm.

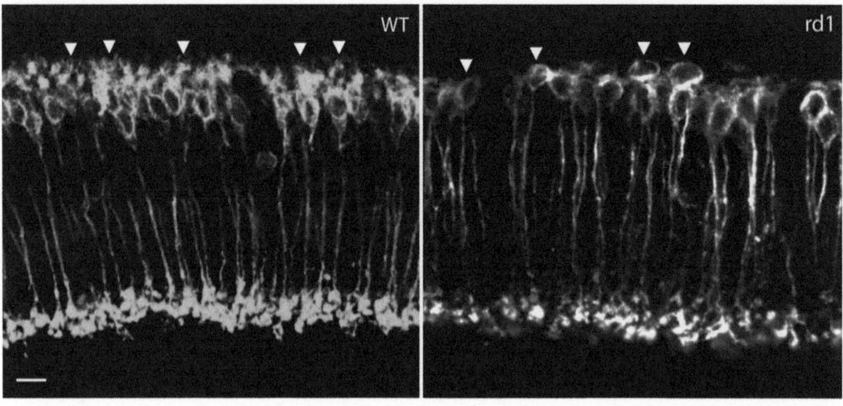

Fig. 2. Remodeling of rod bipolar cells in the *rd/rd* mouse. Cells are immunostained by antibodies against protein kinase C. *WT* normal adult mouse retina. *Triangles* indicate the dendrites of bipolar cells. *Rd1* retina from an *rd/rd* mouse. The dendrites of rod bipolar cells are entirely absent in the *rd/rd* mouse; note the row of "naked" cell bodies at the top of the inner nuclear layer (*triangles*). Scale, 10 μm.

Pan and his colleagues (Bi et al., (28)) used channerhodopsin-2 (Chop2) channels to convert retinal ganglion cells into photosensitive cells in the rd model of photoreceptor degeneration. The protein, originally discovered in green algae, directly forms light-sensitive membrane channels when ectopically expressed in mammalian cells (29).

Bi et al. made intravitreal injection in both normal and *rd1* eyes of an adeno-associated viral (AAV) vectors. The vector was AAV–CAG–Chop2–GFP–WPRE, coding for a Chop2–GFP fusion protein. The vector effectively targeted the ganglion cells and expressed a high-level Chop2 protein in them. The expression of Chop2 after intravitreal injection spread evenly across the retina. From the published images, it appears that 5–10% of all ganglion cells express Chop2. They included both ON and OFF ganglion cells, as well as occasional bipolar, horizontal, and amacrine cells. Not only did Chop2–GFP label the cell membrane of ganglion cell bodies, the entire dendritic arbor and the initial axon of some ganglion cells were also revealed by GFP labeling. The expression of GFP–Chop2 was stable for at least 1 year after the AAV injection, with no evidence of toxicity to the cells.

Extensive electrophysiological experiments were carried out. Initial recordings were from individual dissociated retinal ganglion cells; this is important because it eliminates the possibility that light responses were driven by any possibly surviving photoreceptor cells in the rd retinas. Chop2–GFP positive cells, either retinal ganglion cells or other inner retinal neurons, were identified on the basis of their GFP fluorescence. Light-evoked currents were observed in Chop2–GFP-expressing ganglion cells and other inner neurons. Responses were observed to light stimuli at different wavelengths ranging from 420 to 580 nm, with a peak at 460 nm. This is consistent with the reported peak action spectrum of Chop2 (29). In current clamp, Chop2 drove robust membrane depolarizations in nonspiking neurons and elicited spike firings in spiking neurons, such as the ganglion cells. Even at the optimal wavelength (460 nm) intensities ranging from 2.2×10^{15} to 1.8×10^{18} photos cm^{-2} s^{-1} were required. This was 9 log units brighter than the normal retina's most sensitive (rod-mediated) responses, and at least 5 log units less than the sensitivity of normal cone photoreceptors.

A multielectrode array was used to collect a large sample of recordings of retinal ganglion cells in the whole mounted retina of *rd* mice. Light-evoked spike activities were observed in surviving retinal ganglion cells and persisted even after application of a cocktail of glutamate receptor antagonists, which blocked potential signals from surviving photoreceptors to the ganglion cells. For instance, in the present of CNQX and AP5, which block glutamatergic synapses that would transmit potential signals from surviving photoreceptors to the ganglion cells, light-evoked spike activity was observed in 49 of 58 electrodes, suggesting the spike activity was driven purely and directly by Chop2 protein. (It also confirmed that a significant fraction of all ganglion cells had been transduced and responded to light in the *rd/rd* animals).

Finally, they asked whether the transduced retinal ganglion cells still projected into higher visual centers of the brain via optic nerves, and if so, whether these connections were still functional in the blind mice. Anatomically, by analyzing brain slices of Chop2–GFP-treated *rd1* mice, they confirmed the presence of axonal terminals of GFP-expressing retinal ganglion cells in several regions of the brain, including superior colliculus, dorsal and ventral lateral geniculate nuclei, and optical tract. This demonstrated that the Chop2–GFP-transduced retinal ganglion cells had kept their central projections in the blind mouse. To confirm that these projections were functional, they recorded visual evoked potentials from the visual cortex. Visual evoked potentials were recorded in 9 of 13 Chop2–GFP-treated *rd1* animals. To eliminate any possibility that the visual evoked potentials were driven by residual photoreceptor cells, which would respond at a longer wavelength, the mice also were given light stimuli at 580 nm. No visually evoked potentials were observed in any of the 13 mice tested under this condition. Therefore, Chop2–GFP-transduced retinal ganglion cells indeed are capable of transferring retinal signals to the brain and restoring visually evoked responses in the brain.

2.1. Channelrhodopsin-2 Expressed in Bipolar Cells

A more sophisticated use of Chop2 was reported by Lagali et al. (30), who sought to take advantage of the inner retina's signal processing capabilities, by expressing Chop2 in bipolar cells. While expressing Chop2 in ganglion cells generates a signal that will be transmitted to the brain, they reasoned, that signal – the firing of the ganglion cell – is determined in a fundamentally straightforward way by the opening of the Chop2 cation channel and thus directly limited by the response characteristics of Chop2. In a normal retina, on the other hand, the signal arriving at the retinal ganglion cell is processed by an extensive network of amacrine cells on its way to the ganglion cell. After photoreceptor cells excite bipolar cells, the bipolar cells make outputs not only on retinal ganglion cells but also on amacrine cells. An important role of the amacrine cells is to feed back on the bipolar cell terminals, regulating and adding transience to the output of the bipolar cell. In addition, specialized amacrine cells synapse directly on ganglion cells (amacrine synapses are in fact the majority of all synapses on a retinal ganglion cell) and these shape the responses of the ganglion cells. For example, the output of the bipolar cell depends on the visual context, i.e., on visual stimuli present in the immediate neighborhood of the ganglion cell. To take advantage of these inner retinal circuits, Lagali et al. caused Chop2 to be expressed in bipolar cells.

Lagali et al. added a second interesting wrinkle: they caused Chop2 to be expressed in only a subset of bipolar cells, those that normally respond to light with ON responses. Their reasoning was that expressing Chop2 in all bipolar cells would convert all of the cells into ON cells (since Chop2 is only excitatory – it simply opens a cation channel – when excited by light). Since some bipolar cells are normally OFF cells, expressing Chop2 indiscriminately in all bipolar cells would cause some to be converted from OFF cells to ON cells. This would be mirrored in the responses of the retinal ganglion cells, so that about half of all ganglion cells would be converted from OFF cells to ON cells. Depending on one's assumption about how the brain interprets retinal signals, this might be a problem. Lagali et al. therefore directed the expression of Chop2 specifically to ON bipolar cells.

Again the *rd/rd* strain of mice was studied. In contrast to the other studies reviewed here, Lagali et al. did not attempt to rescue the vision of animals after degeneration of their rods and cones. Instead, they expressed the transgenes by electroporation shortly after birth. They injected subretinally a plasmid that contain a 200-bp promoter sequence of the *Grm6* gene, which encodes the ON bipolar specific glutamate receptor mGluR6, and the sequence for a fusion protein incorporating Chop2 and the yellow fluorescent protein (YFP.) Electroporation was carried out at P_0 or P_1, by passage of capacitive currents through extraocular electrodes after the subretinal injection of plasmid.

The mGluR6 promoter successfully directed expression of the transgene to the ON bipolar cells (Fig. 3). The axonal endings of the Chop2 – expressing cells all terminated in the proximal half of the inner plexiform layer; this sublayer is known to

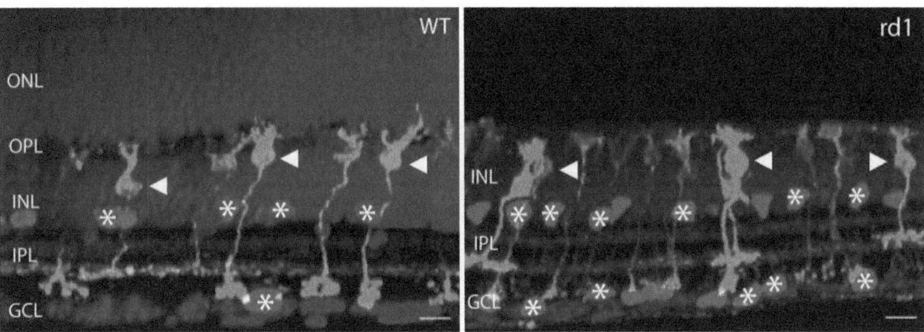

Fig. 3. Expression of Chop2–GFP in normal and degenerated mouse retinas. Many bipolar cells express GFP fluorescence (*triangles*). These bipolar cells send their axonal terminals to the lower half of the inner plexiform layer, indicating that they are ON bipolars. To demarcate the levels of the plexiform layer, the cholinergic amacrine cells (*asterisks*) are also labeled. Scale, 10 μm. By permission from Lagali et al. (30).

specifically contain the endings of the ON bipolar cells. Comparison with the known density of ON bipolar cells revealed that, in the electroporated areas, around 7% of the total population of ON bipolar cells expressed the fusion protein. These included both rod bipolar cells (those normally driven by rod photoreceptors) and cone bipolar cells.

The effectiveness of this manipulation was tested by recording from retinas spread over a multielectrode array, which records the final output of the retina from the retinal ganglion cells. The transduced bipolar cells were able to excite the ganglion cells and, as had been hoped, the ganglion cells showed evidence of the physiological processing known to occur in the inner plexiform layer. For example, there was evidence of center surround antagonism, in which large stimuli give less of a response by the ganglion cell than stimuli that match the size of the ganglion cell's receptive field. Furthermore, the responses of the ganglion cells were relatively transient. Like ganglion cells in wild-type retinas, the response of the ganglion cells did not perfectly mirror the course of the stimulus; it was truncated before the end of stimulation, thus mimicking the normally transient responses of many retinal ganglion cells.

There were certain differences from the normal ganglion cell response. The responses were less vigorous than normal, containing fewer spikes (this was also seen in the experiment of Bi et al., when Chop2 was expressed directly in the retinal ganglion cells). However, this did not prevent substantial and brisk responses recorded from the visual cortex. The signals transmitted to the brain from the ganglion cell had a relatively short latency – in the case illustrated, about 70 ms, which is close to that seen in the wild type. For the (very bright) stimulus tested, the amplitude of the cortical response was about 40% of that seen in the wild type. These findings indicate that the response of the ganglion cells is

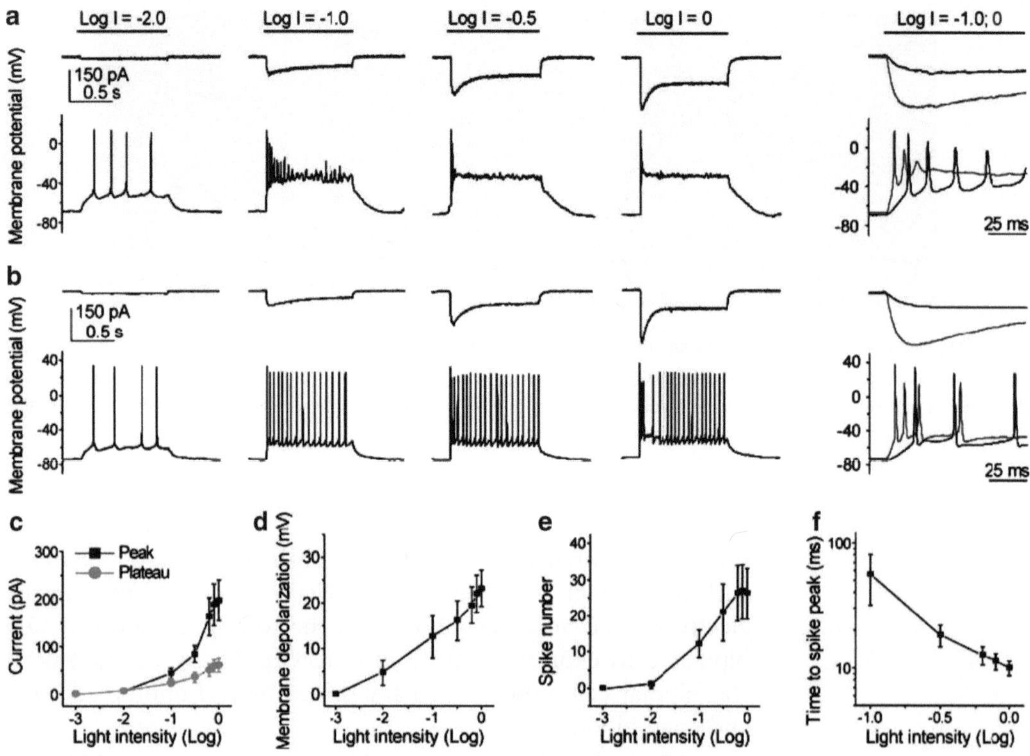

Fig. 4. Responses of retinal ganglion cells to light in an *rd/rd* retina in which Chop2 was expressed in retinal ganglion cells. (**a**, **b**) Show two contrasting types of light response in the Chop2 retina. (**c**–**f**) Intensity – response functions. By permission from Bi et al. (28).

vigorous enough to transmit a robust signal to the brain. The most notable difference between wild type- and Chop2-mediated responses was that the response of the Chop2 expressing retinas was much less sensitive than normal (Fig. 4). Confirming the experiments of Bi et al., the threshold for the Chop2-mediated responses was at least 5 log units higher than that for cone-driven responses in the wild type retinas. As the authors point out, the requirement for unnaturally intense illumination of the retina raises practical issues for using Chop2 in human therapy.

Finally, expressing Chop2 in these retinas restored some simple visual behaviors. When freely moving normal mice are suddenly illuminated, their locomotor behavior changes. They explore their surroundings more, and with a slightly different pattern of starts and stops than in the dark. These behaviors, which are measured by analyzing the locomotor patterns of the mice, were different in untreated *rd/rd* and Chop2 – treated animals, such that the treated animals behaved more like the wild type. A second test was the optomotor response, which is a variant on the familiar optokinetic nystagmus. Chop2 treatment did not restore the eye movements observed in the optokinetic nystagmus, but it did restore a type of following behavior in which the movement of the whole animal is influenced. A freely moving mouse is placed inside a tracking drum and its tendency to circle in one direction or the other is correlated with the rotation of the striped drum. Chop2 treatment restored the tendency of *rd/rd* mice to circle in the direction of movement of the striped drum.

2.2. Melanopsin Expressed in Ganglion Cells

A different approach was taken by Lin et al. (31), who expressed the melanopsin protein in retinal ganglion cells of *rd/rd* mice. This was done in adult animals, in an attempt to rescue vision that had been lost because of degeneration of the photoreceptor cells.

Melanopsin is a light-sensitive protein normally present in a small subset of retinal ganglion cells. These cells, numbering about 600 in the mouse, make a major projection to the intergeniculate leaflet (IGL) and olivary pretectal nucleus (OPN), and to the suprachiasmatic nucleus (SCN), where they have a primary role in synchronizing the circadian oscillator (32). The melanopsin photopigment is an opsin class of G-protein-coupled receptor that uses a *cis*-retinal-based chromophore (10–15, 33). Photostimulation causes isomerization of *cis*-retinal to all-*trans*-retinal and concomitant activation of a G-protein. Although in cultured mammalian cells and in *Xenopus* melanopsin appears to activate pertussis toxin-insensitive $G\alpha_q$ class of G-protein (10–15, 33, 34), there is some evidence that the photopigment can activate other class of G-proteins (13). Members of the $G\alpha_q$ family of G-proteins, $G\alpha_q$, and $G\alpha_{11}$ are functionally redundant (35) and are widely expressed in almost every mammalian tissue (36). The signal transduction pathway downstream of $G\alpha_q$-protein may involve activation of phospholipase-α and dependent intracellular events. The signaling cascade subsequently triggers an increase in intracellular calcium level, both in the ipRGCs (37) and in a heterologous expression system (14). Thus, unlike Chop2, melanopsin signals to the cell membrane with several intervening steps.

An established AAV2 vector injected intravitreally was used to transduce ganglion cells with the melanopsin gene. At ~P80 one of three viral constructs were injected: AAV–Opn4, coding for the melanopsin protein; AAV–Opn4–IRES–cGFP coding for melan opsin protein and enhanced green fluorescent protein; and AAV–eGFP, coding for the green fluorescent protein alone. The vector coding for eGFP alone was used as a sham-injected control.

All of the constructs were expressed in retinal neurons. The cells were distributed more or less uniformly around the retina, with occasional hot spots that were probably near the intravitreal injection site. Retinas injected with AAV–Opn4 had 4,437 ± 1,222 Opn4-expressing cells/retina, almost exactly 10% or the total number of ganglion cells normally present in the mouse (38).

Retinas injected with AAV–Opn4–IRES–eGFP were used to identify the transduced cells for recording. The fluorescence of

GFP was used to identify ganglion cells that had been transduced. This was done during brief illumination by the exciting wavelengths (longer exposure would have bleached the retina's photopigments). After a GFP-expressing cell had been identified, micropipettes for patch recording were guided to that cell under differential interference contrast (DIC) optics using infra red illumination. Once a recording had been established, the retina was stimulated by light. Identifying the cells for recording by their GFP expression ensured that recordings were not carried out from the native melanopsin cells.

The responses had the characteristics previously established for the native melanopsin cells. These are distinguished by long latency and very long persistence – for bright lights, tens of seconds – after termination of the visual stimulus. After recording was complete, the cells were injected with Lucifer Yellow, allowing visualization of their dendritic arbors. The dendritic arbors of the cells expressing melanopsin and GFP had a wide variety of morphologies; these clearly corresponded to many anatomically and functionally distinct types of ganglion cells (39–41). There was no noticeable correlation between the morphology of the cell and the characteristics of the light-induced response; almost all of the diverse physiological types of ganglion cell (ON, OFF, sustained, transient, etc.) responded to light. This result shows that the signaling system that couples melanopsin to membrane depolarization is ubiquitous, or at least very widespread, in retinal ganglion cells; it clearly is not restricted to the native melanopsin-expressing cells. Although there is no certainty that the signaling pathway is identical in every ganglion cell, a functional pathway that can couple activation of melanopsin to a membrane cation channel appears to be present in most types of retinal ganglion cell (as well as in many other neural and nonneuronal cells (10–15, 33)).

The visual abilities of these *rd/rd* mice were also investigated. These ranged from a simple visual reflex to a more complex learned behavior. The first was the pupillary light reflex (PLR). A high-threshold PLR is retained in *rd/rd* mice, mediated by the small complement of native melanopsin cells; sham-injected and untreated mice had a PLR, but it was ~3 log units less sensitive than the PLR in wild type. In *rd/rd* mice with ectopic expression of melanopsin, the PLR was returned to almost the sensitivity observed in mice that had not suffered photoreceptor degeneration (Fig. 4c). This was in fact somewhat surprising, as melanopsin-mediated responses by ganglion cells are generally less sensitive than responses mediated by the rods and cones. However, the responses to light are very much amplified in the melanopsin-transduced ganglion cells and last far longer than normal. This may compensate for a decreased sensitivity. An indication that the pupillary responses were in fact mediated by those responses is that the pupillary responses to light had substantially longer latencies than normal. This is in accord with the long latency of response recorded electrophysiologically in the transduced ganglion cells.

Normal mice avoid open, brightly lit spaces and this innate tendency is the basis of a simple test of their ability to see, similar in some ways to the open field test used by Lagali et al. The test is to give mice a choice between remaining in a bright open field and seeking a dark refuge. The fraction of time spent by the mice in the dark refuge is measured. Untreated *rd/rd* and sham-injected mice behaved randomly, spending ~50% of the time in both chambers. Expression of melanopsin returned the mice lacking rods and cones almost to normal behavior: Mice with normal retinas spent ~80% of the 300-s test interval in the dark, while melanopsin-expressing *rd/rd* mice spent 74%.

Finally, the mice were tested on a task that has a cognitive component, i.e., one in which the mice were required to make a decision based on visual information (42). The test was a classic two choice discrimination alley (Fig. 5a). The mice swam down an alley and had to choose between a bright target, which indicated the location of a safe underwater platform, or a dim one (no platform). For normal mice, this is an easy task: they learned it to greater than 90% accuracy in only a few days. *Rd/rd* mice injected with the AAV–GFP construct did not reach above-chance performance after 8 days of training. *rd/rd* Mice injected with the AAV–Opn4 construct showed a steady improvement, reaching a level of better than 80% correct after the 8-day testing sequence (Fig. 5b).

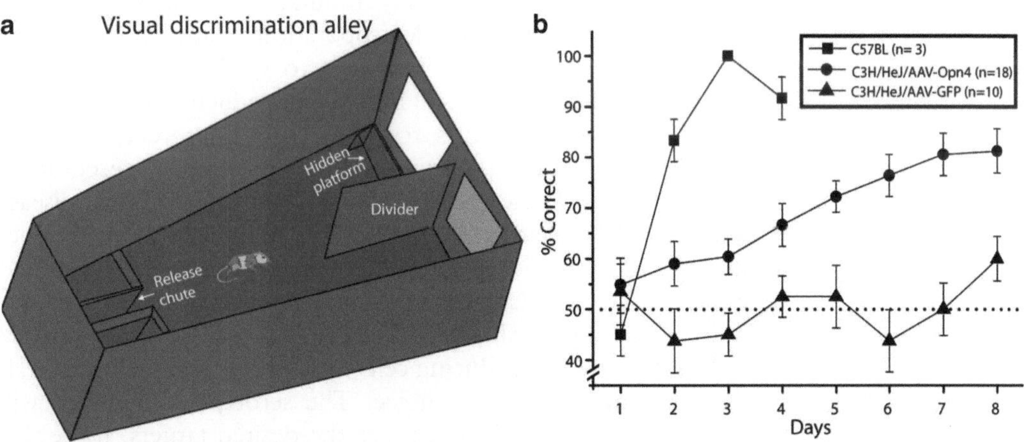

Fig. 5. Rescue of vision in *rd/rd* mice by expression of melanopsin in retinal ganglion cells. (**a**) The mice were required to make a visual discrimination at the end of a water-filled alley; the reward was access to a safe platform, indicated by a bright panel. (**b**) For wild-type mice (*squares*) this is an easy task, learned within a few days. Sham-injected mice (*triangles*) did not progress within the 8-day training cycle. During the same 8 days, melanopsin-treated mice (*circles*) achieved a success rate of >80% correct. By permission from Lin et al. (31).

3. Points of Consensus

Taken together, these results provide unequivocal proof of the principle that expressing a photosensitive protein in the retinal neurons of a retina with photoreceptor degeneration can restore at least simple visual abilities. There is agreement on a number of major points. (1) Channelrhodopsin and melanopsin can readily be expressed in various retinal neurons; (2) neither molecule is highly toxic to the cells, at least in the amounts expressed and for periods of up to 12 months; (3) they can cause retinal ganglion cells to become directly or indirectly responsive to light; (4) the light sensitivity persists for many months. The longest time studied behaviorally appears to be for melanopsin in the ganglion cells, which rescued the PLR for at least 11 months, but the persistence of the Chop2–GFP protein in the bipolar cells suggests that photosensitivity in them also lasted for a year or more – a substantial fraction of the lifetime of a mouse; and (5) the visual signals transmitted to the brain by the transduced retinal neurons can support vision. Several simple visual reflexes were successfully rescued, and in one case a visual behavior that would ordinarily be considered to depend on "conscious" decision-making was enabled.

4. The Future: Technical and Tactical Considerations

A number of decisions will be involved in developing photoreplacement as a practical therapy in humans. The first is choice of the gene therapy vector. Electroporation seems unlikely to be desirable, as is invasive and appears to work well only in very young animals. And the goal must be introduction of genes later in life, after the patient's rod and cone photoreceptors have degenerated. For this purpose, AAV vectors seem at present to be a clear choice. These cause no known human or animal disease and are widely regarded as safe. Because treatment of the eye does not require systemic injection, the risk of transducing unwanted cells is reduced. In fact, AAV2 is currently being used in a phase I human trial of gene therapy in the eye (8, 9). Since we know AAV to be effective in transducing cells with Chop2 and melanopsin, it would seem an obvious choice. The serotype to be used will depend on which retinal cells are the desired targets, to be discussed next. A related and not unimportant issue in choice of serotype is that their tropisms in humans are not well known. Different AAV serotypes preferentially transduce different types of retinal neurons (43). Unfortunately, these tropisms, the mechanisms of which are poorly understood, vary among animal species, so that what is true for the mouse may not be so in primates. Testing in animals with large eyes, including primates, will be required.

The choice of cells to be transduced is also somewhat open. Transducing retinal ganglion cells is a simple and direct approach. These are the cells that transmit a signal to the brain and it seems straightforward to target those cells for introduction of a light-sensitive protein. Several AAV serotypes target the retinal ganglion cells, without need for cell type-specific promoters. Importantly, these cells, in contrast to the bipolar and horizontal cells, appear not to remodel in response to photoreceptor degeneration (2).

On the other hand, transducing bipolar cells has the advantage pointed out by Lagali et al. that shaping and fine-tuning of the Chop2-generated signal can benefit from the synaptic networks of the inner retina. This makes them an appealing target, and Lagali et al. showed directly that at least some of the inner retinal computational functions are carried out when the Chop2-mediated signal is transmitted to the ganglion cell.

Targeting expression to a single class of ON bipolar cells may be an added advantage, which is that it prevents ganglion cells that are normally OFF cells from being turned into ON cells. This sounds on its face like a good thing – and it is an important precedent for cell-type-specific expression of functional genes – but it is not entirely clear how great its real-world importance for vision augmentation would be. The answer ultimately depends on how the brain interprets the output of the retina – whether the brain in some sense "knows" which ganglion cell is supposed to be ON and which is OFF, or instead is more plastic. A careful study in the macaque monkey showed behaviorally that vision is very little compromised in an animal whose retina transmits only OFF signals (the ON system was silenced by intraocular application of APB (44)). If one were to transduce all of the bipolar cells with Chop2 or melanopsin, rather than restricting transduction to the ON cells, then all ganglion cells would presumably become ON cells. It seems quite possible that the brain could put these signals to effective use – indeed; vision might even be improved over the more restricted case, because double the number of ganglion cells could respond to light and this would increase the potential visual resolution.

A final issue in targeting bipolar cells is that they undergo extensive remodeling after degeneration of the photoreceptors. They progressively retract their dendrites and eventually lose all of them (Fig. 2). The retraction of dendrites takes place initially in bipolar cells of the central retina and then gradually advances to those in the peripheral retina. Given that their dendrites are gone, it is very likely that the remodeled bipolar cells downregulate the receptors normally present in them, and that would limit the usefulness of the promoter for the dendritic glutamate receptors (mGluR6) that was used to target Chop2 expression to the ON bipolar cells. Even beyond the issue of specific targeting to one bipolar class, how well the remodeled cells could signal to the ganglion cells remains an open question.

Which protein – channelrhodopsin, melanopsin, or some other – would be most effective? Here, the choice is stark. Chop2, as already noted a membrane-bound molecule that incorporates a cation channel, responds to light with a short initial latency (tens of milliseconds) and a short persistence (tens to a few hundreds of milliseconds). Melanopsin, on the other hand, responds with an initial latency of at least a few hundreds of milliseconds, and its response can outlast the stimulus by tens of seconds – to a bright flash, some transduced cells in the experiments of Lin et al. responded for >100 s. The initial latency would not perhaps be a prohibitive problem, but the long persistence would be a major impediment to realistic visual function.

On the other hand, Chop2 is very insensitive to light. Both of the studies reviewed here estimate that its threshold for responding is approximately 5 log units higher than that of the cones in a normal retina (much less the rods, which are another 3 log units more sensitive.) In effect, Chop2 required light intensities similar to those generated by outdoor sunlight on a summer day. This puts severe limits on how useful Chop2 would be for everyday vision, as it is impractical – if only for the benefit of the patient's friends and family – to light our everyday environment so intensely. It also causes a concern for the safety of the retina, which is famously susceptible to damage by light, especially light at the short wavelengths required for stimulation of channelrhodopsin. Melanopsin, by contrast, functions at intensities approximating those of ordinary indoor room lighting. It is worth noting that Drosophila use a signaling system thought to be identical to that of melanopsin, and their photoresponses are very rapid indeed (45); therefore, slowness may not be intrinsic to the melanopsin system. Thus, Chop2 and melanopsin have contrasting advantages and disadvantages and both seem promising: a high priority experimental task is to discover or engineer a quickly acting melanopsin or a more sensitive channelrhodopsin.

5. Cautious Optimism: How Much Vision to Hope for?

The spatial resolution (acuity) of vision will be limited primarily by the sampling density of the photosensitive elements. The number of cells transduced depends, of course, on the effectiveness of the viral vector and the concentration of viral particles, so that definitive numbers are not yet available. However, it seems from the available evidence in animals that the sampling density of the transduced retina is unlikely to prevent some level of useful vision from being created. In all three of the experiments so far, around 10% of the target cells were transduced. For example, there were 4,437 ± 1,222 ectopically melanopsin-expressing ganglion cells in the retinas of the treated *rd/rd* mice. This amounts to an average of one cell for every 3.75° visual angle, a sampling density that would yield a very coarse visual image but one usable for simple visual tasks.

A small number of careful studies have directly correlated retinal ganglion cell numbers with visual performance in humans (46–49). Although these studies are difficult, the general conclusion seems to be that 30–50% of ganglion cells must be lost before any defect at all is detected by Humphrey perimetry. Many authors have pointed out that there is considerable redundancy in the coverage of the retina by the various types of retinal ganglion cell. From the simple point of view of image processing, this robustness of spatial vision makes sense. To a rough approximation, one might imagine the human retina as a 1-megapixel device (for its 10^6 optic axons). Reducing the pixel density to 10% (as occurs when one switches a digital camera from a resolution of 1,000 × 1,000 pixels to 333 × 333 pixels) allows very clear information to be transmitted. This density of transduced cells is available in animals by the AAV vectors that are available now. Even reducing the pixel density to 1% (10,000 pixels, or by analogy 10,000 transduced ganglion cells) would yield a very grainy image but still a useful one.

A different functional issue is the range of intensities that can be covered by the ectopically expressed protein. A normal retina contains multiple mechanisms of light adaptation, both intrinsic to the rods and cones and in the synaptic circuits of the retina. As a consequence, mammalian retinas can function over most of the ~8 log units of intensity with which the earth is irradiated by light in the visible spectrum, from starlight to sunlight. As was pointed out by Lagali et al., this range-shifting ability is lost for a naked Chop2 molecule in a cell membrane, where the working range is fixed and spans only about 100-fold. This limited range would be a big handicap in everyday life. In fact, it exceeds the variations in light intensity encountered when moving between rooms lit by ordinary household lighting. (The example is for illustrative purposes only, remembering that Chop2 cannot respond at all to light of ordinary indoor intensity.) Perhaps some degree of light adaptation would be added by the inner retinal circuits when Chop2 is expressed in bipolar cells, but this is likely to be minor. Thus, a Chop2-aided retina would presumably need some sort of external aid that kept the retinal input within Chop2's narrow working range – perhaps some sort of microchip-controlled eyeglasses, similar to a virtual reality device and incorporating image intensification and some mechanism for keeping the intensity delivered to the retina centered on the operating range of Chop2. Such a device is not far fetched but would take some engineering and might be cumbersome for the wearer.

In this regard, melanopsin might seem to have an advantage, as the native melanopsin cells have been shown to contain mechanisms for light adaptation (50). However, the range of adaptation (about 1.6 log units) is not great. Furthermore, we do not know that the adaptation mechanism still functions when melanopsin is expressed in neurons that do not normally express it.

The most likely initial use of such a therapy is for patients with late stage retinitis pigmentosa. These patients are densely blind, lacking even vision for large, bright objects. If 10% of the ganglion cells in such a patient became responsive to light, as is readily accomplished in the mouse, it is not overly optimistic to hope that the patient would at the least gain enough vision to orient by visual cues – to locate pieces of furniture, a door frame, and other coarse features of the environment. In fact, these visual tasks are not far removed from the one required of melanopsin-treated mice in the visual discrimination tested by Lin et al. For a densely blind patient, this level of vision would be useful.

It is too early to tell whether or not detailed vision (face recognition) could be achieved using directly photosensitive nonphotoreceptor neurons; that will depend on the numbers of cells that can be transduced, and on the solution found for the problems of sensitivity and timing. If favorable solutions are achieved, vision might be improved not only in retinitis pigmentosa but also for patients suffering from macular degeneration. This is a more challenging task. These patients need help for visual functions more complex than guiding simple ambulation, a capability that macular degeneration patients generally retain by using their intact peripheral retinas. It seems possible from animal experiments that enough cells for an improvement in acuity could be transduced by local subretinal application of a gene therapy vector in the macula. However, this route of delivery has been little studied and many unknowns remain.

This review of the literature was completed in September 2008 and much useful work has been published between then and publication of this volume (early 2011). However, the fundamental issues for photoreplacement therapy – the insensitivity of the channelrhodopsins and slowness of melanopsin – are unchanged.

References

1. Ames A III, Li YY (1992) Energy requirements of glutamatergic pathways in rabbit retina. J Neurosci 12:4234
2. Marc RE, Jones BW, Strettoi E (2003) Neural remodeling in retinal degeneration. Prog Retin Eye Res 22(5):607–655
3. Medeiros NE, Curcio CA (2001) Preservation of ganglion cell layer neurons in age-related macular degeneration. Invest Ophthalmol Vis Sci 424:795–803
4. Bennett J (2000) Gene therapy for retinitis pigmentosa. Curr Opin Mol Ther 2(4):420
5. Campochiaro PA (2002) Gene therapy for retinal and choroidal diseases. Expert Opin Biol Ther 2(5):537–544
6. Sakamoto T, Ikeda Y, Yonemitsu Y (2001) Gene targeting to the retina. Adv Drug Deliv Rev 52(1):93–102
7. Lanyi JK (1986) Halorhodopsin: a light-driven chloride ion pump. Annu Rev Biophys Biophys Chem 15:11–28
8. Dinculescu A, Glushakova L, Min SH, Hauswirth WW (2005) Adeno-associated virus-vectored gene therapy for retinal disease. Hum Gene Ther 16(6):649–663
9. Maguire AM, Simonelli F, Pierce EA et al (2008) Safety and efficacy of gene transfer for Leber's congenital amaurosis. N Engl J Med 358(21):2240–2248
10. Berson DM, Dunn FA, Takao M (2002) Phototransduction by retinal ganglion cells that set the circadian clock. Science 295: 1070–1073
11. Isoldi MC, Rollag MD, Castrucci AM, Provencio I (2005) Rhabdomeric phototransduction initiated by the vertebrate photopigment melanopsin. Proc Natl Acad Sci USA 102(4):1217–1221
12. Melyan Z, Tarttelin EE, Bellingham J, Lucas RJ, Hankins MW (2005) Addition of human melanopsin renders mammalian cells photoresponsive. Nature 433(7027):741–745
13. Newman LA, Walker MT, Brown RL, Cronin TW, Robinson PR (2003) Melanopsin forms a functional short-wavelength photopigment. Biochemistry 42(44):12734–12738
14. Panda S, Nayak SK, Campo B, Walker JR, Hogenesch JB, Jegla T (2005) Illumination of the melanopsin signaling pathway. Science 307(5709):600–604
15. Qiu X, Kumbalasiri T, Carlson SM et al (2005) Induction of photosensitivity by heterologous expression of melanopsin. Nature 433(7027): 745–749
16. Jones BW, Watt CB, Frederick JM et al (2003) Retinal remodeling triggered by photoreceptor degenerations. J Comp Neurol 464:1–16
17. Strettoi E, Pignatelli V (2000) Modifications of retinal neurons in a mouse model of retinitis pigmentosa. Proc Natl Acad Sci USA 97:11020–11025
18. Strettoi E, Porciatti V, Falsini B, Pignatelli V, Rossi C (2002) Morphological and functional abnormalities in the inner retina of the rd/rd mouse. J Neurosci 22:5492–5504
19. Berson EL (1993) Retinitis pigmentosa (The Friedenwald Lecture). Invest Ophthalmol Vis Sci 34:1659–1676
20. Blanks JC, Adinolfi AM, Lolley RN (1974) Synaptogenesis in the photoreceptor terminal of the mouse retina. J Comp Neurol 156(81):81–93
21. Carter-Dawson LD, LaVail MM, Sidman RL (1978) Differential effect of the rd mutation on rods and cones in the mouse retina. Invest Ophthalmol Vis Sci 17(6):489–498
22. Chang GQ, Hao Y, Wong F (1993) Apoptosis: final common pathway of photoreceptor death in rd, rds, and rhodopsin mutant mice. Neuron 11:595–605
23. Farber DB, Flannery JG, Bowes-Rickman C (1994) The rd mouse story: seventy years of research on an animal model of inherited retinal degeneration. Prog Retin Eye Res 13:31–64
24. García-Fernández JM, Jimenez AJ, Foster RG (1995) The persistence of cone photoreceptors within the dorsal retina of aged retinally degenerate mice (rd/rd): implications for circadian organization. Neurosci Lett 187(1):33–36
25. Jiménez AJ, García-Fernández JM, González B, Foster RG (1996) The spatio-temporal pattern of photoreceptor degeneration in the aged rd/rd mouse retina. Cell Tissue Res 284(2):193–202
26. LaVail MM, Matthes MT, Yasumura D, Steinberg RH (1997) Variability in rate of cone degeneration in the retinal degeneration (rd/rd) mouse. Exp Eye Res 65(1):45–50
27. Milam AH, Li ZY, Fariss RN (1998) Histopathology of the human retina in retinitis pigmentosa. Prog Retin Eye Res 17:175–205
28. Bi A, Cui J, Ma YP et al (2006) Ectopic expression of a microbial-type rhodopsin restores visual responses in mice with photoreceptor degeneration. Neuron 50:23–33
29. Nagel G, Szellas T, Huhn W et al (2003) Channelrhodopsin-2, a directly light-gated cation-selective membrane channel. Proc Natl Acad Sci USA 100(24):13940–13945

30. Lagali PS, Balya D, Awatramani GB et al (2008) Light-activated channels targeted to ON bipolar cells restore visual function in retinal degeneration. Nat Neurosci 11(6):667–675
31. Lin B, Koizumi A, Tanaka N, Panda S, Masland RH (2008) Restoration of visual function in retinal degeneration mice by ectopic expression of melanopsin. Proc Natl Acad Sci USA 105:16009–16014
32. Hattar S, Kumar M, Park A et al (2006) Central projections of melanopsin-expressing retinal ganglion cells in the mouse. J Comp Neurol 497(3):326–349
33. Hattar S, Liao HW, Takao M, Berson DM, Yau KW (2002) Melanopsin-containing retinal ganglion cells: architecture, projections, and intrinsic photosensitivity. Science 295:1065–1070
34. Dacey DM, Liao HW, Peterson BB et al (2005) Melanopsin-expressing ganglion cells in primate retina signal colour and irradiance and project to the LGN. Nature 433(7027): 749–754
35. Neves SR, Ram PT, Iyengar R (2002) G protein pathways. Science 296:1636–1639
36. Su AI, Wiltshire T, Batalov S et al (2004) A gene atlas of the mouse and human protein-encoding transcriptomes. Proc Natl Acad Sci USA 101:6062–6067
37. Sekaran S, Sekaran S, Foster RG, Lucas RJ, Hankins MW (2003) Calcium imaging reveals a network of intrinsically light-sensitive inner-retinal neurons. Curr Biol 13:1290–1298
38. Jeon CJ, Strettoi E, Masland RH (1998) The major cell populations of the mouse retina. J Neurosci 18:8936–8946
39. Badea TC, Nathans J (2004) Quantitative analysis of neuronal morphologies in the mouse retina visualized by using a genetically directed reporter. J Comp Neurol 480:331–351
40. Coombs JL, Van Der List D, Chalupa LM (2007) Morphological properties of mouse retinal ganglion cells during postnatal development. J Comp Neurol 503:803–814
41. Kong JH, Fish DR, Rockhill RL, Masland RH (2005) The diversity of ganglion cells in the mouse retina: unsupervised morphological classification and its limits. J Comp Neurol 489(3):293–310
42. Wong AA, Brown RE (2006) Visual detection, pattern discrimination and visual acuity in 14 strains of mice. Genes Brain Behav 5(5):389–403
43. Pang JJ, Lauramore A, Deng WT et al (2008) Comparative analysis of in vivo and in vitro AAV vector transduction in the neonatal mouse retina: effects of serotype and site of administration. Vision Res 48(3):377–385
44. Schiller PH, Sandell JH, Maunsell JH (1986) Functions of the ON and OFF channels of the visual system. Nature 322:824–825
45. Hardie RC, Postma M (2008) Phototransduction in microvillar photoreceptors of drosophila and other invertebrates. In: Masland RH, Albright T (eds) The senses, 1st edn. Academic, Oxford, pp 77–130
46. Harwerth RS, Carter-Dawson L, Shen F, Smith EL III, Crawford ML (1999) Ganglion cell losses underlying visual field defects from experimental glaucoma. Invest Ophthalmol Vis Sci 40(10):2242–2250
47. Kerrigan-Baumrind LA, Quigley HA, Pease ME, Kerrigan DF, Mitchell RS (2000) Number of ganglion cells in glaucoma eyes compared with threshold visual field tests in the same persons. Invest Ophthalmol Vis Sci 41(3):741–748
48. Quigley HA, Addicks EM, Green WR (1982) Optic nerve damage in human glaucoma. III. Quantitative correlation of nerve fiber loss and visual field defect in glaucoma, ischemic neuropathy, papilledema, and toxic neuropathy. Arch Ophthalmol 100(1):135–146
49. Sommer A, Katz J, Quigley HA et al (1991) Clinically detectable nerve fiber atrophy precedes the onset of glaucomatous field loss. Arch Ophthalmol 109(1):77–83
50. Wong KY, Dunn FA, Berson DM (2005) Photoreceptor adaptation in intrinsically photosensitive retinal ganglion cells. Neuron 48:1001–1010
51. Lin B, Masland RH, Strettoi E (2009) Remodeling of cone photoreceptor cells after rod degeneration in rd mice. Exp Eye Res 88(3):589–599

Part III

Molecular Photoswitch Conjugates to Remotely Affect Activity

Chapter 9

Introduction to Part III: Small Molecule Photoswitches

James J. Chambers and Richard H. Kramer

Abstract

In a truly interdisciplinary fashion, organic chemists and molecular biologists have engineered novel systems that allow externally applied light to regulate protein confirmation in living systems. These highly engineered systems typically involve two distinct parts. The first part is a small molecule, photosensitive chromophore that responds to light by changing shape or conformation. At one end of this molecule is usually a ligand or an effector molecule and at the other, sometimes a chemically reactive group that targets a particular amino acid. The second part of these systems is typically a genetically modified protein that has been designed to present a reactive site on the surface of a cell to allow facile chemical coupling of the photoswitch to the protein. These two parts together allow for light to regulate cellular activity by mediating membrane voltage or protein conformation.

Key words: Photoswitches, Azobenzene, Spiropyran, Photochromic molecules

Photosensitive tools for neuronal control can be rationally designed and manufactured through synthetic chemistry. The general strategy is to couple a photoisomerizable molecule (i.e., a "photoswitch") with a ligand- or voltage-gated ion channel or receptor to make it sensitive to light. In theory, the photoswitch can be attached in such a way that a photoisomerization event exerts force on the channel, causing it to open. Alternatively, the photoisomerization event could deliver or remove a ligand from a binding site on the channel or receptor, thereby regulating its activity. In practice, the photoswitchable ligand approach has worked nicely with voltage-gated K^+ channels and glutamate receptors; however, this approach could apply to virtually any ion channel or receptor, so long as there are known ligands that regulate activity.

There are several chemical photoswitches available, but the photoisomerizable small molecule azobenzene has emerged as one of the best suited for biological applications. In darkness, azobenzene exists in a linear *trans* configuration, but 380-nm light promotes isomerization to the bent *cis* configuration which is ~7 Å shorter.

James J. Chambers and Richard H. Kramer (eds.), *Photosensitive Molecules for Controlling Biological Function*, Neuromethods, vol. 55, DOI 10.1007/978-1-61779-031-7_9, © Springer Science+Business Media, LLC 2011

In darkness, the *cis* form relaxes slowly back to the *trans* form (over minutes), but this relaxation can be accelerated by exposure to 500-nm light. Azobenzene compounds are relatively easy to synthesize and decorate, have well-defined geometries, and show high photochemical stability and little photo-induced toxicity.

Erlanger and colleagues were the first to apply the photoisomerizable ligand approach to control the activity of a receptor. They synthesized a water-soluble photoisomerizable molecule, Bis-Q, and a cysteine-reactive derivative, QBr, both of which activate the nicotinic acetylcholine receptor (nAChR) in the *trans* configuration, but not in the *cis* configuration. QBr covalently attaches to the nAChR after reduction of the disulfide bonds between native cysteine residues. QBr was particularly useful for rapidly delivering and removing the ligand to minimize desensitization of the nAChR, enabling detailed study of the mechanisms of receptor activation.

The molecular biology revolution has allowed investigators to take the photoswitchable ligand approach one step further by modifying the protein partner. Instead of relying on native cysteine residues present in the protein, a particular channel or receptor can be targeted for photoswitch attachment by genetically encoding a cysteine into the appropriate location on the protein. The first step is to identify a ligand that can be modified so that it can be conjugated to the azobenzene without losing its ability to bind and regulate channel activity. Structural information about ion channels and receptors can then guide the engineering of the target protein, in particular the position of the cysteine attachment site.

This iterative approach was first used to generate a synthetic photoswitchable azobenzene-regulated K^+ (SPARK) channel. SPARK channels are generated by coupling a photoswitchable ligand, *m*aleimide–*a*zobenzene–*q*uaternary ammonium (MAQ), onto a genetically engineered Shaker K^+ channel. The maleimide is for cysteine tethering, the azobenzene is for photoswitching, and the quaternary ammonium group blocks the pore of the Shaker channel. The channel is blocked only when MAQ is in its extended *trans* form, and not in the shorter *cis* form. MAQ enables control of action potential firing only in neurons that express the cysteine-containing Shaker channel. Visible light blocks SPARK channels, allowing action potential firing. UV light retracts the pore blocker, promoting the flux of potassium ions through the channel, which hyperpolarizes the neuron and inhibits, or silences, natural action potential firing. A mutation that alters the ionic selectivity of the K^+ channel changes the polarity of the effects, enabling depolarization and induction of action potentials with UV light.

Glutamate receptors are another class of ion channels where photoswitchable tethered ligands have proven successful. In 2006, a light-gated ionotropic glutamate receptor (LiGluR) was introduced. This system is based on a tethered glutamate derivative covalently attached to a genetically engineered kainate receptor (iGluR6) via a similar azobenzene tether. In its original embodiment, the tethered neurotransmitter was presented to the clamshell-like binding site in the *cis* configuration (380-nm light) and retracted in the *trans* configuration (500-nm light). Changing the attachment site reversed the polarity, with 500-nm light turning the receptor on and 380-nm light turning it off. LiGluR and its modifications can be employed to control neural activity in vitro and in vivo.

The first-generation light-activated K^+ channel (SPARK) and glutamate receptor (LiGluR) were designed specifically for light-induced neuronal inhibition and excitation. Each of these photoswitch-ready channels was derived from a particular generic channel, chosen because of the availability of prior structure–function information and favorable properties. However, there is no reason why many other K^+ channels and glutamate receptors (or for that matter, any ligand-modulated ion channel) could not be made photoswitchable if a cysteine attachment site were included in the correct position on the channel. Indeed, given sufficient motivation by chemists and neurobiologists, the photoswitchable tethered ligand approach should be applicable to many other types of voltage- and ligand-gated channels.

SPARK channels and LiGluR, like ChR2 and NpHR, are genetically encoded tools and, therefore, can be targeted to particular types of neurons by selective gene expression. The neuronal specificity that comes from genetic targeting is often an advantage, but in some cases, exogenous gene expression is not practical and may not even be desirable (e.g., in humans). This has motivated the development of small molecule photoswitches that can be used in unadulterated tissue and act on native channels or receptors without requiring exogenous gene expression.

An azobenzene-containing photoswitch has also been developed that enables photoregulation of native glutamate receptors. This "reversibly caged" glutamate (Glu-Azo) was shown to act on kainate receptors and reversibly trigger action potential firing in dissociated hippocampal neurons. Although its reversibility might be considered an advantage over classical caged glutamate, its usefulness in brain slices and live animals remains to be demonstrated.

Chapter 10

Photoswitch Design

Andrew A. Beharry and G. Andrew Woolley

Abstract

Photocontrol of protein function with azobenzene-based photoswitches promises to be a powerful tool for probing roles of proteins in vivo. In designing azobenzene-based switches for in vivo use, a number of challenges must be met. In this short review, we highlight progress in meeting some of these challenges. In particular, we focus on recent approaches to achieve (1) a large alteration in conformation and thereby function of azobenzene-modified proteins, (2) long photoswitching wavelengths and long-lived *cis* isomers, and (3) switches that are stable in the reducing intracellular environment.

Key words: Photoswitch, Azobenzene, Light-regulated structure, Photoswitch design

1. Introduction

Elucidating complex protein-mediated biological pathways is a major challenge in chemical biology. Correlating protein function with cellular activity is difficult due to the large number of spatial and temporally regulated cellular processes. For example, a protein may play a particular role in the cell nucleus during a specific growth state. Eliminating or modifying the activity of the protein using genetic tools (e.g., gene knockouts) may not elucidate function since other activities of the protein in other parts of the cell or at other times are also modified. Perturbing protein function using small molecules can permit better temporal and spatial control; however, any lack of specificity makes it difficult to assign the biological consequence to the role of the targeted protein (1). Photocontrol, where the functions of specific proteins can be turned on and off with light, offers an unparalleled level of spatiotemporal control (2, 3).

Engineering a particular target protein so that it is light-sensitive, however, has been a major challenge. An example of a

James J. Chambers and Richard H. Kramer (eds.), *Photosensitive Molecules for Controlling Biological Function*, Neuromethods, vol. 55, DOI 10.1007/978-1-61779-031-7_10, © Springer Science+Business Media, LLC 2011

widely used chemical approach for doing so is the use of caged compounds (2, 4). These small molecule chromophores have been incorporated into proteins to block their function. Absorption of a photon triggers a photolysis reaction which removes the cage and releases the active protein. The *irreversible* photochemistry of these compounds has led to their designation as "phototriggers" (2). That is, once photolysis (or phototriggering) has occurred, the biomolecule can only remain active. Another approach for achieving photosensitivity of a protein is the use of a photoswitch. These molecules undergo inherently *reversible* photochemistry such that the active and inactive states of the biomolecule can be produced in multiple rounds (5). In principle, reversible photochemistry can be used to control cellular activity as a function of time in more complex ways than can be accomplished with irreversible photochemistry. Indeed, a number of naturally occurring, reversible, photoswitchable proteins are known to regulate biochemical processes (6, 7). These proteins contain chromophores that undergo photochemical changes which are then coupled to a protein conformational change. We and others have attempted to mimic this natural photoswitching by incorporating photoisomerizable chromophores site-specifically into proteins (3, 8, 9). The challenges have been to rationally design, synthesize, and apply these small molecule photoswitches effectively such that an alteration in biochemical activity can be achieved within the native cellular environment.

2. Azobenzene-Based Photoswitches

The properties of a range of photoswitches can be found in the book "Molecular Switches" by Ben L. Feringa (10). Perhaps the most widely used class of photoswitches are the azobenzenes (11). Azobenzene can exist in *trans* or *cis* conformations, with the *trans* form more stable by approximately 10 kcal/mol (12, 13). At equilibrium in the dark, azobenzene is essentially all-*trans*. Irradiation at 340 nm produces the *cis* isomer which can revert back to the *trans* state thermally or via irradiation at 450 nm (Fig. 1). The difference in properties between the *trans* and *cis*

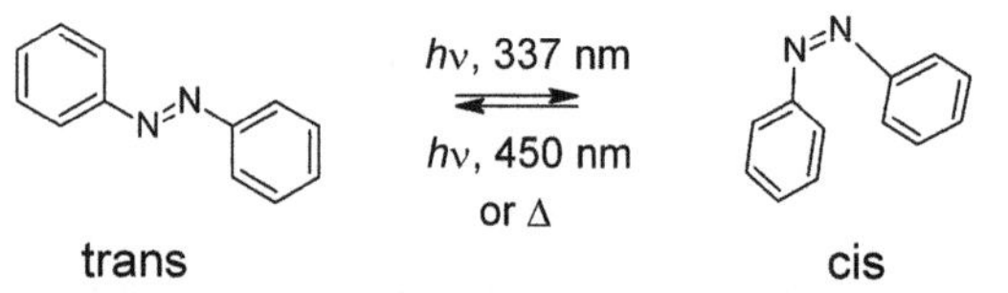

Fig. 1. Photoisomerization of Azobenzene. In the dark, azobenzene exists in the *trans* conformation. Irradiation at 337 nm produces the *cis* isomer. Thermal relaxation or irradiation at 450 nm regenerates the *trans* isomer.

isomers makes azobenzene attractive for photocontrolling biomolecules. For example, the *cis* form has a bent shape with a shorter end-to-end distance than the extended *trans* form. Also, the *trans* form has a dipole moment of nearly zero, while the *cis* has a dipole moment of approximately 3 Debye. Large fold-changes in the amount of each isomer can be produced photochemically with high quantum yields and (for many derivatives) with minimal photobleaching. In addition, the photoisomerization event takes place on a picosecond time scale, allowing study of light-induced dynamics of the attached biomolecule with good temporal separation from the switch.

In principle, the *trans–cis* isomerization of azobenzene can be used to directly affect the structure/activity of a protein. For example, light-sensitive enzymes have been generated via nonspecific labeling of protein lysine residues with an amine-reactive azobenzene derivative (14). Azobenzene has also been incorporated site-specifically near an enzyme active site by fragment complementation (15) or non-natural amino acid mutagenesis (16, 17). In all these cases, the observed effects are not easy to predict, difficult to interpret in structural terms and only modest changes in activity were produced. Others have embarked on a more structure-based approach for using azobenzene to alter activity. Initial investigations exploited the shape change of azobenzene to design free photoswitchable inhibitors of enzymes and ion channels (18, 19). Later, light-switchable ion channels were developed by tethering an azobenzene-containing ligand near the ligand-binding site (20).

In some cases, linking both ends of azobenzene to a target biomolecule can have a large effect on protein conformation and activity. Moroder and colleagues incorporated azobenzene within the backbone of cyclic peptides (21–23). The *cis* form, having a shorter end-to-end distance, promoted formation of a beta-type turn, while the *trans* form led to a series of unfolded conformations. Although this is a powerful approach for altering secondary structure of one common protein motif, backbone incorporation is not straightforward with larger proteins. We have focused on developing a strategy for photocontrolling proteins via side chain reactive switches that are intramolecularly cross-linked after protein synthesis. Below, we outline several criteria which need to be met for the activity of a protein to be effectively altered and for the switch to operate inside cells.

3. Desirable Properties for Biological Photoswitches

To achieve effective photocontrol of protein function in vivo: (1) the isomerization of the photoswitch must be effectively coupled to a protein conformational change. More specifically, the protein must be inactive in one isomeric state of the switch and active in the other. (2) Since the absorption spectra of *trans* and *cis* isomers overlap extensively, irradiation typically produces photostationary states that are ~80% *cis* or ~95% *trans* (15). Thermal relaxation, however, produces 100% of the *trans* isomer and is therefore preferred in order to generate larger changes in *cis* isomer content. Since unmodified azobenzene relaxes on a timescale of days at room temperature, derivates with faster relaxation rates may be required if thermal relaxation is used to reset the switch. (3) The photoswitch must undergo photochemistry at wavelengths compatible with cells and tissues (>350 nm). Longer absorption wavelengths will reduce the degree of light scattering allowing for deeper penetration within tissues, in addition to avoiding absorption by other biomolecules (e.g., NADH). Since unmodified azobenzene absorbs in the UV region (~340 nm), red-shifting its absorption wavelength is desirable. (4) The azobenzene-based photoswitch must be stable in the cellular environment. That is, it must not be metabolized or degraded in some way, once introduced. (5) Finally, the ease of synthesis and subsequent conjugation to the protein is also a consideration. The remainder of this review will focus on how we and others have tried to meet the criteria outlined above to photocontrol the structure and function of proteins using azobenzene-based photoswitches.

4. Photo controlling α-Helices

Due to the widespread occurrence of α-helices in proteins and the large body of knowledge concerning helix stability and conformational dynamics, we focused on photocontrolling this element of protein secondary structure (8). We designed azobenzene cross-linkers containing chloroacetamide groups to allow for selective modification of cysteine residues in the presence of other reactive side-chains (24, 25). Cysteine residues occur relatively rarely in proteins and can be incorporated site-selectively by standard mutagenesis techniques. Sulfonate groups were added to confer water solubility on the switch, allowing for cross-linking reactions to be carried out under aqueous conditions (26). The switch was designed to be symmetrical so that only one species is formed upon cross-linking. We note that using maleimide groups for Cys reactivity can lead to various diastereomeric products. Finally, a minimum number of single bonds were used to link the azo moiety and the peptide backbone to effectively couple the isomerization event to a conformational change.

Starting from commercially available 2,5-diaminobenzenesulfonic acid, the 2-amino group can be selectively acetylated. Oxidation to the azo gives the acetylated azobenzene photoswitch

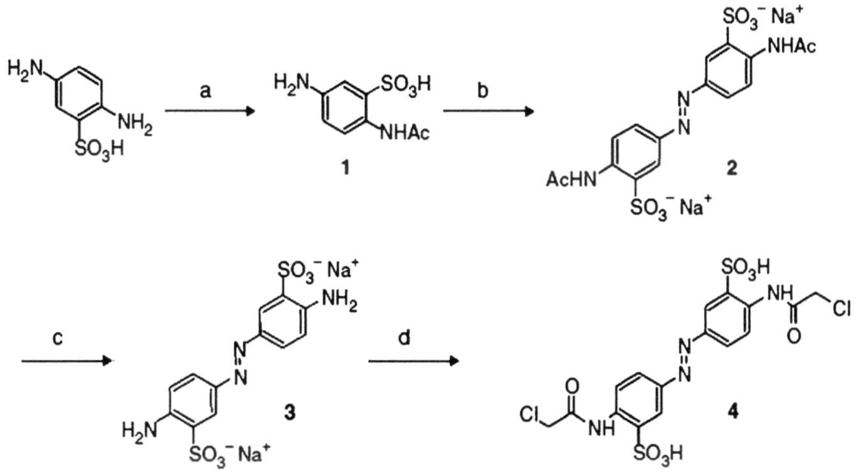

Fig. 2. Synthesis of a water soluble thiol-reactive diamidoazobenzene. (**a**) glacial acetic acid and acetic anhydride, (**b**) water, sodium carbonate followed by sodium hypochlorite, (**c**) water and HCl followed by NaOH, and (**d**) chloroacetic acid and chloroacetic anhydride.

(2); deprotection and chloroacetylation of the para-amino groups yields the thiol-reactive sulfonated diamidoazobenzene (25, 26) (Fig. 2).

Using molecular dynamics simulations, the distributions of end-to-end distances were calculated for the *cis* and *trans* forms of this cross-linker. Using this information, model helical peptides were designed with pairs of Cys residues spaced such that one isomeric form of the linker would stabilize or "fit" the structure of the helix (27). For example, with cysteines spaced i, $i+7$ residues apart on the peptide, the *cis* form end-to-end distance range (11–15 Å) is compatible with helical structure. In contrast, the *trans* form has an end-to-end distance range (19–23 Å) that is too extended to stabilize the helical conformation for this Cys spacing. *Trans* to *cis* photoisomerization at 365 nm thus leads to an increase in helical content, whereas thermal *cis* to *trans* relaxation in the dark causes a decrease in helix content (27). A longer cysteine spacing (i.e. i, $i+11$) can produce the opposite effect where the *trans* (dark-adapted) form of the cross-linker stabilizes helical structure while the *cis* form decreases it (28). A detailed conformational analysis revealed that photocontrol of helix content does not involve specific interactions between the switch and the peptide, making this approach general for a variety of peptide sequences (27, 29). Time-resolved ORD and IR measurements revealed that azobenzene isomerization occurred on a picosecond timescale (similar to unmodified azobenzene), whereas peptide folding/unfolding occurred on a 100 ns to 1 μs timescale (similar to an unperturbed helical peptide) (30, 31). Thus, the linker can be viewed as simply corralling the intrinsic conformational dynamics of the peptide, which can be predicted by comparing end-to-end distance ranges. This approach has been used to reversibly photocontrol the binding of α-helical peptides to proteins (32) and DNA (33, 34).

5. Spectral Tuning and Altering Thermal Relaxation

The thermal relaxation of diamidoazobenzene from *cis* to *trans* occurs with a half-life of ~20 min at room temperature (24). The half-life of the *cis* isomer is one important aspect in determining the practical usefulness of the switch. If one wishes to carry out detailed structural studies of the attached biomolecule, a longer half-life of the *cis* isomer may be required. On the other hand, if one wishes to produce a pulsed conformational change, then rapid return to the *trans* state is desired. Altering the nature of the substituents on the aromatic rings can substantially affect the thermal relaxation rate. The effects of substituents can be rationalized by considering their effects on the degree of N–N single bond character of the azo moiety. Although the mechanism of isomerization has been debated (i.e., rotation or inversion) (13, 35–37), an increase in resonance or dipole character is found to lower the activation barrier for *cis* to *trans* thermal isomerization. For example, to enhance the lifetime of the *cis* isomer, the diamido moiety at the para positions was replaced by an alkyl sp^3 carbon ($\tau_{1/2}$ 43 h). To shorten the lifetime of the diamido switch, carbamate ($\tau_{1/2}$ 80 s) or urea groups ($\tau_{1/2}$ 11 s) were incorporated at the para positions (38) (Fig. 3).

Substituents on the phenyl rings of azobenzene also have a strong influence on the absorption maximum. As the degree of electron donation at ortho and para positions is increased, the absorption maximum red-shifts (39). The excited state has been proposed to have similar dipole character to the transition state for thermal relaxation (40). To increase the switching wavelengths of azobenzene-based photoswitches, we designed a para-substituted diethylaminoazobenzene (41). This exhibited a λ_{max} of ~480 nm, a substantial red-shift compared to unmodified azobenzene, and a half-life for thermal relaxation of ~50 ms at room temperature. A piperazine moiety was then incorporated to conformationally constrain the switch (42) (Fig. 4). Photoswitching in this case occurred at ~400–450 nm and the thermal relaxation rate was on the timescale of a few seconds. Twisting of the piperazine ring relative to the phenyl ring to reduce steric interactions between the α-methylene and ortho hydrogens on the piperazine and phenyl rings is likely responsible for the hypsochromic shift and slower rate of relaxation compared to the acyclic counterpart (42).

Although a photoswitch with fast thermal relaxation may serve as a useful tool to photocontrol the function of a protein

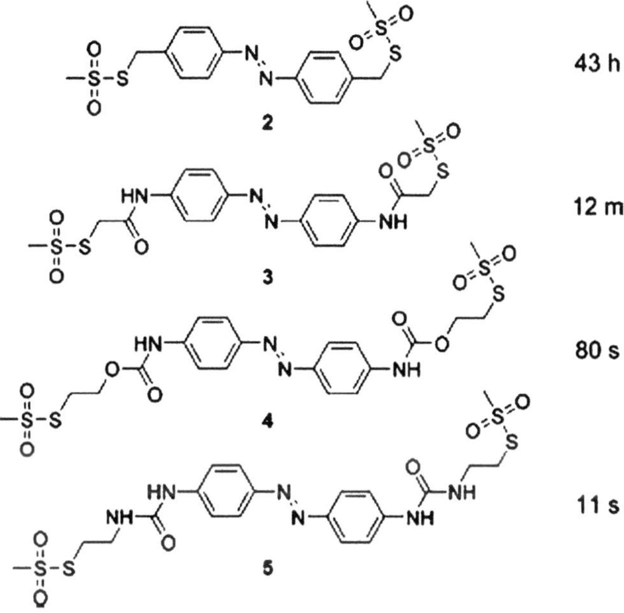

Fig. 3. Azobenzene series with different *cis* thermal stabilities. By varying the degree of single bond character of the azo bond, the half-life of the *cis* isomer can be varied. An increase in electron donation of these para-substituted compounds, results in a shorter half-life of the *cis* form.

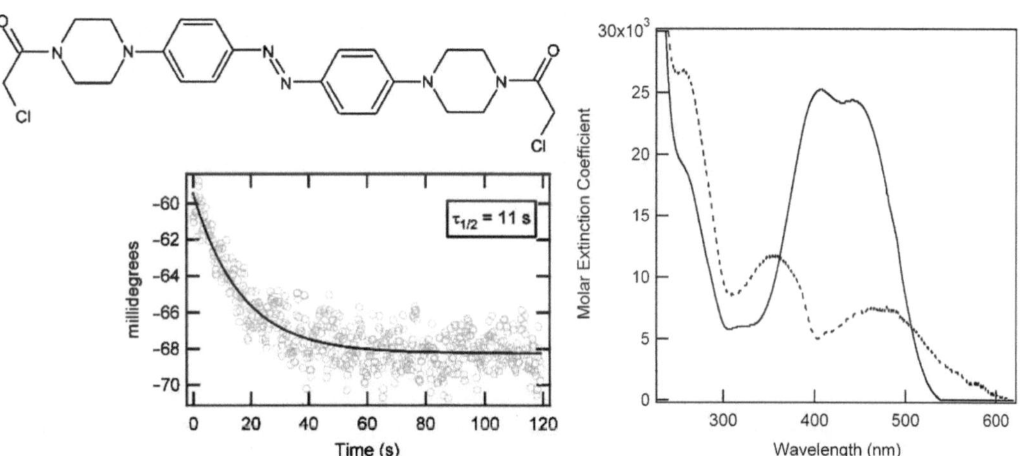

Fig. 4. A switch for rapid photocontrol of helix content. *Right*, 100% *trans* (*solid*) and calculated 100% *cis* (*dashed*) absorption spectrum of switch cross-linked to a model peptide. *Left*, time-dependent circular dichroism change at 225 nm after 400 nm irradiation. An increase in helical content with time is observed as the switch relaxes back from *cis* to *trans*. Solution: 50% methanol/50% sodium phosphate pH 7. Reproduced from (42) by permission of The Royal Society of Chemistry.

that is involved in a fast biochemical process, there are a few practical issues associated with having a thermally unstable *cis* form. First, to compete with the thermal process, higher intensity light may be required for *trans* to *cis* photoisomerization and this high intensity light may be harmful to the cells or tissues under study. Second, characterization of isomerization induced conformational changes requires instruments with high time resolution (e.g., to measure rapid changes in helical content).

In an attempt to produce longer wavelength photoswitches that showed relatively slow thermal relaxation, we designed a series of ortho-substituted diamino azobenzenes (43). The degree of electron donation was varied by incorporating acyclic or cyclic amines. The presence of a 6-membered ring (e.g., piperidine or piperazine) introduces steric interactions that lead to loss of sp2 character of the donating nitrogen as discussed above. In contrast, 5-membered rings (e.g., pyrrolidine) or no ring at all (e.g., dimethylamine) relieves the steric clash and leads to enhanced delocalization. Diamido moieties were retained at the para positions to preserve the end-to-end distance change that was known to cause substantial structural changes in model peptides. This series of compounds could be produced using one synthetic route, with the key step being the use of silver oxide in acetone to produce the azo species (Fig. 5). Blue (450 nm), cyan (480 nm), and green (530 nm) absorbing switches were synthesized (43). The blue-absorbing, water soluble (piperazino) switch retained switching wavelengths like its para-substituted counterpart but with a ~30-fold slower thermal relaxation rate. Several factors may account for this: (1) the ortho groups may be packing closely around the azo group forming a hydrophobic cage stabilizing the *cis* structure, (2) the packing around the azo group may also form a local nonpolar environment that will destabilize the dipolar transition state, and/or (3) the ortho groups in general cause substantial twisting of the benzene rings in the C–N = N–C plane

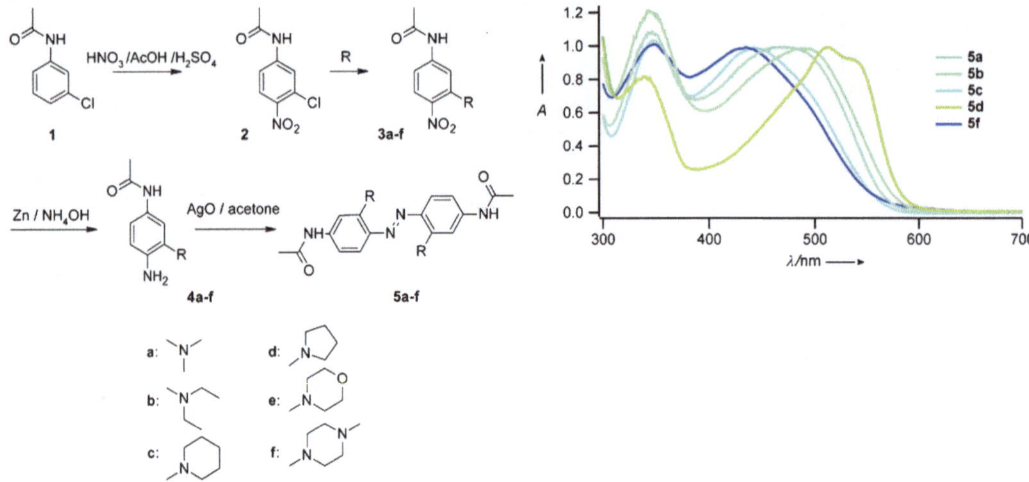

Fig. 5. An ortho-substituted-diaminoazobenzene series. *Left*, synthesis of the ortho-amino series. *Right*, UV/Vis spectra of the ortho-amino series in 70/30 acetonitrile/phosphate buffer pH 7. Varying the electron-donating power of the amino groups has a strong effect on the absorption spectra. Reproduced from (43) with permission.

of the ground state *trans* isomer. Ring twisting could exist in the transition state, thereby having a destabilizing effect. The latter point is consistent with molecular models of the *trans* isomer, a low molar extinction coefficient (relative to their para-substituted counterparts) and studies by Nishimura and colleagues who proposed a similar twisting effect for methyl groups at the ortho positions (44). By combining enhanced delocalization and steric effects, a fine balance between producing a substantial amount of *cis* isomer (~50%) with low power illumination and absorption at relatively long wavelengths can be achieved. Photoswitches that absorb at different regions in the spectrum may allow for multiple components in a biochemical system to be turned on and off with different colors of light.

6. Stability in a Cellular Environment

To date only extracellular targets have been successfully modified in vivo. In such cases achieving target specificity is the major consideration. Intracellular targets require, in addition, the photoswitch to be membrane permeable as well as to be stable to the intracellular reducing environment. Typically, this reducing environment is maintained by the tripeptide glutathione present in its reduced form at ~1–10 mM concentrations (45). Cross-linkers bearing disulfide linkages to proteins will be reductively cleaved rapidly under such conditions, thus stable linkages, such as thioethers are preferred. Glutathione is also known to reduce azo groups under certain conditions (46). Moroder and colleagues have demonstrated that an azobenzene-based peptide photoswitch was susceptible to reduction by glutathione (47). The mechanism of reduction appears to involve attack of the thiol group of glutathione on the azo double bond to form a sulfenyl hydrazide adduct. This species can react with a second molecule of glutathione to produce oxidized glutathione and the reduced hydrazo compound as products. The rates of these reactions will depend on the glutathione concentration, the peptide concentration as well as the redox potential of the cross-linker.

Diaminoazobenzenes do not seem to be susceptible to glutathione reduction when incubated with excess amounts for an overnight period under physiological conditions in vitro (42). We found that photoswitching persisted for at least 3 h after this incubation period. This resistance can be attributed to the electron-rich character of this photoswitch. With other types of photoswitches, catalysis of thermal *cis* to *trans* isomerization rather than reduction by glutathione have been reported (48). In this case, the thiolate is thought to reversibly add to the N=N double bond, transiently reducing bond order and freeing rotation.

Of course, a variety of other in vivo pathways may lead to loss of azobenzene photoswitching. An encouraging indication of azobenzene intracellular stability is the report by Bose et al. (16). Using in vivo nonsense codon suppression, azobenzene was introduced as an amino acid (i.e., *p*-phenylazo-phenylalanine) into the *Escherichia coli* catabolite activator protein. Successful production of the modified protein indicates that the azo amino acid was stable in vivo. Additionally, Fischer and colleagues recently reported photoswitchable versions of cyclosporine A bearing diamidoazobenzene photoswitches. These compounds were shown to be resistant to glutathione reduction in both isomeric forms and remained photoswitchable in human whole blood lysates (49).

7. Increasing the Degree of Structural Change

Azobenzene photoswitching would be most effective if the *trans* or *cis* isomer could completely unfold/inactivate or fold/activate the attached protein. The degree of peptide/protein structural change will depend, in general, on the magnitude of the end-to-end distance change of the azo compound. However, simply extending the end-to-end distance of the molecule does not necessarily imply a larger distance change will occur upon isomerization. Azobenzenes cross-linked through amide, amine, carbamate or urea linkages extend the end-to-end distance of the molecule, but their flexibilities reduced the effective change in distance upon isomerization (38). For example, with the 4,4′-diamidoazobenzene switch series the *cis* form has an end-to-end distance distribution of ~11–15 Å and ~19–23 Å for the *trans* form. A minimum distance change of only 4 Å is possible due to the flexibility of the *trans* and *cis* isomers. In some cases both isomers may have conformers that share end-to-end distances, as observed for the diaminoazobenzene series (41, 42). In such cases, isomerization may lead to small or even no conformational changes, unless the cysteine spacings are carefully optimized to fit the small range of nonoverlapping conformers.

Extending the end-to-end distance of the molecule to induce a large conformational change requires a highly rigid structure. Moroder and colleagues introduced an azobenzene-based cross-linker bearing rigid alkyne units at each end that was successfully used to photocontrol the conformation of collagen peptide sequences (50, 51). Although their alkyne units increased the length of the cross-linker, the presence of methylene groups permitted some flexibility. Liskamp and colleagues also reported photoswitchable azobenzene derivatives bearing rigid spacers (52). Standaert and Park aimed at maximizing biomolecular

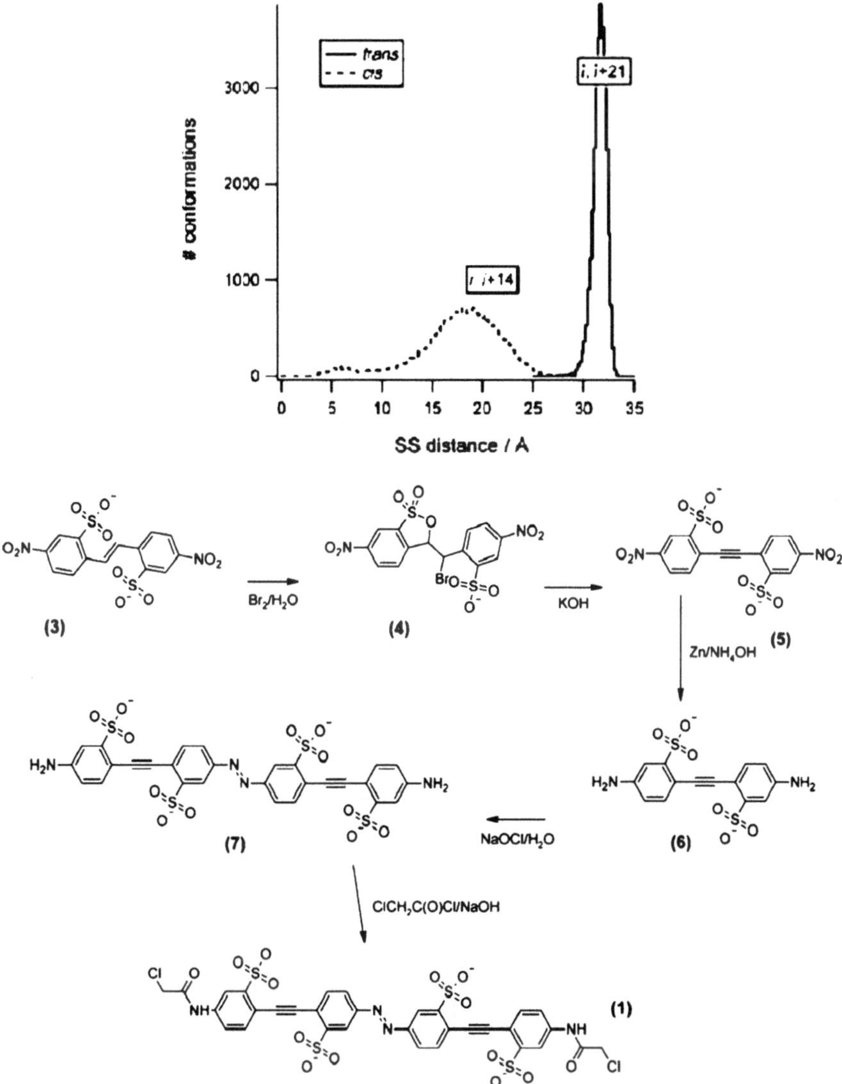

Fig. 6. Synthesis of a long, rigid photoswitch. *Bottom*, synthetic route. *Top*, end-to-end distances for *cis* and *trans* isomers calculated from molecular dynamics simulations. Distances between Cys side-chains spaced (i, $i+x$) in an α-helix are shown. Reproduced from (53) with permission.

conformational changes by designing azobenzene-based photoswitches incorporating extended biphenyl units with a variety of substitution patterns (48). In particular, one designated mpABC exhibited an end-to-end distance change of 13 Å when modeled as part of a peptide backbone. Although the photochemical properties of mpABC were determined, the conformational effects on peptides are yet to be tested experimentally.

We have synthesized and characterized a long rigid photoswitchable cross-linker containing an oligo (phenylene ethynylene) scaffold (53). Following bromination of compound (3) (Fig. 6), treatment with potassium hydroxide with slight heating yields the alkyne moiety. Subsequent reduction of the nitro groups to amines followed by oxidation yields the photoswitchable rigid azobenzene compound (7).

We estimated the end-to-end distance ranges of the *trans* and *cis* isomers of this cross-linker using molecular dynamic simulations. The distance for the *trans* form ranged from 30 to 33 Å, while the *cis* range was 13–24 Å, so that a minimum distance change of ~6 Å is possible – significantly greater than that of the diamidoazobenzene photoswitch series. Irradiation at 400 nm promoted formation of the *cis* isomer, which had a half-life of ~2 h at room temperature in aqueous solution (53). Interestingly, at higher temperatures irradiation produced a greater percentage of the *cis* isomer. The primary effect of increased temperature, as discussed above is to increase the thermal *cis* to *trans* relaxation rate, so that, typically, a lower fraction of *cis* would be formed at higher temperatures for a given irradiation intensity. In this case, however, increased temperature appears to affect the photochemical quantum yields in a manner that produces more *cis* isomer at higher temperatures. Perhaps, isomerization in this case creates a requirement for the rapid movement of many atoms during the short lifetime of the excited state. This, in turn, lowers the quantum yield for isomerization and perhaps increases its sensitivity to temperature.

8. Summary

This short review has attempted to summarize the motivating principles for photoswitch design and to highlight progress in producing photoswitches with properties ideally suited for photocontrol of biological systems. Clearly, there is considerable scope for improvement in terms of creating switches with longer wavelength absorption profiles and greater end-to-end distance changes. Despite a large number of variations on the core structure, the versatility of the azobenzene chromophore has not yet been exhausted. Indeed, bridged azobenzenes have recently been reported in which the *cis* isomer is the more stable form in the dark (54). With improvements in computational methods, switches with desired properties may also be designed in silico (13, 37). These photoswitches are likely to find diverse applications as control elements in a range of biological settings.

References

1. Alaimo PJ, Shogren-Knaak MA, Shokat KM (2001) Chemical genetic approaches for the elucidation of signaling pathways. Curr Opin Chem Biol 5:360–367
2. Goeldner M, Givens R (eds) (2005) Dynamic studies in biology: phototriggers, photoswitches and caged biomolecules. Wiley–VCH, Weinheim, Germany

3. Gorostiza P, Isacoff EY (2008) Optical switches for remote and noninvasive control of cell signaling. Science 322:395–399
4. Ellis-Davies GC (2007) Caged compounds: photorelease technology for control of cellular chemistry and physiology. Nat Methods 4:619–628
5. Willner I, Willner B (1993) Chemistry of photobiological switches. In: Morrison H (ed) Biological applications of photochemical switches, vol 2. Wiley, Toronto, pp 1–110
6. van der Horst MA et al (2005) From primary photochemistry to biological function in the blue-light photoreceptors PYP and AppA. Photochem Photobiol Sci 4:688–693
7. Hellingwerf KJ (2002) The molecular basis of sensing and responding to light in microorganisms. Antonie Leeuwenhoek 81:51–59
8. Woolley GA (2005) Photocontrolling peptide alpha helices. Acc Chem Res 38:486–493
9. Renner C, Moroder L (2006) Azobenzene as conformational switch in model peptides. Chembiochem 7:868–878
10. Feringa B (ed) (2001) Molecular switches. Wiley–VCH, Weinheim, Germany
11. Rau H (1990) Photoisomerization of azobenzenes. In: Rabek JF (ed) Photochemistry and photophysics, vol 2. CRC, Boca Raton, FL, pp 119–141
12. Dias AR et al (1992) Enthalpies of formation of cis-azobenzene and trans azobenzene. J Chem Thermodyn 24:439–447
13. Crecca CR, Roitberg AE (2006) Theoretical study of the isomerization mechanism of azobenzene and disubstituted azobenzene derivatives. J Phys Chem A 110:8188–8203
14. Willner I, Rubin I (1996) Control of the structure and functions of biomaterials by light. Angew Chem Int Ed Engl 35:367–385
15. James DA, Burns DC, Woolley GA (2001) Kinetic characterization of ribonuclease S mutants containing photoisomerizable phenylazophenylalanine residues. Protein Eng 14:983–991
16. Bose M, Groff D, Xie J, Brustad E, Schultz PG (2006) The incorporation of a photoisomerizable amino acid into proteins in E. coli. J Am Chem Soc 128:388–389
17. Muranaka N, Hohsaka T, Sisido M (2002) Photoswitching of peroxidase activity by position-specific incorporation of a photoisomerizable non-natural amino acid into horseradish peroxidase. FEBS Lett 510:10–12
18. Westmark PR, Kelly JP, Smith BD (1993) Photoregulation of enzyme activity – photochromic, transition-state-analog inhibitors of cysteine and serine proteases. J Am Chem Soc 115:3416–3419
19. Krouse ME, Lester HA, Wassermann NH, Erlanger BF (1985) Rates and equilibria for a photoisomerizable antagonist at the acetylcholine receptor of Electrophorus electroplaques. J Gen Physiol 86:235–256
20. Banghart M, Borges K, Isacoff E, Trauner D, Kramer RH (2004) Light-activated ion channels for remote control of neuronal firing. Nat Neurosci 7(12):1381–1386
21. Behrendt R et al (1999) Photomodulation of the conformation of cyclic peptides with azobenzene moieties in the peptide backbone. Angew Chem Int Ed 38:2771–2774
22. Renner C, Cramer J, Behrendt R, Moroder L (2000) Photomodulation of conformational states. II. Mono- and bicyclic peptides with (4-aminomethyl)phenylazobenzoic acid as backbone constituent. Biopolymers 54: 501–514
23. Renner C, Kusebauch U, Loweneck M, Milbradt AG, Moroder L (2005) Azobenzene as photoresponsive conformational switch in cyclic peptides. J Pept Res 65:4–14
24. Kumita JR, Smart OS, Woolley GA (2000) Photo-control of helix content in a short peptide. Proc Natl Acad Sci USA 97:3803–3808
25. Zhang Z, Burns DC, Kumita JR, Smart OS, Woolley GA (2003) A water-soluble azobenzene cross-linker for photocontrol of peptide conformation. Bioconjug Chem 14:824–829
26. Burns DC, Zhang F, Woolley GA (2007) Synthesis of 3, 3′-bis(sulfonato)-4, 4′-bis (chloroacetamido)azobenzene and cysteine cross-linking for photo-control of protein conformation and activity. Nat Protoc 2:251–258
27. Burns DC et al (2004) Origins of helix–coil switching in a light-sensitive peptide. Biochemistry 43:15329–15338
28. Flint DG, Kumita JR, Smart OS, Woolley GA (2002) Using an azobenzene cross-linker to either increase or decrease peptide helix content upon trans-to-cis photoisomerization. Chem Biol 9:391–397
29. Kumita JR, Flint DG, Smart OS, Woolley GA (2002) Photo-control of peptide helix content by an azobenzene cross-linker: steric interactions with underlying residues are not critical. Protein Eng 15:561–569
30. Bredenbeck J, Helbing J, Kumita JR, Woolley GA, Hamm P (2005) Alpha-helix formation in a photoswitchable peptide tracked from picoseconds to microseconds by time-resolved IR spectroscopy. Proc Natl Acad Sci USA 102:2379–2384

31. Chen E, Kumita JR, Woolley GA, Kliger DS (2003) The kinetics of helix unfolding of an azobenzene cross-linked peptide probed by nanosecond time-resolved optical rotatory dispersion. J Am Chem Soc 125:12443–12449
32. Kneissl S, Loveridge EJ, Williams C, Crump MP, Allemann RK (2008) Photocontrollable peptide-based switches target the anti-apoptotic protein Bcl-xL. Chembiochem 9: 3046–3054
33. Guerrero L et al (2005) Photochemical regulation of DNA-binding specificity of MyoD. Angew Chem Int Ed Engl 44:7778–7782
34. Guerrero L, Smart OS, Woolley GA, Allemann RK (2005) Photocontrol of DNA binding specificity of a miniature engrailed homeodomain. J Am Chem Soc 127:15624–15629
35. Chang CW, Lu YC, Wang TT, Diau EW (2004) Photoisomerization dynamics of azobenzene in solution with S1 excitation: a femtosecond fluorescence anisotropy study. J Am Chem Soc 126:10109–10118
36. Tiago ML, Ismail-Beigi S, Louie SG (2005) Photoisomerization of azobenzene from first-principles constrained density-functional calculations. J Chem Phys 122:094311
37. Dokic J et al (2009) Quantum chemical investigation of thermal cis-to-trans isomerization of azobenzene derivatives: substituent effects, solvent effects, and comparison to experimental data. J Phys Chem A 113:6763–6773
38. Pozhidaeva N, Cormier ME, Chaudhari A, Woolley GA (2004) Reversible photocontrol of peptide helix content: adjusting thermal stability of the cis state. Bioconjug Chem 15:1297–1303
39. Rau H (1990) Photochromism. In: Durr H, Bouas-Laurent H (eds) Molecules and systems. Elsevier, Amsterdam, pp 165–192
40. Nishimura N, Tanaka T, Asano M, Sueishi Y (1986) A volumetric study on the thermal cis-to-trans isomerization of 4-(dimethylamino) 4′-nitroazobenzene and 4,4′-bis(dialkylamino) azobenzenes: evidence of an inversion mechanism. J Chem Soc Perkin Trans II, 1839–1845
41. Chi L, Sadovski O, Woolley GA (2006) A blue-green absorbing cross-linker for rapid photoswitching of peptide helix content. Bioconjug Chem 17:670–676
42. Beharry AA, Sadovski O, Woolley GA (2008) Photo-control of peptide conformation on a timescale of seconds with a conformationally constrained, blue-absorbing, photo-switchable linker. Org Biomol Chem 6:4323–4332
43. Sadovski O, Beharry AA, Zhang F, Woolley GA (2009) Spectral tuning of azobenzene photoswitches for biological applications. Angew Chem Int Ed Engl 48:1484–1486
44. Nishimura N et al (1976) Thermal cis-to-trans isomerization of substituted azobenzenes II. Substituent and solvent effects. Bull Chem Soc Jpn 49:1381–1387
45. Lopez-Mirabal HR, Winther JR (2008) Redox characteristics of the eukaryotic cytosol. Biochim Biophys Acta 1783:629–640
46. Kosower EM, Kanety-Londner H (1976) Glutathione. 13. Mechanism of thiol oxidation by diazenedicarboxylic acid derivatives. J Am Chem Soc 98:3001–3007
47. Boulegue C, Loweneck M, Renner C, Moroder L (2007) Redox potential of azobenzene as an amino acid residue in peptides. Chembiochem 8:591–594
48. Standaert RF, Park SB (2006) Abc amino acids: design, synthesis, and properties of new photoelastic amino acids. J Org Chem 71:7952–7966
49. Zhang Y, Erdmann F, Fischer G (2009) Augmented photoswitching modulates immune signaling. Nat Chem Biol 5:724–726
50. Kusebauch U, Cadamuro SA, Musiol HJ, Moroder L, Renner C (2007) Photocontrol of the collagen triple helix: synthesis and conformational characterization of bis-cysteinyl collagenous peptides with an azobenzene clamp. Chemistry 13:2966–2973
51. Kusebauch U et al (2006) Photocontrolled folding and unfolding of a collagen triple helix. Angew Chem Int Ed Engl 45:7015–7018
52. Kuil J, van Wandelen LT, de Mol NJ, Liskamp RM (2009) Switching between low and high affinity for the Syk tandem SH2 domain by irradiation of azobenzene containing ITAM peptidomimetics. J Pept Sci 15:685–691
53. Zhang F, Sadovski O, Woolley GA (2008) Synthesis and characterization of a long, rigid photoswitchable cross-linker for promoting peptide and protein conformational change. Chembiochem 9:2147–2154
54. Siewertsen R et al (2009) Highly efficient reversible Z-E photoisomerization of a bridged azobenzene with visible light through resolved S(1)(npi*) absorption bands. J Am Chem Soc 131:15594–15595

Chapter 11

Photoswitchable Voltage-Gated Ion Channels

Doris L. Fortin and Richard H. Kramer

Abstract

Multiple strategies enabling the control of cellular function with light have been developed. These strategies include the expression of intrinsically photosensitive proteins and the use of photosensitive molecules that target native or exogenously expressed proteins. In particular, the use of small molecules containing a photoisomerizable moiety, such as azobenzene, enables the photosensitization of proteins that would otherwise be light insensitive. Photosensitivity is targeted to the protein of interest by connecting the photoisomerizable moiety to a specific agonist or antagonist. Two classes of azobenzene-containing photoswitches have been developed for exogenously expressed or endogenous voltage-gated K^+ channels. In both cases, the photoswitch molecule consists of an azobenzene linked to a pore-blocking quaternary ammonium ion. Addition of a maleimide group to the photoswitch has enabled covalent attachment of the photoswitch molecule to a genetically engineered cysteine on the surface of a modified Shaker K^+ channel, allowing light to regulate action potential firing in transfected neurons treated with the photoswitch. Replacing the maleimide with different chemical groups eliminates the requirement for a genetically engineered cysteine, allowing regulation of endogenously expressed K^+ channels in treated cells. The modular nature of the photoswitch molecule allows flexibility in the design of each functional group, yielding a combinatorial toolkit for optical regulation of genetically engineered or native proteins that enables optical control of a variety of physiological functions.

Key words: Azobenzene, Photoswitch, Chemical modification, Orthogonal, Actuator, Genetic targeting, Exogenous expression, Neuronal firing

1. Introduction

Neurons and other excitable cells harbor ion channels that are activated by voltage, ligands, temperature, mechanical force, but not by light. Consequently, experimental manipulation of the nervous system *in vitro* and *in vivo* has relied heavily on the application of electrical or chemical stimuli that target these proteins. The response of the system can then be recorded using electrical or optical methods at the level of single cells or populations of cells. Although

James J. Chambers and Richard H. Kramer (eds.), *Photosensitive Molecules for Controlling Biological Function*, Neuromethods, vol. 55, DOI 10.1007/978-1-61779-031-7_11, © Springer Science+Business Media, LLC 2011

much has been learned from the use of electrical stimulation and/or chemical perfusion devices, the recent development of light-based actuators to manipulate the function of neuronal circuits is leading to a renaissance in the field of functional neuroanatomy (1–3). Optical stimulation offers several advantages over electrical and chemical stimulation. Light can be projected with great spatial precision on subcellular structures, single cells, or many cells, contiguous or not, simultaneously. During the course of an experiment, light can be redirected between different cells or groups of cells, enabling control of complex networks. Light can be turned on and off precisely, offering exceptional temporal precision for the control of cellular activity. Because of its ability to penetrate tissue, it may be possible to stimulate cells optically from afar, minimizing invasiveness and potential damage to the structure under study. Light may thus act as a "remote control" for neuronal activity. But how can light be used to manipulate the activity of neurons that have no endogenous photoresponsive proteins?

Several strategies have been described that enable light to control cellular activity (Fig. 1). These include photolysis of caged molecules and heterologous expression of intrinsically photosensitive proteins and will be described only briefly in this chapter since they have been reviewed elsewhere (2–4) and are also described in other sections of this book. This chapter will focus on a third, opto-chemical strategy: the design of small photoisomerizable molecules that impart light sensitivity onto genetically modified or native voltage-gated ion channels. Their use to control the activity of neurons will also be described.

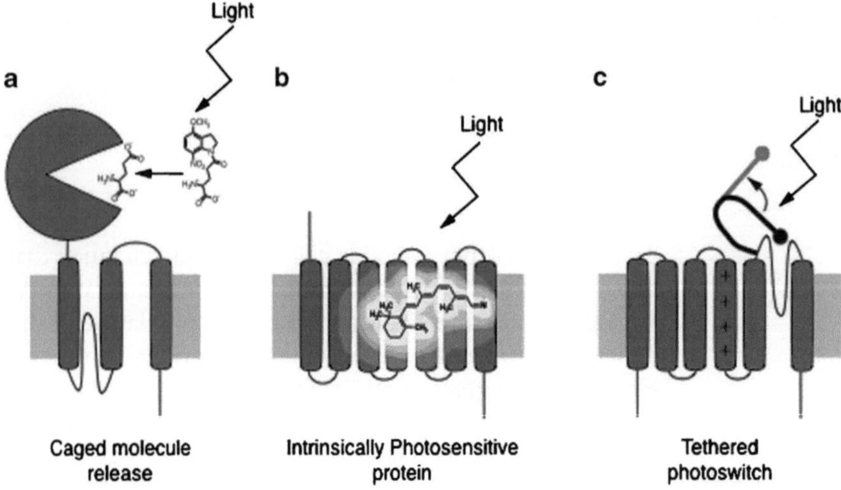

Fig. 1. Optical methods for cellular stimulation. (**a**) Photolysis of a caged molecule liberates the active form of the chemical, which can act on its target protein. (**b**) Intrinsically photosensitive proteins contain a natural chromophore that isomerizes upon exposure to light resulting in protein activation. (**c**) Proteins can be photosensitized by application of a molecular photoswitch. Subsequent photoisomerization of the photoswitch controls protein function.

2. Optical Methods for Cellular Stimulation

2.1. Caged Molecules

2.1.1. Control of Native Proteins

One popular approach devised to control cellular activity has been to conjugate bioactive molecules of interest to photolabile protective groups, in effect "caging" the molecule to prevent its action. Upon exposure to light, the protective group is photolyzed and the previously inert molecule is liberated, becoming free to mediate its biological function(s) (Fig. 1a). Various molecules have been caged successfully including neurotransmitters, nucleotides, peptides, and enzymes (4). Because light triggers uncaging, the bioactive molecule can be released precisely in space and time and quantitatively by grading the light intensity. Glutamate uncaging in the nervous system accurately mimics the kinetics of synaptic transmission (5, 6) and has been extensively used to map neuronal circuits (7–12). It has been estimated that the spatial resolution of one-photon glutamate uncaging is ~10 μm (13). Given that many neurons express glutamate receptors and the liberated glutamate can diffuse in the extracellular space, it may be difficult to restrict stimulation with glutamate uncaging to the targeted area. Two-photon laser photostimulation dramatically increases the resolution of uncaging to the submicron level, allowing photostimulation of individual synapses (14). In addition, complex stimulation patterns can now be generated using digital micromirrors (15), acoustico-spatial deflectors (6) or liquid-crystal spatial light modulators (16) enabling tailored excitation at multiple sites to mimic more faithfully normal neuronal activity.

2.1.2. Control of Exogenously Expressed Proteins

A creative twist to the uncaging method described above involves the generation of caged molecules specific for proteins not normally expressed in the cells or tissues under study. Non-native ion channels are expressed heterologously, via transfection or transgenesis, to confer sensitivity to the caged molecule (17). Thus, both the stimulus and the susceptibility to the stimulus are restricted locally using photolysis and limited expression, respectively, minimizing off-target action by the released molecules. Activation of heterologously expressed $P2X_2$ receptors by photolysis of caged ATP has been used successfully to activate specific populations of neurons and induce a variety of behaviors in *Drosophila* (17). Photolysis of caged ATP can also induce action potential firing in cultured mammalian neurons transfected with the $P2X_2$ receptor (17). Likewise, photolysis of caged capsaicin induces action potential firing in neurons transfected with the capsaicin receptor, TRPV1 (17). The recent generation of a knock-in mouse conditionally expressing TRPV1 (18) opens the possibility of refining the expression pattern of the nonnative receptor for photostimulation of specific neuronal populations using caged capsaicin.

2.2. Photosensitive Proteins

To circumvent the limitations associated with freely diffusible molecules, novel methods for optical stimulation have been developed that rely on photosensitive proteins. Two broad classes of photosensitive proteins have been described: intrinsically photosensitive proteins and those rendered photosensitive by treatment with an exogenous chemical.

2.2.1. Rhodopsins

Intrinsically photosensitive proteins can be genetically introduced into neurons to impart light sensitivity. These proteins include Channelrhodopsin-2 (ChR2) (19), melanopsin (20), and the ChARGe system (consisting of arrestin-2, rhodopsin, and a G-protein α-subunit) (21) for induction of action potential firing, whereas halorhodopsin (22, 23) and the vertebrate rhodopsin (RO4) (24) inhibit action potential firing. These proteins, broadly called rhodopsins, contain the chromophore retinal that photoisomerizes upon light absorption (25) (Fig. 1b). Thus, in addition to exogenous protein expression, retinal must be present in sufficient amount to impart light sensitivity. For instance, it is necessary to feed retinal to Drosophila and *Caenorhabditis elegans* to generate functional, light-sensitive ChR2 protein (26, 27). Retinal addition may also be required for prolonged experiments using RO4 in the chick spinal cord preparation (24). In contrast, the mammalian nervous system contains sufficient endogenous retinal to support ChR2, and by extension halorhodposin, function (28–31).

Intrinsically photosensitive proteins are not natively expressed in neurons and thus must be introduced using transfection methods or expressed transgenically in intact organisms (reviewed in (1)). Targeted expression of photosensitive proteins can be achieved using stereotactic viral injection (32), cell-type specific viruses (33–35) or cell-specific promoters (36). Alternatively, sparse expression of the photosensitive proteins, such as is the case under the control of the Thy-1 promoter (37), can be used to restrict the number of photosensitive cells. Targeted or restricted expression of the photosensitive protein results in genetically encoded specificity: only a subset of neurons can respond to light. This feature may be essential for deciphering the contribution of different types of neurons to the function of the nervous system. In addition, circuit diagrams may be assembled more easily by stimulating sparse, genetically specified neurons expressing photosensitive proteins and recording the response of their target cells, possibly in an all-optical system using calcium or voltage-sensitive dye imaging (6, 15, 23, 38). Recent years have seen a veritable explosion of papers describing the use of ChR2 to map connectivity in brain slices and anesthetized animals as well as to induce behavior in model organisms and awake animals (26–31, 39–43). Intrinsically photosensitive proteins have been reviewed elsewhere (2, 3) as well as in Sect. 2.2 and will not be discussed further here.

2.2.2. Caged Proteins

Early efforts to control protein function with light have relied on "bulk caging" specific amino acids in a purified protein (44). Photoregulation of protein activity is then studied in an *in vitro* assay or in cells after microinjection of the caged protein. Alternatively, the introduction of unnatural, caged amino acids by the nonsense suppression method can also be used to photosensitize proteins (45). Nonsense suppression requires genetic manipulation of the DNA to introduce a stop codon at the position of interest in the protein under study, followed by its heterologous expression in target cells. A suppressor tRNA specific for the stop codon and acylated with the unnatural, caged amino acid is also prepared and introduced into cells. After successful generation and expression of the caged protein, light can be used to trigger the photolysis of the caged moiety enabling control of protein function. A major difficulty in generating caged proteins is the identification of an amino acid residue that once caged, will enable regulation of protein function with light. Although this may be conceptually feasible for enzymes, it may be more difficult to predict for proteins such as ion channels that rely on large, complex domains for function. Nonetheless, caged amino acids have been introduced in K^+ channels to modulate their function with light. For example, the regulation of ion conduction through the inward rectifier K^+ channel Kir2.1 was assessed by replacing a specific tyrosine (Y242) with caged tyrosine (Y(ONB)) using the nonsense suppression method. Uncaging led to a rapid decrease in Kir2.1 current. Unexpectedly, this decrease in current was due to the endocytosis of Kir2.1 (46), a mechanism of channel regulation not previously appreciated from conventional mutagenesis studies (47). The ability to generate a properly localized Kir2.1 arrested at a specific step in a cellular pathway followed by time-resolved uncaging was crucial in revealing the contribution of channel trafficking in the regulation of ion conduction.

The generation of caged proteins can be technically challenging, which may explain why this technique remains somewhat underexploited. Other factors may also limit their use in biological systems. For instance, certain caging groups require long irradiation times that may not always be compatible with biological function. Long irradiation time and high light intensity may result in protein backbone cleavage, a feature that has been exploited in *s*ite-specific, *n*itrobenzyl-*i*nduced *p*hotochemical *p*roteolysis (SNIPP) of proteins that contain 2-(nitrophenyl)-glycine (Npg) (48). The microenvironment of the caged residue, such as local pKa, steric hindrance, or presence of tryptophan in the vicinity, may influence the extent of uncaging (49). Thus, it may not always be possible to predict the properties of the caged residue or the efficiency of uncaging, at least when the modified residue is buried deep inside the protein. Although irradiation of caged proteins permits time-resolved studies, uncaging is a one-way process that consumes the starting material and cannot be reversed. In contrast, the development and use of photosensitive molecules that can be reversibly manipulated with light would open new frontiers in the study of protein and cell function.

2.2.3. Photosensitive Proteins by Chemical Treatment

The key step in imparting photosensitivity to an otherwise light-insensitive target protein is chemical treatment with a synthetic light-sensitive gate, called a photoswitch (Fig. 1c). Photoswitches contain a photosensitive group, such as azobenzene (50), spiropyran (51), or hemithioindigo (52), whose properties change upon illumination (Fig. 2). Retinal, which is an essential component of rhodopsins (25), can be considered a natural photoswitch, although it is not always necessary to add the molecule exogenously to obtain functional rhodopsins. Upon illumination with the appropriate wavelength, spiropyran undergoes ring opening, increasing its polarity, whereas azobenzene, retinal and hemithioindigo undergo a *cis–trans* isomerization, changing their length and geometry. These molecular changes can be exploited to trigger biophysical and cellular events. Various photoisomerizable molecules have been used to generate photoswitches that regulate the activity of ion channels (53–59), but this review will focus on azobenzene-containing photoswitches for the control of ion channels.

3. Reversible Photocontrol of Genetically Engineered Ion Channels

3.1. Azobenzene

Azobenzene can be repetitively photoswitched with a high quantum yield and little to no photobleaching, making it ideal for use in photoswitch molecules (60). The two stable azobenzene isomers, *cis* and *trans*, differ in length by as much as 7 Å, allowing azobenzene to be used as a bistable mechanical lever that maneuvers a functional group in one of two positions, for example, on the surface of a protein (Fig. 1c). Because it is more stable thermodynamically, the *trans*, or extended, form of azobenzene predominates after prolonged incubation in the dark. Exposure to near UV light (~360–380 nm) triggers photoisomerization to the *cis* state whereas the reverse *cis* to *trans* conversion is accelerated by visible light (>460 nm) (Fig. 2a). Light-induced conversion between the two isomers is extremely rapid, occurring on the picosecond time scale (61–63). In contrast, the rate of thermal relaxation from *cis*- to *trans*-azobenzene varies widely, from seconds to days, often depending on the nature of chemical substitution on the azobenzene core (64). Because the *cis* and *trans* states of azobenzene have overlapping absorption spectra, complete photoconversion to either state is not possible with light. Instead, at each wavelength, a photostationary state is achieved with

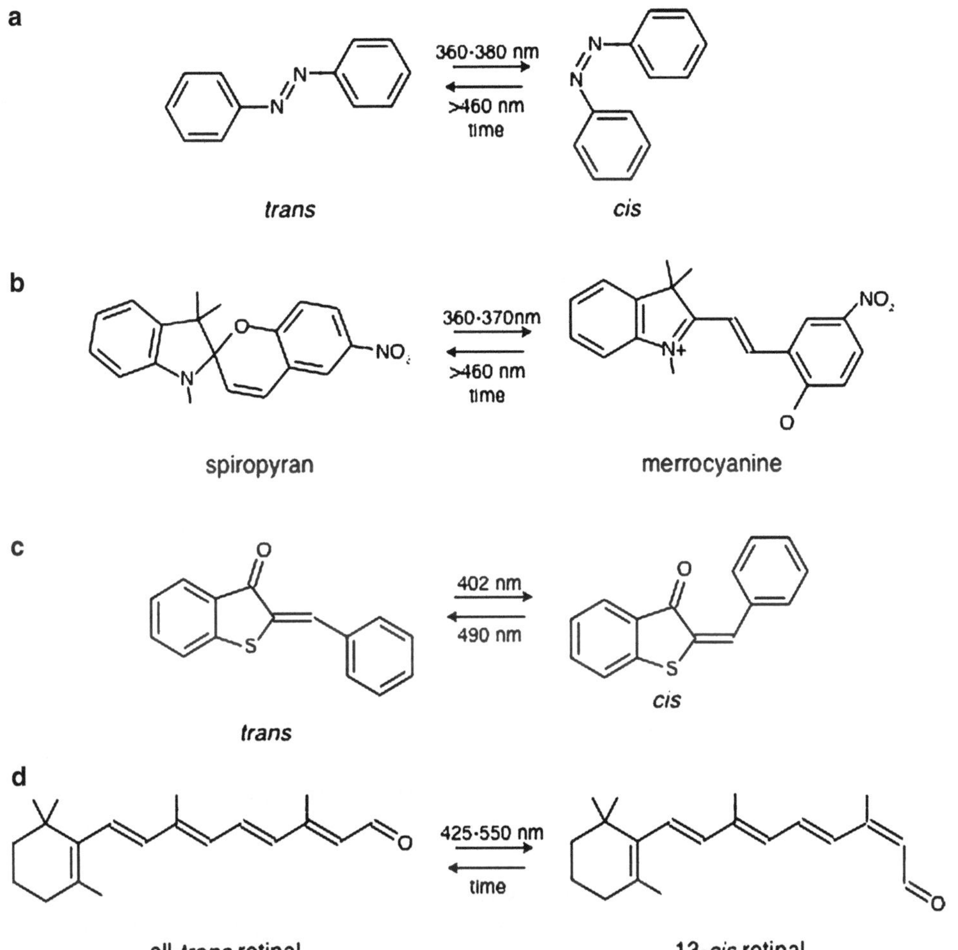

Fig. 2. Natural and synthetic photoswitch molecules used in light-regulated proteins. (**a**) Azobenzene undergoes *trans* to *cis* isomerization after illumination with 360–380 nm light. Visible light above 460 nm light returns the molecule to the *trans* state. Thermal relaxation from *cis* to *trans* also occurs with time in darkness. (**b**) Spyropyran undergoes ring opening to generate the more polar merrocyanine when illuminated with 360–370 nm light. Darkness and illumination with wavelengths above 460 nm favor the less polar spiropyran configuration. (**c**) Hemithioindigo undergoes *trans* to *cis* isomerization after illumination with 402 nm light whereas the *cis* state can be restored with 490 nm light or darkness. (**d**) In Channelrhodopsin-2, the naturally occurring photoswitch retinal is covalently attached and adopts the all-*trans* configuration at rest. Exposure to 425–450 nm light induces isomerization to 13-*cis*-retinal. In all cases, light-induced conformational changes in the photoswitch can be harnessed for the control of protein activity and biological function.

different relative amounts of *cis* and *trans* isomers. Thus, ~85% of azobenzene molecules in a population will be converted to *cis* by illumination with the optimal wavelength (~380 nm). Similarly, under visible light (~500 nm), ~90% of the molecules will adopt the *trans* conformation and near full conversion (>99%) can be reached after prolonged incubation of azobenzene molecules in darkness (64).

3.2. Acetylcholine Receptor

The nicotinic acetylcholine receptor (nAchR) is a pentameric protein that consists of a cation channel coupled to an extracellular ligand-binding site. Because of its essential role at the neuromuscular junction and the availability of electric organs from electric fish, which provided an accessible source of the receptor, the nAchR emerged as a model ligand-gated ion channel (65). Erlanger and colleagues were the first to implement the use of azobenzene-containing molecules to regulate the activity of an ion channel (66), generating a soluble photoisomerizable agonist for the nAchR, Bis-Q. When applied to the electric eel electroplaque under visible light, Bis-Q is predominantly in its *trans* form and acts as a full agonist of the nAchR. Exposure to 330 nm light induces isomerization of Bis-Q to its *cis* configuration, which shows little to no agonist activity toward the receptor. Silman and Karlin describe a similar photoisomerizable agonist, QBr, which covalently attaches to a previously reduced surface cysteine on the nAchR. Tethering of QBr, unlike tethering of conventional agonists ((67) and see below), enables regulation of nAchR with light. Specifically, photoisomerization of tethered QBr to the *trans* configuration results in activation of the nAchR while subsequent isomerization to *cis* minimizes inactivation or desensitization, greatly facilitating kinetic studies. For instance, rapid photoisomerization of tethered QBr uncovered the existence of a rate-limiting step for nAchR activation, distinct from ligand binding (68).

3.3. SPARK

As exemplified by the development of Bis-Q and QBr, there has been a long-standing interest in generating light-regulated ion channels to gain insight into the mechanism of action of the channels *in vitro* and in a cellular context. However, because ion channels play critical roles in a variety of biological processes, such as neuronal signaling, muscle contraction, and immune signaling, light-regulated ion channels also provide a means to manipulate these cellular functions. We have successfully used azobenzene-based photoswitches to impart light sensitivity onto voltage-gated K^+ channels (54, 55, 69).

3.3.1. Voltage-Gated K^+ Channels

Voltage-gated K^+ channels are composed of four six-transmembrane domains subunits, termed S1–S6, arranged in a structure that displays four-fold symmetry. Each channel subunit can be roughly divided into two basic parts: a central pore region formed by the S5–S6 region of each subunit surrounded by four voltage-sensing domain (S1–S4). At rest, an activation gate, located deep in the pore region, is closed, preventing the flow of K^+ ions. Changes in membrane potential result in the movement of several voltage-sensing arginines, located in S4, followed by a conformational change within the channel that opens the activation gate allowing ion flow (70). Voltage-gated K^+ channels play a crucial role in regulating cellular excitability by setting the resting membrane potential, contributing to the action potential waveform, and regulating firing frequency.

3.3.2. Molecular Tape Measures

Nearly all voltage-gated K^+ channels are blocked by the binding of quaternary ammonium ions, such as tetraethylammonium (TEA), to an extracellular site in the pore region, although the affinity of the pore for the blocker can vary by three orders of magnitude (71). The affinity of voltage-gated K^+ channels for externally applied TEA is generally defined by a single amino acid at the entrance of the pore (position 449 in *Drosophila* Shaker). Replacement of the naturally occurring threonine at this position with an aromatic residue (T449Y or T449F) enhances TEA affinity of Shaker by as much as 40-fold (72, 73).

Taking advantage of the surface accessibility of the TEA-binding site on voltage-gated K^+ channels, Blaustein and colleagues designed a series of compounds to map the distance between the pore and specific residues on the surface of the channel (74). These "molecular tape measures" consisted of a quaternary ammonium ion and a cysteine-reactive maleimide group, separated by a flexible poly-glycine-based linker of variable length. Individual cysteines were introduced at different test positions in a Shaker variant containing the high affinity external TEA-binding site mutation T449F (72). Exposure to a given "molecular tape measure" leads to the tethering of the molecule, via its maleimide moiety, to the cysteine engineered on the surface of the channel. If the flexible linker is long enough, the TEA moiety is positioned in the channel pore, irreversibly blocking ion conduction (Fig. 3a). However, when the distance between the TEA-binding site and the cysteine optimally match the length of the "molecular tape measure," covalent tethering is accelerated by affinity labeling (75, 76) such that the reversible binding of TEA to the pore brings the maleimide in close proximity to its target cysteine (77). Several cysteine positions located in the S1–S2, S3–S4, and S5–P loops were mapped using the "molecular tape measure" approach. In most cases, the level of irreversible block was minimal with very short tethers but increased sharply when a critical length was reached. The minimal length at which ~100% irreversible channel block occurs corresponds to the distance between the TEA-binding site and the introduced cysteine. For example, the distance between E422C, located with the S5–P loop segment of Shaker was estimated to be ~15 Å, longer than the distance previously estimated from the electrostatics of TEA binding (10 Å for a neighboring residue) (78) but only slightly shorter than estimates obtained from the crystal structure of KcsA (16–18 Å) (79) and from the electrostatics of charybdotoxin binding (17 Å) (80). The "molecular tape measure" approach is not limited to obtaining static measurement of distances on the surface of proteins but

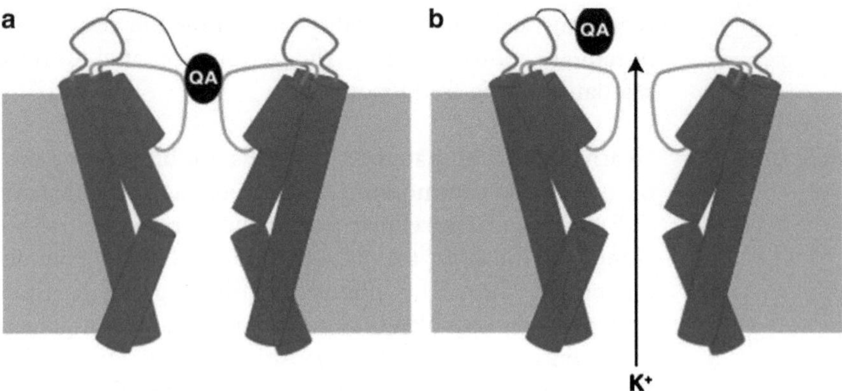

Fig. 3. Molecular tape measures to map protein surfaces. Molecular tape measures for Shaker consist of a quaternary ammonium (QA) ion and a cysteine-reactive maleimide group, separated by a flexible linker of variable length. Exposure to a given molecular tape measure leads to the tethering of the molecule, via its maleimide moiety, to a cysteine engineered on the surface of the channel. The position of the engineered cysteine and the length of the linker determine whether tethering results in channel block (**a**) or not (**b**). Using this method, the distances between the QA-binding site and individual test sites on the Shaker channel surface were estimated.

has also been used to explore the range of motion experienced during channel gating by the Shaker voltage sensor. For instance, tethering of a 45 Å long Gly_7TEA to cysteines located in the S3–S4 region is not affected by channel gating suggesting that those positions do not experience a large motion during the channel gating cycle (81). The distances and maps obtained from the studies by Blaustein and colleagues have served as landmarks in the assessment of the first crystal structure from a voltage-gated K^+ channel (82, 83) and continue to guide structure/function studies of voltage-gated ion channels.

A critical feature of the "molecular tape measure" approach is the ability to determine the effective end-to-end distance of the molecule. However, there are several sources of potential errors in assigning distances using "molecular tape measures" (76). First, if the "molecular tape measures" contain a flexible linker, they will exist as a population with a Gaussian distribution of lengths, which on average will be different than the extended length of the molecule (84). In addition, linker flexibility may impart further uncertainty to the measured distances. For instance, in the case of Shaker, blockers longer than the minimal length for ion channel block are still effective because they accommodate the increase in linker length by folding about rotatable bonds (77). Indeed when designing multivalent ligands, it may be useful to use flexible linkers longer than the optimal distance between individual binding sites to ensure multivalent binding (85). Similarly, flexibility of a target protein may also accommodate increasing linker length. This may be particularly relevant when mapping distances between residues located in flexible regions of proteins, such as those in the extracellular loops of K^+ channels. Thus, the use of a series of molecules containing flexible linkers of different lengths combined with molecules based on rigid linkers, which will have a more narrow length distribution, may allow a more precise assignment of distances between two particular sites.

3.3.3. SPARK Design

We chose the Shaker K^+ channel as a first target for photosensitization because of the abundant biophysical and structural information available about this channel. Specifically, the position of amino acid E422, located in the S5-P loop, was estimated to be ~15–18 Å from the TEA-binding site (74, 79, 80). In experiments using Shaker E422C, a 4 Å difference in the length of tethered blockers (11 Å vs. 15 Å) made the distinction between ineffective or complete block of the channel (74). Taking this into consideration, a new chemical gate was designed to regulate the activity of a variant Shaker channel using light (54). The chemical gate, termed MAQ, consists of a *M*aleimide moiety for cysteine tethering, a photoisomerizable *A*zobenzene, and a *Q*uaternary ammonium group to block the channel (Fig. 4a). The system is designed such that once MAQ is tethered to its target channel, a Shaker variant containing the E422C mutation, the blocking group can reach the pore of the channel only when the azobenzene is in its *trans* (long; ~17 Å), but not its *cis* (short; ~10 Å) state, resulting in a *S*ynthetic *P*hotoisomerizable *A*zobenzene-*R*egulated *K*$^+$ (SPARK) channel (Fig. 4b). Illumination with the appropriate wavelength extends or retracts the blocker from the pore, regulating whether the channel conducts K^+ ions or not.

3.3.4. Photoregulation of SPARK

The effects of MAQ were first tested on *Xenopus laevis* oocytes expressing a variant of Shaker E422C containing the Δ6-46 mutation to abolish fast inactivation (86). Channel block developed slowly upon application of MAQ to outside-out oocyte membrane patches. Importantly, channel block persisted after washout of MAQ, suggesting that covalent tethering of the photoswitch had occurred. Subsequent exposure to 380 nm light relieved some of the channel block, whereas 500 nm light or prolonged incubation in darkness restored channel block (Fig. 4c). Taking advantage of the photostationary state established at different wavelengths of light, the relative amounts of *cis* and *trans* isomers were varied to grade the amount of current passing through SPARK (54). The dynamic range of light-regulated SPARK activity was maximal when a mutation for low affinity TEA binding (T449V) (73) was also present in SPARK, potentially because of the high local TEA concentration at the pore after covalent tethering of the photoswitch.

SPARK was next tested in cultured hippocampal neurons to determine whether light-regulated channels could control

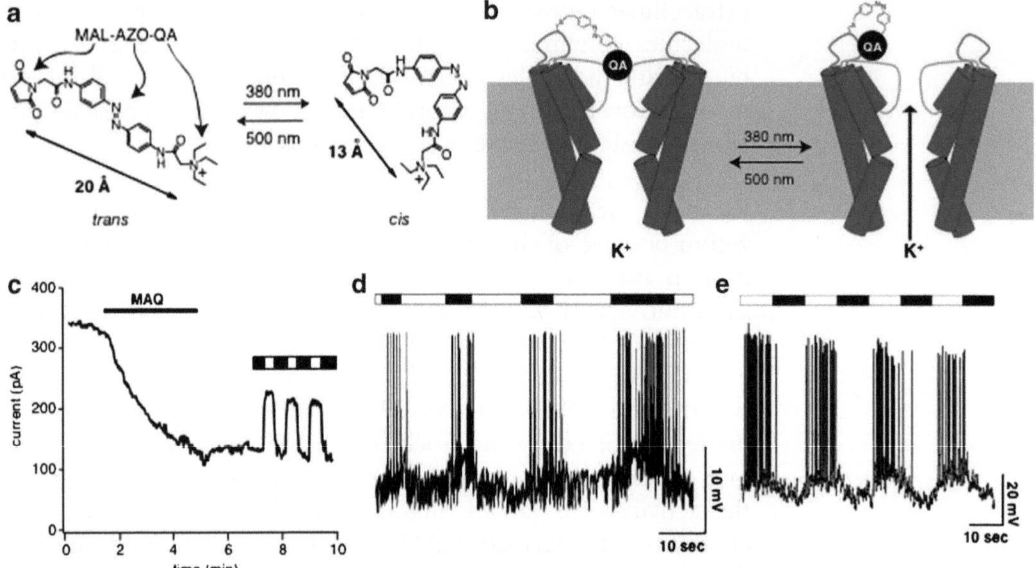

Fig. 4. SPARK channels to regulate neuronal activity in genetically specified cells. (**a**) MAQ consists of a maleimide group (MAL) and a quaternary ammonium (QA) group separated by a photosensitive azobenzene moiety (AZO). Azobenzene isomerizes to the *cis* form upon illumination with 380 nm light. The *trans* form is restored upon illumination with 500 nm light. (**b**) Tethering MAQ to a genetically engineered cysteine (E422C) in a modified Shaker channel enables photoregulation of ion conduction. Under 500 nm light, MAQ is in the *trans* form, the QA group reaches the pore of the channel and blocks ion flow. Photoisomerization to the *cis* state with 380 nm light removes the blocker from the pore, allowing ion flow. (**c**) MAQ application blocks ion conduction in an outside-out patch obtained from an oocyte expressing the modified E422C Shaker channel. Ion conduction remains blocked even after washing, indicating that covalent tethering has occurred. Illumination with 380 nm light restores some of the current, whereas 500 nm light blocks ion flow. (**d**) Expression of the modified E422C Shaker in neurons results in a large resting K^+ conductance that silences spontaneous activity. Treatment with MAQ and subsequent illumination with 500 nm light restores spontaneous activity. (**e**) Modifying the ion selectivity of Shaker to allow Na^+ flow results in the generation of a depolarizing channel that can be photoregulated after treatment with MAQ. In this case, illumination with 380 nm light induces action potential firing that is terminated upon illumination with 500 nm light and channel block. Panel (**e**) is reproduced with permission from reference (69).

neuronal activity. The channel was further modified to ensure that the photoswitch, not the voltage sensor, would be the primary regulator of gating by introducing a mutation (L366A) that shifts voltage-dependent activation to hyperpolarized potentials (87). Expression of this Shaker variant (Δ6-46; L366A; E422C; T449V) results in a high resting K^+ conductance that can be photoregulated upon application of MAQ. Thus, photoisomerization of MAQ to the *cis* state opens the chemical gate, triggering a K^+ current that hyperpolarizes the membrane potential and silences neuronal activity. Subsequent illumination with 500 nm light induces closure of the chemical gate, allowing action potential firing (54) (Fig. 4d). As expected, light had no effect on MAQ-treated neurons that did not express Shaker Δ6-46; L366A;

E422C; T449V. The SPARK channel therefore provides an orthogonal means to regulate the activity of neurons using light.

3.3.5. Modifying SPARK – Covalent Attachment

The introduction of cysteine residues has been used extensively for the covalent attachment of reactive molecules, including photoswitches, to proteins. Alternatively, several chemical labeling approaches have been developed to target distinct genetically encoded protein tags (88–95), greatly expanding the toolkit for orthogonal labeling and regulation of proteins. For example, novel functionalities can be installed onto surface proteins containing a 15 amino acid-long biotin acceptor peptide by a reengineered *Escherichia coli* biotin ligase (93) or using a six amino acid consensus sequence that directs the introduction of an aldehyde group into the protein by a formyl-generating enzyme and subsequent incubation with aldehyde-specific reagents (90). Another approach for the photoregulation of proteins involves the introduction of unnatural amino acids that are themselves photoisomerizable (96, 97), although it is unclear whether photoswitching these amino acids will affect protein function as dramatically as exogenously applied photoswitches that covalently attach to their target protein (49).

3.3.6. Modifying SPARK – Channel Biophysics

By including previously characterized mutations (98) and taking into account the three-dimensional structure of the K^+ channels (99, 100), it is possible to tailor SPARK for particular biological applications. For instance, a single amino acid change in the pore region (V443Q) converts the normally K^+ selective Shaker into a nonspecific cation channel by changing the Na^+:K^+ permeability ratio from <0.02 to 0.70 (101). Engineering this mutation in the Shaker Δ6-46, L366A, E422C, T449V context enables optochemical gating of a ***depolarizing*** conductance after MAQ treatment (69). Thus, illumination with 380 nm light triggers action potential firing in neurons expressing this depolarizing SPARK (D-SPARK), which can then be halted by illumination with 500 nm light (Fig. 4e). The availability of two light-regulated channels with opposite polarity is of particular interest for use in retinal ganglion cells, which typically fire at light onset (ON cells) or light offset (OFF cells). Using the same wavelengths of light, it would be possible to photoregulate SPARK and D-SPARK expressed in distinct retinal ganglion cells to elicit opposite effects on the membrane potential, mimicking OFF and ON cells, respectively.

3.3.7. Modifying SPARK Channels – Localization

Voltage-gated channels require specific targeting sequences to reach their proper subcellular locations. Mammalian homologues of Shaker, the Kv1 family of channels, normally localize to the cell body and axon of neurons by virtue of a conserved N-terminal sequence known as the T1 domain (102, 103). Localization of

SPARK to the cell body and axon is adequate for the photocontrol of spontaneous action potential firing, the output function of the neuron (54). However, to shunt the inputs to a neuron, it may be more efficient to hyperpolarize the membrane potential specifically in dendrites. To enhance dendritic hyperpolarization, it may be necessary to retarget SPARK to dendrites by adding a 16 amino acid-long dileucine-containing motif, isolated from the Kv4.2 channel. This motif is sufficient to relocalize a model membrane protein, CD4, as well as channels from the Shaker family to the somatodendritic compartment (104). A more synaptic localization may also be achieved by "grafting" a PDZ-binding domain onto SPARK (105).

In most neurons, action potentials are initiated in the initial segment of the axon, from which they propagate to activate downstream neurons. Subsequent depolarization of the cell body is thought to play a role in information coding whereas depolarization of the dendrites may contribute to synaptic plasticity (106, 107). In addition, it has been proposed that back-propagation of the action potential into the dendrites may relay information about axonal output back to synapses (107). Thus, targeting of D-SPARK to specific compartments may allow photocontrol of different aspects of neuronal function. Dendritic localization of D-SPARK can be achieved as described above using the Kv4.2 dileucine motif or a PDZ-binding domain. Likewise, membrane proteins can be targeted to the initial segment of the axon using the cytoplasmic loop linking domains II and III of voltage-gated Na^+ channels (108, 109). Photocontrol of D-SPARK localized to the axon initial segment will mimic the way spikes are initiated *in vivo* and will likely induce "normal" downstream events such as synaptic plasticity.

3.3.8. Modifying SPARK – Photoswitch Characteristics

Open SPARK channels revert to their closed state within minutes in the dark because the *cis* configuration of MAQ relaxes spontaneously to the *trans* configuration (54). Illumination with 380 nm light will keep the chemical gate opened, but prolonged illumination may not always be desirable. Thermal stability, that is the rate of spontaneous relaxation from *cis* to *trans* in darkness, can be varied (from milliseconds to days) by altering the position and nature of substituents on the aromatic rings of the azobenzene moiety (64, 110, 111). In particular, addition of methyl groups to the *ortho* position of the benzene rings dramatically enhances thermal stability of MAQ in solution (112). The generation of a more stable *cis*-MAQ would be potentially useful for the long-term manipulation of neuronal activity in situations when continuous illumination is not possible.

Another noteworthy characteristic of azobenzene photoswitches is their spectral sensitivity. Currently, MAQ undergoes maximal photoisomerization to the *cis* state around 380 nm, a near-UV wavelength. Because light can be scattered by living tissue, it may be desirable to shift the action spectra of the photoswitch toward longer wavelengths. Chemical modification of the azobenzene moiety may enable changes in the absorption spectra of the molecule, although this could also alter the thermal stability of the photoswitch (110). Likewise, it may be useful for use in deep tissue to generate a modified azobenzene photoswitch with a more significant 2-photon cross-section. Although the 2-photon cross-section of several azobenzene-containing molecules has been investigated (113), nonlinear optics have not yet been used successfully to photoregulate channels modified with azobenzene-containing photoswitches.

The modular nature of azobenzene-based photoswitch molecules enables flexible design of each functional group, yielding a combinatorial toolkit for the optical regulation of proteins. Modification of the azobenzene moiety can lead to changes in the properties of the photoswitch, such as thermal stability and action spectra. In addition, the choice of reactive group and ligand determines which protein is targeted by the photoswitch. Since the description of SPARK, the photoswitch approach has been applied to the regulation of endogenous voltage-gated K^+ channels to enable optical regulation of cellular excitability (55), which will be described in the next section. Imparting photocontrol to the voltage-gated K^+ channels was facilitated by the extensive knowledge of the biophysics and structure of these channels. Likewise, a photoswitchable ionotropic glutamate receptor was designed rationally (56), based on the X-ray structure of the ligand-binding domain of iGluR6 bound to an agonist (114). A cysteine was positioned on the surface of iGluR6 such that the photoswitch MAG (*m*aleimide-*a*zobenzene-*g*lutamate) reaches the ligand-binding site and activates the receptor only in its *cis* configuration. The resulting light-activated glutamate receptor, LiGluR, reversibly depolarizes cells and promotes neuronal firing (56, 115, 116).

4. Reversible Photocontrol of Native Ion Channels

Existing light-regulated ion channels such as ChR2, SPARK, and LiGluR can impart light sensitivity to neuronal firing, but only if their gene is first introduced into the cell of interest followed by sufficient expression of the protein on the plasma membrane. Genes encoding light-activated proteins can be introduced transgenically into some organisms, but this may perturb the development, as well as function of the cells expressing the genes. Alternatively, exogenous gene expression can be achieved using chemical or viral-mediated transfection but this can be slow, nonuniform, and nontrivial for some cells or in some organisms. For the treatment of human diseases, gene therapy holds great promise but suffered some setbacks recently, at least in the case of methods based on viral gene delivery vehicles (117).

We have described a new method based on photoswitchable molecules that confer light sensitivity to proteins without requiring genetic engineering and subsequent exogenous gene expression (55). This approach can be used to photosensitize endogenous proteins and control their activity in genetically unadulterated cells or tissues, simply by bathing them for a relatively short time (minutes) in a solution containing the photoswitch molecule. We have designed photoswitchable molecules that specifically target voltage-gated K^+ channels with a pore-blocking TEA and also contain an electrophilic group for attachment to the channel as well as a photoswitchable azobenzene moiety. Similar to the previously described MAQ, these photoswitches block target channels only in their extended, or *trans*, configuration, but not in their short, *cis* form. However, photoswitch-treated endogenous channels differ in several ways from SPARK channels. Whereas the SPARK system involves exogenous expression of an extensively modified Shaker K^+ channel to enable opto-chemical control, the new photoswitches act on endogenous K^+ channels that have no introduced cysteine and no mutations to modify gating. Hence these photoswitches define a new, one-component system for conferring light-sensitivity.

4.1. Affinity Labeling

Since its original description over 50 years ago, the concept of affinity labeling has contributed significantly to our understanding of how proteins, and in particular enzymes and receptors, carry out their biological functions (75). Traditional affinity labels, also known as active site-directed ligands, first bind reversibly to their binding site and then, by virtue of a highly reactive group, become attached covalently to the target protein within or outside the ligand-binding site. Affinity labels have been used to identify amino acids involved in ligand binding and catalysis as well as to map distances between a ligand-binding site and selected residues in a variety of proteins. In addition, affinity labels can activate or inhibit protein function since covalent attachment results in persistent occupancy of the ligand-binding site (67, 75). Affinity labels traditionally consist of a reactive electrophile such as maleimide or iodoacetamide, for attachment to highly nucleophilic amino acids, such as cysteine (118). However, it can be difficult to affinity label proteins that lack an appropriately positioned cysteine or contain multiple cysteines. Recent work has begun exploring the use of alternative electrophiles for covalent attachment to the surface of proteins that does not rely on cysteines. For instance, epoxide-containing probes have been used to affinity label human carbonic anhydrase II (119, 120). Mass spectrometry analysis revealed that the epoxide-containing probe labeled a single histidine residue located outside the active site (119), confirming that epoxide is a suitable electrophile to label proteins on residues other than cysteines. Expanding on the affinity labeling of human carbonic anhydrase II, Harvey and Trauner modified the epoxide-containing probe to include a photoisomerizable azobenzene moiety, allowing them to modulate carbonic anhydrase II activity using light (121). Other electrophilic moieties, including acrylamide, chloroacetamide and unsaturated ketone, have been successfully used to label cysteines and other amino acid residues *in vitro* (122, 123) and may also be suitable for affinity labeling.

4.2. Photoisomerizable Molecules for K^+ Channels Control

4.2.1. Shaker

A variety of electrophiles were used in the generation of photoswitch molecules for native K^+ channels, including epoxide, chloroacetamide and acrylamide, generating EAQ, CAQ, and AAQ, respectively. As for the previously described MAQ, these molecules consist of a K^+ channel-specific pore-blocking quaternary ammonium, a photoisomerizable azobenzene and a reactive electrophile. We first tested AAQ, CAQ, and EAQ on cells heterologously expressing a Shaker channel engineered to contain an appropriately positioned nucleophilic amino acid (E422C). Addition of AAQ, CAQ, or EAQ to Shaker E422C photosensitized the channel, enabling regulation of ion flow with light. Importantly, photosensitivity could also be imparted onto a Shaker lacking the cysteine substitution. Given that a wild-type Shaker channel became photosensitive after treatment with AAQ, CAQ, or EAQ, we tested if other TEA-sensitive K^+ channels could also be photosensitized (55). HEK293 cells were transfected with various K^+ channels before treatment with AAQ. Several K^+ channels became light sensitive, including Kv1.2, Kv1.3, Kv1.4, Kv2.1, and Kv4.2. Light sensitivity was less pronounced for Kv3.3 and BK channels, whereas Kv3.1 exhibited no detectable photosensitivity under the same AAQ treatment conditions. Importantly, neither voltage-gated Na^+ channels (Nav1.2) nor voltage-gated Ca^{2+} channels (L-type) became light sensitive after AAQ treatment.

4.2.2. Endogenous K^+ Channels

To ultimately test the effectiveness of AAQ in photosensitizing endogenous K^+ channels, we recorded voltage-gated outward K^+ currents from cultured hippocampal neurons before and after AAQ treatment. AAQ application and subsequent exposure to 500 nm light blocked a large fraction of voltage-gated K^+ currents. Photoisomerization of AAQ to its *cis* configuration with 380 nm relieved K^+ channel block, restoring K^+ currents to levels indistinguishable from those before treatment with AAQ (Fig. 5a). Thus, treatment of neurons with AAQ enables photocontrol of K^+ channels without generating a population of irreversibly blocked channels. Voltage-gated K^+ channels could be similarly photosensitized by treatment with CAQ. In contrast, treatment

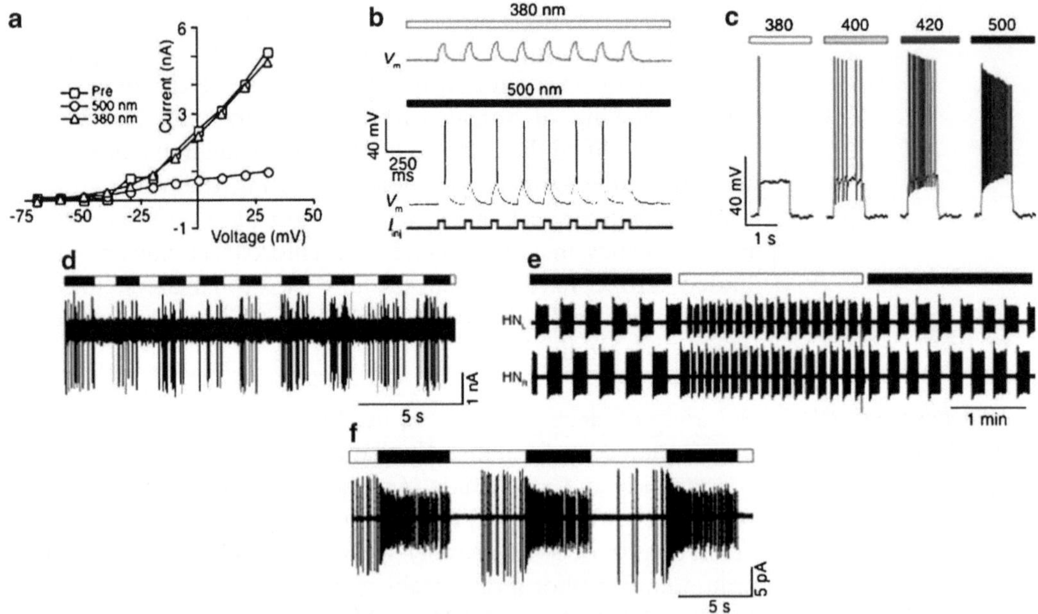

Fig. 5. Photoswitches for native K+ channels are generated by replacing the cysteine-specific maleimide different chemical groups. (**a**) Treatment of cultured hippocampal neurons with AAQ and subsequent illumination with 500 nm light blocks a large fraction of the outward current measured with whole-cell recording (*circles*). Illumination with 380 nm light restores outward current (*triangles*) to pre-treatment levels (*squares*). (**b**) After photoswitch treatment, current injections that do not result in membrane potential deflections large enough to trigger action potential firing under 380 nm light reliably elicit spikes in 500 nm light. (**c**) Different numbers of action potentials are fired in response to a given current injection when the extent of K+ channel block is modified by setting the relative proportion of *cis* and *trans* isomers with different wavelengths of light. (**d–f**) Photoswitches for native K+ channels enable photoregulation of neuronal activity in freshly obtained tissue without genetic manipulation. Extracellular, loose-patch recordings obtained from a cerebellar basket cell (**d**), the leech heart central pattern generator interneurons (**e**), and a retinal ganglion cell (**f**) show that neuronal activity is increased when K+ channels are blocked under 500 nm light (*black bars*) and decreased under 380 nm light (*white bars*).

with MAQ, which has a cysteine-specific electrophile, did not result in photosensitization of neuronal K+ channels. Despite having a reactive electrophile and causing K+ channel blockade in their thermally relaxed state, AAQ and CAQ did not have obvious deleterious effects on cultured hippocampal neurons, up to several hours after treatment. However, EAQ caused significant cell death in cultured hippocampal neurons and was not characterized further.

4.2.3. Using AAQ to Modulate Neuronal Excitability

Like other light-activated ion channels, optical control of AAQ-modified K+ channels can alter the frequency of action potential firing in neurons. For example, photoisomerization of AAQ to its *trans* configuration with continuous 500 nm light increases firing frequency of neurons firing action potentials under 380 nm light. Likewise, action potentials can be triggered in neurons close to their firing threshold by photoisomerizing AAQ from *cis* to *trans* using 500 nm light. However, unlike ChR2 and LiGluR, which encode nonspecific cation channels that directly depolarize the membrane potential, AAQ targets voltage-gated K^+ channels, which serve as a "brake" rather than a direct trigger of activity. Thus, although light can induce action potential firing in AAQ-treated neurons, the temporal fidelity is lower than with other light-activated ion channels that directly depolarize the membrane. Nonetheless, photocontrol of AAQ-modified channels occurs rapidly, resulting in modulation of neuronal firing within hundreds of milliseconds after the onset of illumination, with the exact delay varying with light intensity. Because AAQ targets native ion channels that control cellular excitability, it is possible to alter the threshold for action potential firing in treated neurons using light. Thus, current injections (simulating synaptic inputs) that fail to elicit action potential firing under 380 nm light reliably induce firing under 500 nm light (Fig. 5b). In addition, different wavelengths of light can be used to control the relative proportion of *cis* and *trans* photoisomers of azobenzene, resulting in graded K^+ channel block and fine control over cellular excitability. For example, injection of depolarizing current in AAQ-treated neurons induces an increasing number of action potentials as the extent of K^+ channel block is varied with different wavelengths of light (Fig. 5c). Rather than being a simple digital I/O device to induce action potential firing, AAQ acts as an analog device for the graded control of cellular excitability.

Another noteworthy property of azobenzene-containing photoswitches is the stability of the *trans* isomer. Flashes of 500 nm light followed by darkness result in sustained control of firing that long outlived the light stimulus. Likewise, because *cis*-AAQ thermally relaxes over a period of several minutes, 380 nm light flashes (1 s/min) are sufficient to maintain AAQ-modified channels >90% unblocked. The persistent nature of both AAQ photoisomers eliminates the need for continuous illumination during photocontrol of modified channels and thus minimizes photodamage to the photoswitch and target cells. This feature differentiates AAQ-mediated optical control from regulation by glutamate uncaging and ChR2, which generally affects neurons only during the illumination period. Mutations have recently been identified that extend the lifetime of the ChR2 open state, generating a channel that remains open after the light stimulus has ceased. The opened ChR2 can then be turned off by a different wavelength of light. Introduction of these mutations in ChR2 thus results in a bistable depolarizing channel that can be used for sustained depolarization of cells (124).

4.2.4. Using AAQ in Intact Circuits

To photomodulate endogenous K^+ channels in intact tissue, an additional requirement is that the photoswitch molecule and light reach the cells of interest. We first evaluated AAQ-mediated optical control of basket cell excitability using the loose-patch recording configuration in freshly obtained cerebellar slices. Recordings were obtained in the presence of AMPA, NMDA, and $GABA_A$ receptor antagonists to isolate the basket cells from the rest of the circuit. As expected, unblocking K^+ channels with 360 nm light reduced basket cell firing, whereas blocking K^+ channels with 500 nm light promoted firing (Fig. 5d). These results demonstrate that the penetration of AAQ and the delivery of light are not significantly impeded in brain tissue. Since Purkinje neurons receive inhibitory inputs from basket cells, photoregulation of basket cell firing may be expected to modulate their firing pattern. Indeed, in the context of intact synaptic transmission, Purkinje neurons increased their firing rate when illuminated with 360 nm light and decreased their firing rate under 500 nm light. Thus, the effects of AAQ were overlaid on normal tissue circuitry, enabling regulation of the whole neuronal pathway.

We also tested AAQ on the medicinal leech *Hirudo medicinalis*, a system where techniques for the introduction of foreign genes, required to express genetically encoded light-activated channels, are not widely used. We obtained extracellular recordings from the heart central pattern generator interneurons (HN cells) that control the contraction of the heart by bursting in alternation. Modeling studies have predicted changes in the bursting pattern of HN cells upon K^+ channel modification (125). FMRFamine, an important neuropeptide in the leech, regulates the burst period of HN cells, possibly by modulating voltage-gated K^+ currents (126). By allowing specific and reversible photoregulation of K^+ channels, our opto-chemical approach provides a means to assess the contribution of K^+ channels in the bursting pattern of HN cells. Consistent with modeling studies predictions, unblocking AAQ-modified K^+ channels with 380 nm light decreased the burst period of HN cells, whereas 500 nm light extended the period (Fig. 5e). The ability to photoregulate neurons in the leech heartbeat central pattern generator demonstrates that AAQ-mediated photosensitization is a powerful approach to control K^+ channels and electrical activity in an intact neural circuit without genetic modification.

The installation of light sensitivity on neurons may enable the artificial input of information downstream from sites of damage or degeneration. Additionally, because it targets endogenous channels, AAQ may be useful when the introduction of foreign genes via gene therapy is not practical or deemed unsafe. A particularly relevant tissue for AAQ-mediated optical regulation of excitability is the retina, the sole part of the nervous system that is exposed to light *in vivo*. For example, the loss of visual function caused by degeneration of rods and cones could be alleviated by treatments that impart light sensitivity to downstream neurons that are normally light insensitive. We tested whether AAQ treatment could impart photosensitivity on retinal ganglion cells (RGCs), which relay information from the retina to the brain. Loose-patch recordings obtained from RGCs after AAQ treatment showed that their firing increased in 500 nm light and decreased in 380 nm light, owing to the block and unblock of AAQ-modified K^+ channels (Fig. 5f). Our data raise the possibility that AAQ treatment, along with an appropriate optical system, may be used as an alternative to multielectrode-based retinal prosthetic devices (127) to restore visual function in retinas with damaged or degenerated rod and cone photoreceptors.

4.2.5. Cell-Specific Photosensitization

Photoswitches such as AAQ target intrinsic cellular proteins and thus impart photosensitivity on all treated cells, as long as they express the photoswitch target. This is in contrast to genetically encoded light-activated proteins whose expression can be restricted to defined subpopulations of neurons. Widespread photosensitivity may facilitate functional analysis of processes that involve the coordinated firing of multiple cells. However, if regulation of particular cells is desired, there are three ways to limit AAQ-mediated photosensitization. First, the photoswitch molecules can be applied locally so that only a restricted cell or group of cells becomes photosensitized. We also found that illumination of a subpopulation of cells with 380 nm light during AAQ treatment prevents their photosensitization, providing a second means to restrict installation of light sensitivity. Third, after AAQ treatment, light of the appropriate wavelength can be projected locally to regulate the photoswitch in individual cells or group of cells. The key asset of the photoswitch for endogenous proteins approach is that light sensitivity can be installed onto freshly obtained tissue, unadulterated by exogenous gene expression and possible developmental consequences of ectopic protein expression.

4.2.6. Mechanism of AAQ Action

We originally envisioned the following events upon treatment with AAQ: first, the QA binds the pore of the channel, slowing its departure from the vicinity of channel and increasing the local effective concentration of the reactive moiety. Second, covalent attachment occurs if the channel possesses a nucleophilic amino acid located at the appropriate distance from the QA-binding site (~17 Å). We attempted to identify the site of attachment of AAQ by mutating single candidate amino acids on the surface of the Shaker channel to alanine, a nonreactive amino acid. These mutant channels were expressed in HEK293 cells, treated with AAQ and photosensitization quantified. No single mutation resulted in the generation of an AAQ-insensitive channel. During the course of these studies, we also found that changing the affinity of Shaker for external TEA block, by mutating a single amino acid (72, 128), did not influence the rate or extent of AAQ-mediated photosensitization. This is consistent with the broad range of affinities for external TEA block exhibited by different AAQ-sensitive K^+ channels (55). A photoswitch in which the reactive acrylamide was replaced with a nonreactive acetyl group also photosensitized Shaker expressed in HEK293 cells, albeit with lower potency. Covalent attachment is thus not necessary for efficient photosensitization of K^+ channels. Then how does AAQ confer persistent (at least several hours) photosensitivity to endogenous neuronal channels? Although the mechanism of action of AAQ and related photoswitches is still under investigation, the persistence of photosensitization indicates that if the molecule does not covalently attach to the channel, it may nonetheless become trapped in protein crevices or permeate through the membrane, accumulating within the membrane itself or the intracellular space and resulting in long-lasting photosensitization.

5. Summary and Perspectives

Several approaches have been described to impart light sensitivity to proteins and cellular functions. Intrinsically photosensitive proteins such as ChR2 and halorhodopsin exhibit high fidelity and rapid kinetics for switching on and off and have been used to interrogate neural circuits *in vitro* and control animal behavior (26–31, 39–43). We have developed a different method for imparting light sensitivity to heterologously expressed proteins, based on synthetic photoisomerizable molecules. A covalent attachment site, such as a cysteine, is first added to the protein of interest enabling the attachment of a photoswitch. Because of the modular nature of the photoswitch and the possibility to genetically modify the target protein, this approach has tremendous flexibility and in theory can be applied to many receptors, ion channels, and other signaling proteins. The versatility of the photoswitch approach has already been demonstrated with the photosensitization of different classes of proteins including a voltage-gated K^+ channel (54) and an ionotropic glutamate receptor (56). Extension of this approach to other proteins is likely forthcoming based on the increasing availability of structural data and extensive knowledge of pharmacology for ion channels and receptors.

When the introduction and expression of foreign genes is not possible or prohibitively difficult, small molecule uncaging can be used to activate cell surface receptors and regulate cellular function. However, photorelease is irreversible and liberated photoproducts can diffuse away from the uncaging spot resulting in unintended activation of receptors on untargeted cells. In addition, prolonged uncaging can lead to local depletion of the caged molecule. Small, photoisomerizable molecules that target native proteins circumvent the potential pitfalls associated with caged molecules, allowing for persistent on and off photoregulation of protein function. For instance, photochromic ligands enable control of protein function through a reversible change in the shape or polarity of a bistable photoswitch, such as azobenzene (66, 68, 129). Like caged molecules, photochromic ligands need to be continuously present in the extracellular space to regulate their target proteins. However, they are not consumed during the process of photoregulation and enable on/off control of protein function. Expanding on the design of a genetically encoded light-regulated K^+ channel, termed SPARK (54), we generated photoswitch molecules to target endogenous channels. Treatment with these photoswitches successfully imparted light sensitivity to K^+ channels expressed natively in neurons, enabling photocontrol of neuronal excitability in dissociated cultures and semi-intact neuronal circuits (55).

It is likely that all three methods surveyed here will continue to find use in the study of proteins and cell physiology. An important variable in the photoregulation of proteins and cellular activity not discussed in depth here is the ability of light to reach the cell(s) of interest. The development of portable epifluorescence (130), one-photon (131, 132) and two-photon (133) microendoscopes has enabled visualization of cells embedded deep in tissue that were previously inaccessible to *in vivo* imaging (130) and could be adapted for photoregulation of proteins *in vivo*. An optical neural interface has recently been developed to control neuronal activity with ChR2 in mouse brain (134). Based on the success of photosensitization methods for the control of neuronal activity in model systems and organisms, it is conceivable that these methods could be used as medical tools for the noninvasive input of information into the nervous system. The ability to photoregulate neuronal activity downstream from sites of injury or degeneration could help restore the function of the affected neuronal pathway, heralding a new era in molecular medicine (135, 136).

Acknowledgment

This research was supported by grants from the Lawrence Berkeley National Laboratory and the National Institutes of Health.

References

1. Luo L, Callaway EM, Svoboda K (2008) Genetic dissection of neural circuits. Neuron 57:634–660
2. Sjulson L, Miesenbock G (2008) Photocontrol of neural activity: biophysical mechanisms and performance in vivo. Chem Rev 108:1588–1602
3. Zhang F, Aravanis AM, Adamantidis A, de Lecea L, Deisseroth K (2007) Circuit-breakers: optical technologies for probing neural signals and systems. Nat Rev Neurosci 8:577–581
4. Ellis-Davies GC (2007) Caged compounds: photorelease technology for control of cellular

chemistry and physiology. Nat Methods 4:619–628

5. Callaway EM, Yuste R (2002) Stimulating neurons with light. Curr Opin Neurobiol 12:587–592
6. Shoham S, O'Connor DH, Sarkisov DV, Wang SS (2005) Rapid neurotransmitter uncaging in spatially defined patterns. Nat Methods 2:837–843
7. Bureau I, Shepherd GM, Svoboda K (2004) Precise development of functional and anatomical columns in the neocortex. Neuron 42:789–801
8. Callaway EM, Katz LC (1993) Photostimulation using caged glutamate reveals functional circuitry in living brain slices. Proc Natl Acad Sci USA 90:7661–7665
9. Dalva MB, Katz LC (1994) Rearrangements of synaptic connections in visual cortex revealed by laser photostimulation. Science 265:255–258
10. Shepherd GM, Pologruto TA, Svoboda K (2003) Circuit analysis of experience-dependent plasticity in the developing rat barrel cortex. Neuron 38:277–289
11. Shepherd GM, Stepanyants A, Bureau I, Chklovskii D, Svoboda K (2005) Geometric and functional organization of cortical circuits. Nat Neurosci 8:782–790
12. Shepherd GM, Svoboda K (2005) Laminar and columnar organization of ascending excitatory projections to layer 2/3 pyramidal neurons in rat barrel cortex. J Neurosci 25:5670–5679
13. Dodt H, Eder M, Frick A, Zieglgansberger W (1999) Precisely localized LTD in the neocortex revealed by infrared-guided laser stimulation. Science 286:110–113
14. Matsuzaki M, Ellis-Davies GC, Nemoto T, Miyashita Y, Iino M, Kasai H (2001) Dendritic spine geometry is critical for AMPA receptor expression in hippocampal CA1 pyramidal neurons. Nat Neurosci 4:1086–1092
15. Wang S, Szobota S, Wang Y et al (2007) All optical interface for parallel, remote, and spatiotemporal control of neuronal activity. Nano Lett 7:3859–3863
16. Lutz C, Otis TS, Desars V, Charpak S, Digregorio DA, Emiliani V (2008) Holographic photolysis of caged neurotransmitters. Nat Methods 5:821–827
17. Zemelman BV, Nesnas N, Lee GA, Miesenbock G (2003) Photochemical gating of heterologous ion channels: remote control over genetically designated populations of neurons. Proc Natl Acad Sci USA 100:1352–1357
18. Arenkiel BR, Klein ME, Davison IG, Katz LC, Ehlers MD (2008) Genetic control of neuronal activity in mice conditionally expressing TRPV1. Nat Methods 5:299–302
19. Boyden ES, Zhang F, Bamberg E, Nagel G, Deisseroth K (2005) Millisecond-timescale, genetically targeted optical control of neural activity. Nat Neurosci 8:1263–1268
20. Melyan Z, Tarttelin EE, Bellingham J, Lucas RJ, Hankins MW (2005) Addition of human melanopsin renders mammalian cells photoresponsive. Nature 433:741–745
21. Zemelman BV, Lee GA, Ng M, Miesenbock G (2002) Selective photostimulation of genetically chARGed neurons. Neuron 33:15–22
22. Han X, Boyden ES (2007) Multiple-color optical activation, silencing, and desynchronization of neural activity, with single-spike temporal resolution. PLoS ONE 2:e299
23. Zhang F, Wang LP, Brauner M et al (2007) Multimodal fast optical interrogation of neural circuitry. Nature 446:633–639
24. Li X, Gutierrez DV, Hanson MG et al (2005) Fast noninvasive activation and inhibition of neural and network activity by vertebrate rhodopsin and green algae channelrhodopsin. Proc Natl Acad Sci USA 102:17816–17821
25. Siebert F (2003) Retinal Proteins. In: Durr H, Bouas-Laurent H (eds) Photochromism: Molecules and Systems, 2nd edn. Elsevier, New York, pp 756–792
26. Schroll C, Riemensperger T, Bucher D et al (2006) Light-induced activation of distinct modulatory neurons triggers appetitive or aversive learning in Drosophila larvae. Curr Biol 16:1741–1747
27. Nagel G, Brauner M, Liewald JF, Adeishvili N, Bamberg E, Gottschalk A (2005) Light activation of channelrhodopsin-2 in excitable cells of Caenorhabditis elegans triggers rapid behavioral responses. Curr Biol 15:2279–2284
28. Adamantidis AR, Zhang F, Aravanis AM, Deisseroth K, de Lecea L (2007) Neural substrates of awakening probed with optogenetic control of hypocretin neurons. Nature 450:420–424
29. Arenkiel BR, Peca J, Davison IG et al (2007) In vivo light-induced activation of neural circuitry in transgenic mice expressing channelrhodopsin-2. Neuron 54:205–218
30. Huber D, Petreanu L, Ghitani N et al (2008) Sparse optical microstimulation in barrel cortex drives learned behaviour in freely moving mice. Nature 451:61–64
31. Lagali PS, Balya D, Awatramani GB et al (2008) Light-activated channels targeted to

ON bipolar cells restore visual function in retinal degeneration. Nat Neurosci 11:667–675

32. Zlokovic BV, Apuzzo ML (1997) Cellular and molecular neurosurgery: pathways from concept to reality – part II: vector systems and delivery methodologies for gene therapy of the central nervous system. Neurosurgery 40:805–812, discussion 812–813
33. Verma IM, Weitzman MD (2005) Gene therapy: twenty-first century medicine. Annu Rev Biochem 74:711–738
34. Jang JH, Lim KI, Schaffer DV (2007) Library selection and directed evolution approaches to engineering targeted viral vectors. Biotechnol Bioeng 98:515–524
35. Kuhlman SJ, Huang ZJ (2008) High-resolution labeling and functional manipulation of specific neuron types in mouse brain by Cre-activated viral gene expression. PLoS ONE 3:e2005
36. Heintz N (2004) Gene expression nervous system atlas (GENSAT). Nat Neurosci 7:483
37. Caroni P (1997) Overexpression of growth-associated proteins in the neurons of adult transgenic mice. J Neurosci Methods 71:3–9
38. Gobel W, Helmchen F (2007) In vivo calcium imaging of neural network function. Physiology (Bethesda) 22:358–365
39. Bi A, Cui J, Ma YP et al (2006) Ectopic expression of a microbial-type rhodopsin restores visual responses in mice with photoreceptor degeneration. Neuron 50:23–33
40. Petreanu L, Huber D, Sobczyk A, Svoboda K (2007) Channelrhodopsin-2-assisted circuit mapping of long-range callosal projections. Nat Neurosci 10:663–668
41. Wang H, Peca J, Matsuzaki M et al (2007) High-speed mapping of synaptic connectivity using photostimulation in Channelrhodopsin-2 transgenic mice. Proc Natl Acad Sci USA 104:8143–8148
42. Douglass AD, Kraves S, Deisseroth K, Schier AF, Engert F (2008) Escape behavior elicited by single, channelrhodopsin-2-evoked spikes in zebrafish somatosensory neurons. Curr Biol 18:1133–1137
43. Liewald JF, Brauner M, Stephens GJ et al (2008) Optogenetic analysis of synaptic function. Nat Methods 5:895–902
44. Marriott G, Roy P, Jacobson K. (2003) Preparation and light-directed activation of caged proteins. Methods Enzymol. 360:274–288
45. Dougherty DA (2000) Unnatural amino acids as probes of protein structure and function. Curr Opin Chem Biol 4:645–652
46. Tong Y, Brandt GS, Li M et al (2001) Tyrosine decaging leads to substantial membrane trafficking during modulation of an inward rectifier potassium channel. J Gen Physiol 117:103–118
47. Wischmeyer E, Doring F, Karschin A (1998) Acute suppression of inwardly rectifying Kir2.1 channels by direct tyrosine kinase phosphorylation. J Biol Chem 273:34063–34068
48. England PM, Lester HA, Davidson N, Dougherty DA (1997) Site-specific, photochemical proteolysis applied to ion channels in vivo. Proc Natl Acad Sci USA 94:11025–11030
49. Loudwig S, Bayley H (2005) Light activated proteins: an overview. In: Goeldner M, Givens R (eds) Dynamic studies in biology: phototriggers, photoswitches and caged biomolecules. Wiley–VCH, Weinheim, Germany, pp 253–304
50. Rau H (2003) Azo Compounds. In: Durr H, Bouas-Laurent H (eds) Photochromism: Molecules and Systems, 2nd edn. Elsevier, New York, pp 165–192
51. Guglielmetti R (2003) 4n+2 Systems: Spiropyrans. In: Durr H, Bouas-Laurent H (eds) Photochromism: Molecules and Systems, 2nd edn. Elsevier, New York, pp 314–466
52. Cordes T, Heinz B, Regner N et al (2007) Photochemical Z→E isomerization of a hemithioindigo/hemistilbene omega-amino acid. Chemphyschem 8:1713–1721
53. Banghart MR, Volgraf M, Trauner D (2006) Engineering light-gated ion channels. Biochemistry 45:15129–15141
54. Banghart M, Borges K, Isacoff E, Trauner D, Kramer RH (2004) Light-activated ion channels for remote control of neuronal firing. Nat Neurosci 7:1381–1386
55. Fortin DL, Banghart MR, Dunn TW et al (2008) Photochemical control of endogenous ion channels and cellular excitability. Nat Methods 5:331–338
56. Volgraf M, Gorostiza P, Numano R, Kramer RH, Isacoff EY, Trauner D (2006) Allosteric control of an ionotropic glutamate receptor with an optical switch. Nat Chem Biol 2:47–52
57. Kocer A, Walko M, Meijberg W, Feringa BL (2005) A light-actuated nanovalve derived from a channel protein. Science 309:755–758
58. Lougheed T, Borisenko V, Hennig T, Ruck-Braun K, Woolley GA (2004) Photomodulation of ionic current through hemithioindigo-modified gramicidin channels. Org Biomol Chem 2:2798–2801

59. LoudwigS, BayleyH (2006) Photoisomerization of an individual azobenzene molecule in water: an on-off switch triggered by light at a fixed wavelength. J Am Chem Soc 128: 12404–12405
60. Renner C, Moroder L (2006) Azobenzene as conformational switch in model peptides. Chembiochem 7:868–878
61. Lednev IK, Ye T-Q, Hester RE, Moore JN (1996) Femtosecond time-resolved UV-visible absorption spectroscopy of *trans*-azobenzene in solution. J Phys Chem 100:13338–13341
62. Nagele T, Hoche R, Zinth W, Wachtveitl J (1997) Femtosecond photoisomerization of cis-azobenzene. Chem Phys Lett 272: 489–495
63. Ihalainen JA, Bredenbeck J, Pfister R et al (2007) Folding and unfolding of a photoswitchable peptide from picoseconds to microseconds. Proc Natl Acad Sci USA 104:5383–5388
64. Pozhidaeva N, Cormier ME, Chaudhari A, Woolley GA (2004) Reversible photocontrol of peptide helix content: adjusting thermal stability of the cis state. Bioconjug Chem 15:1297–1303
65. Lefkowitz RJ (2004) Historical review: a brief history and personal retrospective of seven-transmembrane receptors. Trends Pharmacol Sci 25:413–422
66. Bartels E, Wassermann NH, Erlanger BF (1971) Photochromic activators of the acetylcholine receptor. Proc Natl Acad Sci USA 68:1820–1823
67. Silman I, Karlin A (1969) Acetylcholine receptor: covalent attachment of depolarizing groups at the active site. Science 164:1420–1421
68. Lester HA, Krouse ME, Nass MM, Wassermann NH, Erlanger BF (1980) A covalently bound photoisomerizable agonist: comparison with reversibly bound agonists at Electrophorus electroplaques. J Gen Physiol 75:207–232
69. Chambers JJ, Banghart MR, Trauner D, Kramer RH (2006) Light-induced depolarization of neurons using a modified Shaker K(+) channel and a molecular photoswitch. J Neurophysiol 96:2792–2796
70. Tombola F, Pathak MM, Isacoff EY (2006) How does voltage open an ion channel? Annu Rev Cell Dev Biol 22:23–52
71. Kavanaugh MP, Hurst RS, Yakel J, Varnum MD, Adelman JP, North RA (1992) Multiple subunits of a voltage-dependent potassium channel contribute to the binding site for tetraethylammonium. Neuron 8:493–497
72. Heginbotham L, MacKinnon R (1992) The aromatic binding site for tetraethylammonium ion on potassium channels. Neuron 8:483–491
73. MacKinnon R, Yellen G (1990) Mutations affecting TEA blockade and ion permeation in voltage-activated K+channels. Science 250:276–279
74. Blaustein RO, Cole PA, Williams C, Miller C (2000) Tethered blockers as molecular 'tape measures' for a voltage-gated K+channel. Nat Struct Biol 7:309–311
75. Wold F (1977) Affinity labeling – an overview. Methods Enzymol 46:3–14
76. Kramer RH, Karpen JW (1998) Spanning binding sites on allosteric proteins with polymer-linked ligand dimers. Nature 395:710–713
77. Blaustein RO (2002) Kinetics of tethering quaternary ammonium compounds to K(+) channels. J Gen Physiol 120:203–216
78. Bretschneider F, Wrisch A, Lehmann-Horn F, Grissmer S (1999) External tetraethylammonium as a molecular caliper for sensing the shape of the outer vestibule of potassium channels. Biophys J 76:2351–2360
79. Doyle DA, Morais Cabral J, Pfuetzner RA et al (1998) The structure of the potassium channel: molecular basis of K+ conduction and selectivity. Science 280:69–77
80. MacKinnon R, Miller C (1989) Mutant potassium channels with altered binding of charybdotoxin, a pore-blocking peptide inhibitor. Science 245:1382–1385
81. Darman RB, Ivy AA, Ketty V, Blaustein RO (2006) Constraints on voltage sensor movement in the shaker K+channel. J Gen Physiol 128:687–699
82. Jiang Y, Lee A, Chen J et al (2003) X-ray structure of a voltage-dependent K+channel. Nature 423:33–41
83. Laine M, Papazian DM, Roux B (2004) Critical assessment of a proposed model of Shaker. FEBS Lett 564:257–263
84. Green NS, Reisler E, Houk KN (2001) Quantitative evaluation of the lengths of homobifunctional protein cross-linking reagents used as molecular rulers. Protein Sci 10:1293–1304
85. Krishnamurthy VM, Semetey V, Bracher PJ, Shen N, Whitesides GM (2007) Dependence of effective molarity on linker length for an intramolecular protein-ligand system. J Am Chem Soc 129:1312–1320
86. Hoshi T, Zagotta WN, Aldrich RW (1990) Biophysical and molecular mechanisms of

Shaker potassium channel inactivation. Science 250:533–538
87. Lopez GA, Jan YN, Jan LY (1991) Hydrophobic substitution mutations in the S4 sequence alter voltage-dependent gating in Shaker K+channels. Neuron 7:327–336
88. Marks KM, Nolan GP (2006) Chemical labeling strategies for cell biology. Nat Methods 3:591–596
89. Chen I, Ting AY (2005) Site-specific labeling of proteins with small molecules in live cells. Curr Opin Biotechnol 16:35–40
90. Carrico IS, Carlson BL, Bertozzi CR (2007) Introducing genetically encoded aldehydes into proteins. Nat Chem Biol 3:321–322
91. Adams SR, Campbell RE, Gross LA et al (2002) New biarsenical ligands and tetracysteine motifs for protein labeling in vitro and in vivo: synthesis and biological applications. J Am Chem Soc 124:6063–6076
92. Griffin BA, Adams SR, Tsien RY (1998) Specific covalent labeling of recombinant protein molecules inside live cells. Science 281:269–272
93. Chen I, Howarth M, Lin W, Ting AY (2005) Site-specific labeling of cell surface proteins with biophysical probes using biotin ligase. Nat Methods 2:99–104
94. Keppler A, Gendreizig S, Gronemeyer T, Pick H, Vogel H, Johnsson K (2003) A general method for the covalent labeling of fusion proteins with small molecules in vivo. Nat Biotechnol 21:86–89
95. Keppler A, Pick H, Arrivoli C, Vogel H, Johnsson K (2004) Labeling of fusion proteins with synthetic fluorophores in live cells. Proc Natl Acad Sci USA 101:9955–9959
96. Bose M, Groff D, Xie J, Brustad E, Schultz PG (2006) The incorporation of a photoisomerizable amino acid into proteins in E. coli. J Am Chem Soc 128:388–9
97. Juodaityte J, Sewald N (2004) Synthesis of photoswitchable amino acids based on azobenzene chromophores: building blocks with potential for photoresponsive biomaterials. J Biotechnol 112:127–138
98. Jan LY, Jan YN (1992) Structural elements involved in specific K+channel functions. Annu Rev Physiol 54:537–555
99. Long SB, Campbell EB, Mackinnon R (2005) Crystal structure of a mammalian voltage-dependent Shaker family K+channel. Science 309:897–903
100. Long SB, Tao X, Campbell EB, MacKinnon R (2007) Atomic structure of a voltage-dependent K+channel in a lipid membrane-like environment. Nature 450:376–382
101. Heginbotham L, Lu Z, Abramson T, MacKinnon R (1994) Mutations in the K+channel signature sequence. Biophys J 66:1061–1067
102. Gu C, Jan YN, Jan LY (2003) A conserved domain in axonal targeting of Kv1 (Shaker) voltage-gated potassium channels. Science 301:646–649
103. Rivera JF, Chu PJ, Arnold DB (2005) The T1 domain of Kv1.3 mediates intracellular targeting to axons. Eur J Neurosci 22:1853–1862
104. Rivera JF, Ahmad S, Quick MW, Liman ER, Arnold DB (2003) An evolutionarily conserved dileucine motif in Shal K+channels mediates dendritic targeting. Nat Neurosci 6:243–250
105. Prybylowski K, Chang K, Sans N, Kan L, Vicini S, Wenthold RJ (2005) The synaptic localization of NR2B-containing NMDA receptors is controlled by interactions with PDZ proteins and AP-2. Neuron 47:845–857
106. Bean BP (2007) The action potential in mammalian central neurons. Nat Rev Neurosci 8:451–465
107. Sjostrom PJ, Rancz EA, Roth A, Hausser M (2008) Dendritic excitability and synaptic plasticity. Physiol Rev 88:769–840
108. Garrido JJ, Giraud P, Carlier E et al (2003) A targeting motif involved in sodium channel clustering at the axonal initial segment. Science 300:2091–2094
109. Lemaillet G, Walker B, Lambert S (2003) Identification of a conserved ankyrin-binding motif in the family of sodium channel alpha subunits. J Biol Chem 278:27333–27339
110. Chi L, Sadovski O, Woolley GA (2006) A blue-green absorbing cross-linker for rapid photoswitching of peptide helix content. Bioconjug Chem 17:670–676
111. Forber CL, Kelusky EC, Bunce NJ, Zerner MC (1985) Electronic Spectra of cis- and trans-Azobenzenes: Consequences of Ortho substitution. J Am Chem Soc 107:5884–5890
112. Banghart MR, Mourot A, Fortin DL, Yao JZ, Kramer RH, Trauner D. (2009) Photochromic blockers of voltage-gated potassium channels. Angew Chem Int Ed Engl 248:9097-9101
113. Leonardo De Boni LMSCZCRM (2005) Degenerate two-photon absorption spectra in azoaromatic compounds. Chemphyschem 6:1121–1125
114. Mayer ML (2005) Crystal structures of the GluR5 and GluR6 ligand binding cores: molecular mechanisms underlying kainate receptor selectivity. Neuron 45:539–552

115. Gorostiza P, Volgraf M, Numano R, Szobota S, Trauner D, Isacoff EY (2007) Mechanisms of photoswitch conjugation and light activation of an ionotropic glutamate receptor. Proc Natl Acad Sci USA 104:10865–10870
116. Szobota S, Gorostiza P, Del Bene F et al (2007) Remote control of neuronal activity with a light-gated glutamate receptor. Neuron 54:535–545
117. Edelstein ML, Abedi MR, Wixon J (2007) Gene therapy clinical trials worldwide to 2007 – an update. J Gene Med 9:833–842
118. Brinkley M (1992) A brief survey of methods for preparing protein conjugates with dyes, haptens, and cross-linking reagents. Bioconjug Chem 3:2–13
119. Chen G, Heim A, Riether D et al (2003) Reactivity of functional groups on the protein surface: development of epoxide probes for protein labeling. J Am Chem Soc 125:8130–8133
120. Wakabayashi H, Miyagawa M, Koshi Y, Takaoka Y, Tsukiji S, Hamachi I (2008) Affinity-labeling-based introduction of a reactive handle for natural protein modification. Chem Asian J 3:1134–1139
121. Harvey JH, Trauner D (2008) Regulating enzymatic activity with a photoswitchable affinity label. Chembiochem 9:191–193
122. Levitsky K, Boersma MD, Ciolli CJ, Belshaw PJ (2005) Exo-mechanism proximity-accelerated alkylations: investigations of linkers, electrophiles and surface mutations in engineered cyclophilin-cyclosporin systems. Chembiochem 6:890–899
123. Weerapana E, Simon GM, Cravatt BF (2008) Disparate proteome reactivity profiles of carbon electrophiles. Nat Chem Biol 4:405–407
124. Berndt A, Yizhar O, Gunaydin LA, Hegemann P, Deisseroth K (2009) Bi-stable neural state switches. Nat Neurosci 12(2):229–234
125. Hill AA, Lu J, Masino MA, Olsen OH, Calabrese RL (2001) A model of a segmental oscillator in the leech heartbeat neuronal network. J Comput Neurosci 10:281–302
126. Nadim F, Calabrese RL (1997) A slow outward current activated by FMRFamide in heart interneurons of the medicinal leech. J Neurosci 17:4461–4472
127. Lakhanpal RR, Yanai D, Weiland JD et al (2003) Advances in the development of visual prostheses. Curr Opin Ophthalmol 14:122–127
128. MacKinnon R, Heginbotham L, Abramson T (1990) Mapping the receptor site for charybdotoxin, a pore-blocking potassium channel inhibitor. Neuron 5:767–771
129. Volgraf M, Gorostiza P, Szobota S, Helix MR, Isacoff EY, Trauner D (2007) Reversibly caged glutamate: a photochromic agonist of ionotropic glutamate receptors. J Am Chem Soc 129:260–261
130. Flusberg BA, Cocker ED, Piyawattanametha W, Jung JC, Cheung EL, Schnitzer MJ (2005) Fiber-optic fluorescence imaging. Nat Methods 2:941–950
131. Jung JC, Schnitzer MJ (2003) Multiphoton endoscopy. Opt Lett 28:902–904
132. Flusberg BA, Nimmerjahn A, Cocker ED et al (2008) High-speed, miniaturized fluorescence microscopy in freely moving mice. Nat Methods 5:935–938
133. Helmchen F, Fee MS, Tank DW, Denk W (2001) A miniature head-mounted two-photon microscope. high-resolution brain imaging in freely moving animals. Neuron 31:903–12
134. Aravanis AM, Wang LP, Zhang F et al (2007) An optical neural interface: in vivo control of rodent motor cortex with integrated fiberoptic and optogenetic technology. J Neural Eng 4:S143–S156
135. Airan RD, Hu ES, Vijaykumar R, Roy M, Meltzer LA, Deisseroth K (2007) Integration of light-controlled neuronal firing and fast circuit imaging. Curr Opin Neurobiol 17:587–592
136. Kramer RH, Chambers JJ, Trauner D (2005) Photochemical tools for remote control of ion channels in excitable cells. Nat Chem Biol 1:360–365

Chapter 12

Optical Manipulation of Protein Activity and Protein Interactions Using Caged Proteins and Optical Switch Protein Conjugates

Yuling Yan and Gerard Marriott

Abstract

One of the major challenges in biology is to understand better the molecular regulation of signaling pathways that control complex cellular processes such as motility and proliferation. The experimental approaches employed in such investigations must gather information on changes in protein structure and function over a hierarchy of biological organization, spanning the nanometer dimensions of single protein complexes to micron-sized lamellipodia. These structural and organizational changes take place over correlated timescales that span milliseconds or less for the formation of protein complexes, to minutes for movements of an entire cell. Caged effector molecules and site-selective introduction of photochromic probes within biomolecules have been used as part of an approach for optical control of biomolecular interactions and activities within cells. The photochromic spiropyran-containing molecules undergo rapid and reversible, optically driven transitions between a colorless spiro state and a brightly colored merocyanine state.

Key words: Molecular probes, Photochromism, Spiro compounds, Merocyanine, Pyrimidinones

1. Introduction

A major challenge in molecular and cellular biology is to understand the molecular regulation of signaling pathways that control complex cellular processes such as motility and proliferation (1–3). The experimental approaches employed in such investigations must gather information on changes in protein structure and function over a hierarchy of biological organization, spanning the nanometer dimensions of single protein complexes to micron-sized lamellipodia. These structural and organizational changes take place over correlated timescales that span milliseconds or less for

James J. Chambers and Richard H. Kramer (eds.), *Photosensitive Molecules for Controlling Biological Function*, Neuromethods, vol. 55, DOI 10.1007/978-1-61779-031-7_12, © Springer Science+Business Media, LLC 2011

the formation of protein complexes, to minutes for movements of an entire cell. Moreover, control of protein activity and protein complex formation during cell signaling pathways is usually reversible with many reactions being confined to specific loci (illustrated in Fig. 1a). One could argue then that a complete understanding of the control of cell signaling pathways will require a multiscale, kinetic analysis of reactions underlying protein activity and complex formation, ranging from single interacting proteins to ensemble populations within living cells. These studies require advances in multiscale image microscopy of living cells and tissue, including improved methods for super-resolution imaging (4–7), and integration of techniques used in chemical relaxation kinetics (8) for image-based analysis of protein activity (9). Kinetic imaging of proteins and their complexes in living cells could yield information on sequence of protein interactions within a signaling pathway and measure the rate constants for these interactions as schematized in Fig. 1b. These investigations require a means to perturb specifically the reaction under study and then to monitor its return to the new equilibrium state. For studies of proteins within living cells, these pulse-probe studies must be conducted within an image and recorded with high spatial and temporal resolution (10).

There are many different types of rapid perturbation techniques one can use to perturb the concentration of a molecular species involved in a chemical reaction (11), including specific

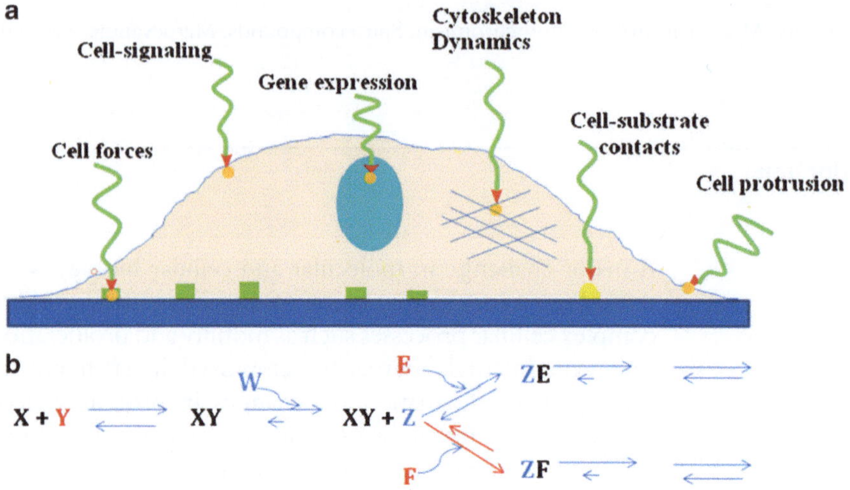

Fig. 1. (**a**) Complex biological processes such as those indicated in the figure are often triggered by external chemical or physical cues and are then regulated at specific loci within the cell by signaling pathways that control the interactions and activities of multiple proteins. (**b**) A schematic representation of a reversible signaling pathway involving the formation and disassembly of protein complexes (such as XY) and ligand–protein complexes (such as ZE and ZF) and protein activity (such as enzymatic transformation of W to Z). Optical manipulation of photoresponsive proteins (such as those highlighted in *red*) provides opportunities to study the sequence of molecular events associated with these pathways and determine kinetic rate constants for reversible reactions in the pathway.

proteins. Since we are interested in perturbing specific proteins within a micro–nanoscopic sample, for example, within a compartment of a living cell, we will neither consider perturbations based on physical properties such as temperature or pressure, nor those that generate a concentration jump using rapid flow. The most common approach to change the concentration or activity of a protein in a cell is based on controlling the level of a gene encoding that protein using transfection of the gene of interest, or by using the RNAi technique to control the level of the mRNA transcript. However, these perturbations usually require hours to days to take effect and cannot control the protein concentration at a specific locus in a cell. This limitation is significant because many signaling pathways remain active for tens of seconds or less, and are often confined to specific sites in the cell (12, 13). The requirement for rapid and localized perturbations to study kinetic properties of a particular reaction in a cell mostly limits the means to control the level of a particular species in a protein-based reaction to optical perturbations. Moreover, the trend of studying signaling in the context of higher organizational levels of cells, such as within living tissue or animals versus noninteracting cells grown on a coverslip, requires a means to trigger the protein perturbation rapidly and reversibly with high spatial resolution within a three-dimensional living sample. Clearly, these latter demands limit the nature of the perturbing source of electromagnetic radiation, which can be delivered rapidly and with high spatial resolution for cells grown in culture or for those deep within tissue using two-photon excitation with near-infrared light (14).

2. Optical Control of Protein Interactions and Protein Activity

In our view, the best general approach to control protein activity or protein complexes rapidly is based on optical control of small molecule actuators that are directly or indirectly attached to the protein of interest. Optical perturbations can be used to bring about the activation or inactivation of specific protein activity or complex formation in a sample, with impressive resolution approaching a few microseconds and submicron (12, 15–19). Optical control of the binding or activity of a photoresponsive protein or protein conjugate may be irreversible in the case of caged proteins (20, 21), or reversible in the case of genetically engineered photoresponsive ion channels and protein conjugates of optical switch probes (18, 19, 22, 23). Here we discuss two different optical approaches to manipulate protein interactions in conjugates of proteins with small molecule actuators. We limit our discussion to in vitro and in vivo studies on protein conjugates of the 2-nitrophenyl (caged) and the nitrospirobenzopyran (optical switch) groups, which can be used, respectively, to control protein interactions irreversibly or reversibly.

2.1. Irreversible Optical Control: Caged Compounds and Caged Proteins

In a now classic study, Jack Kaplan (15) introduced the light-directed activation of caged compounds technique for rapid triggering of protein substrates from inactive, photolabile precursors. In particular, he showed that photoactivation of ATP from caged ATP could be used to trigger the activity of Na:K pumps in native erythrocyte membrane preparations. This probe was also used as part of an approach to derive kinetic parameters for specific reactions in the actomyosin crossbridge cycle (reviewed by (24)). Others advanced the caged compound approach by developing caged derivatives of protein-binding ligands including caged glutamate, caged cyclic AMP, and second messengers such as caged calcium and caged IP_3 (25–27). It is important to recognize that the amount of the photoactivated ligand or substrate required to trigger the reaction can be quite high and may limit the effectiveness of the technique. For example, in the case of myosin II, the k_m for ATP is ~60 μM (16), and so ideally one would want to generate at least twice as much ATP in the sample from the caged ATP. This is not easily achieved using single pulses of near-ultraviolet light. In particular, the quantum yield for the uncaging of 2-nitrophenyl adducts of ATP is low (<0.1; (24)), which requires a far higher concentration of caged ATP (up to 10 mM). Consequently, the high absorption of these samples requires using a caged group with a low extinction coefficient (500 M^{-1} cm^{-1} at 320 nm (24)). Moreover, at a concentration of 10 mM, caged ATP acts as a competitive inhibitor for myosin II (28), a complication that must be accounted for when analyzing the kinetics of myosin II activity in the sample. Moreover, the high energy of the near-ultraviolet light used for these single pulse perturbations is not always compatible with that used in studies on living cells or tissue.

On the contrary, the amount of energy required to manipulate a protein using caged probes can be significantly reduced by removing the caged group from an inactive protein conjugate (20). We originally defined a caged protein as a conjugate whose activity is inactivated by chemical modification with one or more caged groups (29). Removal of caged groups from a protein conjugate is achieved by using near-ultraviolet light (typically 365 nm), generating a native protein whose activity is often restored to preconjugation levels (16). For example, the polymerization activity of G-actin was shown to be inhibited by modifying up to four lysine residues with 6-nitroveratryloxycarbonyl chloride (NVOC-Cl (10)). Ultraviolet irradiation of the caged actin conjugate liberated the native, polymerization-competent G-actin, as well as carbon dioxide and 3,4-dimethoxy-2-nitrosobenzaldehyde as photoproducts, the latter being scavenged by mercaptans (15). It is important to subject the caged and uncaged forms of the protein to rigorous biochemical and spectroscopic characterization in order to show that the caged conjugate is stable, and that the irradiation, the uncaging reaction, or photoproducts do not inactivate the protein of interest or other proteins in the sample (15). Studies from our group and others have shown that caged proteins are effective molecular tools for in vitro and in vivo control of protein activity and protein interactions (12, 16, 17, 20, 30, 31). Two examples of the light-directed activation of caged protein technique are presented below.

2.2. In Vitro Studies

Specific modification of the cysteine-374 residue within the SH1-helix of myosin II with a thiol-reactive caged group (bromoethyl-4,5-dimethoxynitrobenzene) leads to a dramatic decrease in the ability of the caged myosin conjugate to support movement of actin filaments (32). The mechanism of inhibition does not involve the ATPase activity, which actually increases in the conjugate. Rather the presence of the caged group prevents the reversible helix–coil transition within the SH1 helix (Fig. 2a) such

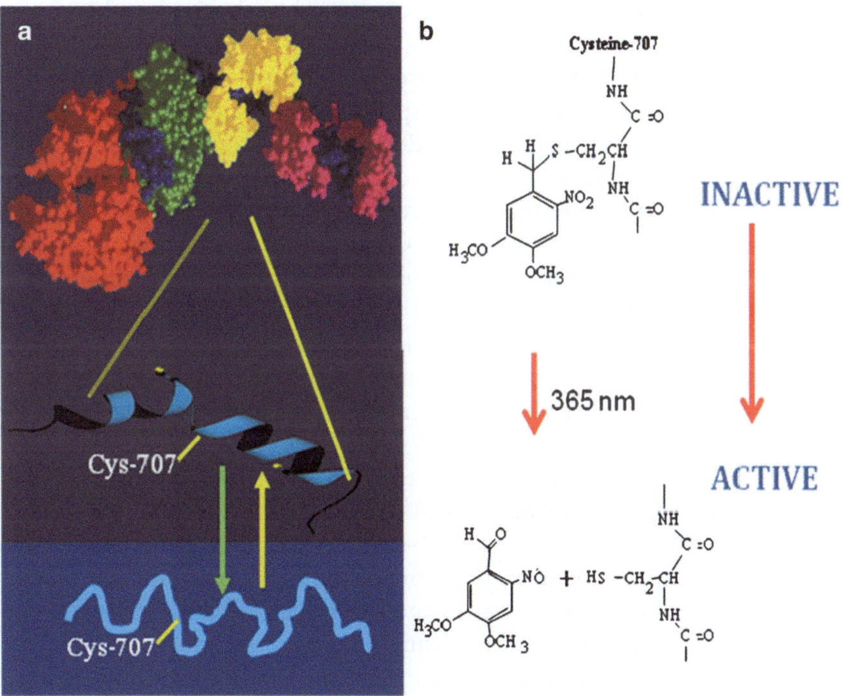

Fig. 2. (**a**) Conformational transition in myosin II involving a reversible helix–coil transition in the SH1 helix that links events associated with the binding and hydrolysis of ATP with the movement of the lever arm. (**b**) Modification of cysteine 374 in the SH1 helix with 1-bromomethyl-4,5-dimethoxynitrobenzene completely inhibits the movement of actin filaments as assayed in an in vitro motility assay (16). This inhibition is dependent on the chemical modification of cysteine-374 and is believed to be linked to the prevention of the helix–coil or the coil–helix transition in SH1. Removal of the caged group using a pulse of 365 nm regenerates the native myosin II and fully restores the motility of actin filaments.

that the ATPase activity is uncoupled from the subsequent conformational transition of the lever arm that leads to the power stroke (16). This inhibitory uncoupling is rapidly reversed, however, upon removal of the caged group using near-ultraviolet light, which generates native myosin II (Fig. 2b) that moves actin filaments at the same velocity as control myosin II (16). The caged group is removed from the myosin conjugate with a time constant of about 22 ms as measured by the rate of decrease of the aci-nitro intermediate, the rate-limiting step in the cleavage of the thioether bond. These studies highlight several advantages in using a caged protein compared to a caged substrate. First, the amount of uncaging required for the kinetic analysis is spectacularly less than that used with the caged ATP approach. In particular, we use a monolayer of caged myosin attached to the surface of the coverslip and can conduct the assay within a 10-μm field of view at a video rate, which of course requires far less energy for uncaging compared to a caged substrate. This feature also allows us to optimize some spectroscopic properties of the caged group for studies using biological samples, including the use of the 4,5-dimethoxy-substituted 2-nitrophenyl caged group, which has a tenfold higher extinction coefficient and far more red-shifted absorption spectrum compared to the 2-nitrophenyl group.

2.3. In Vivo Studies

We used the light-directed activation of caged protein technique (20) in a multiscale spatio-temporal, dynamic analysis of cell signaling pathways underlying cell motility (12). In this study, a simple amino-reactive caged reagent (NVOC-Cl) was used to label essential lysine residues of thymosin β4 (Tβ4), a small polypeptide that buffers G-actin in cells. A generalized depiction of this study is shown in Fig. 3, where Tβ4 is shown as the actin-binding protein (ABP). Caged (NVOC-conjugated) Tβ4 has a greatly reduced affinity for the conjugate of G-actin; therefore, once loaded into living and motile cells, it distributes uniformly, unable to bind to actin. However, irradiation of caged Tβ4 within a subregion of the cell using a microsecond pulse of near-ultraviolet light removes the caged group, thereby elevating the concentration of active Tβ4 in that region. This perturbation increases the amount of Tβ4–G-actin complex and thereby reduces the amount of G-actin for polymerization at that site. The kinetics of this process was imaged and quantified indirectly by imaging the behavior of the leading edge and the motile response to the perturbation. Thus the localized photoactivation of Tβ4 at the wing of a leading edge in a motile keratocyte reduced the rate of actin polymerization at that site and was accompanied by a decrease in protrusive activity at that site. On the contrary, the rate of actin filament dynamics for other regions of the leading edge was unaffected, as seen by normal and continued membrane protrusion. The balance of protrusive normal and inhibited force generation

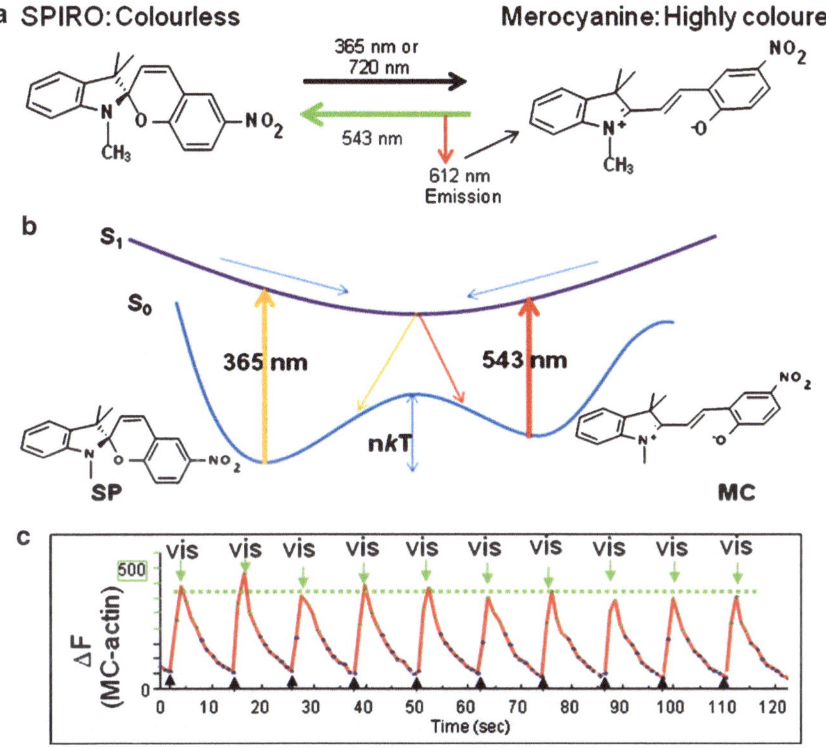

Fig. 3. (**a**) Chemical structures of NitroBIPS and the light-driven transitions between the SP and the MC state. The SP to MC transition is driven by excitation of the SP state using 365- or 720-nm (two-photon) light, while the MC to SP transition can be rapidly driven by 543- or 960-nm (two-photon) light. The excited MC state is depopulated via two competing processes, namely photochemistry that ultimately forms the colorless SP state or by return to the MC ground state with emission of fluorescence. The MC-fluorescence signal provides a simple and sensitive means to readout the state of the SP/MC switch in a sample. (**b**) Potential energy curve for reversible SP/MC transitions in NitroBIPS. SP is thermodynamically favored at 20°C. Cleavage of the spiro bond forms MC. Transitions occur in the ground or excited states, with the latter occurring at a rate of $>10^8$ faster. (**c**) Optical switching between the SP and MC states as seen through the change in the MC-fluorescence signal. The SP to MC transition is triggered quantitatively using a single pulse of 365-nm light (*black arrow*). Irradiation of the MC state with visible light leads to the formation of the SP state or else return to the MC ground state with red fluorescence. The efficiency of these excited-state processes is governed by their respective quantum yields. This underlies the constant profile of the MC-fluorescence signal for each cycle of optical switching study.

along the leading edge caused the cell to pivot at the site of Tβ4 uncaging. The G-actin–Tβ4 complex returns (within 10 s) to a new equilibrium concentration primarily as a result of the diffusion of free Tβ4 from the irradiated region. The rate of actin filament dynamics and protrusive activity is thereby restored to uniform levels along the leading edge, and the cell resumes its unidirectional motile behavior. These multiscale studies suggest that a persistent and polarized cellular response involving reactions between diffusible proteins requires a sustained generation of the active form of the protein at the site of activation.

As discussed earlier, successful application of the light-directed activation of caged protein technique in living cells requires key control experiments to show, for example, that photoproducts generated during the excited uncaging reaction do not interfere with the reaction under study or the health of the cell. The uncaging technique is not well-suited particularly for studies within tissue and within living organisms because of the need to use near-ultraviolet light, which has poor transmittance in biological material. Unfortunately, the 2-nitrophenyl caging group has a poor two-photon absorption cross-section, while other caged groups with better two-photon excitation properties are cleaved via a triplet-state, or possibly even a radical-based, reaction (33). For the time being, then, the light-directed activation of caged protein technique (29) will most likely be limited to studies of proteins within living cells grown on a coverslip or exposed on the surface of a tissue (12, 17).

3. Reversible Control of Protein Interactions and Protein Activity

Optical approaches to control protein activity in cells, tissue, and organisms *reversibly* include the use of genetically engineered, photoresponsive ion channels and their conjugates with small molecule actuators (18, 19). These methods, while providing reversible and direct control of protein molecules, require genetic manipulation of the cell or animal to introduce the engineered ion channel proteins (18). In the Volgraf et al. (19) study, reversible optical manipulation of the active and inactive states of the ion channel also requires attaching a photoresponsive actuator such as azobenzene to a specific cysteine residue on the ion channel. In all cases studied to date, reversible manipulation of the two states of these photoresponsive channel proteins is realized by using one-photon excitation. This can prove to be limiting, especially when using cells within tissue, which has poor transmittance in the near-UV region. Consequently, these studies are usually performed on photoresponsive ion channel proteins expressed in cells in culture, exposed on the surface of the tissue or else by using an excitation device implanted within the animal or tissue (18).

3.1. Optical Switches

In the following section, work from our group is described that shows how small molecule actuators based on a class of molecular optical switch can be used to modulate specific dipolar interactions on proteins rapidly and reversibly. Since studies using optical switches on proteins are fairly recent and few and far between (23, 34–36), we find it necessary to provide a detailed account of their spectroscopic and photochemical properties and to illustrate using model systems how these properties can be best exploited to control the activity and interactions of optical switch protein conjugates reversibly. Optical switches (23) are small molecule actuators that can undergo quantitative transitions between two distinct structural states via excited- and/or ground-state reactions (Fig. 3a, b). Optical switches have been studied by chemists for many decades, and there are several different classes including the dihydro-indolizines, azobenzenes, diarylethanes, chromenes, napthopyrans, spiropyrans (NitroBIPS), spirooxazines (naphthoxazine), and fulgides (reviewed in (37)). The spectroscopic and photophysical properties of optical switch probes in these experiments have largely been determined within organic solvents, and only rarely within aqueous environments. This disparity is largely based on the fact that most optical switches are insoluble in water, or else because the quantum yield for excited-state transitions between the two states is too low (38). Moreover, some classes of photochromic probes are not strictly reversible, and typically involve an irreversible side reaction that depletes the photochromic species (37).

Perhaps the best suited optical switch probes for studies on biomolecules are the nitrospirobenzopyrans (NitroBIPS), naphthoxazines, and diarylethanes (23, 34, 39). In designing an optical switch for applications with proteins or other biomolecules, it becomes necessary to characterize carefully the spectroscopic and photochemical properties of the free probe and its protein conjugate, and then to optimize these properties perhaps by selective labeling to a unique residue in the protein (23, 40). Until recently, however, there were few reactive forms of optical switches for labeling biomolecules and most were linked randomly to the protein through carboxyl-activated *N*-hydroxysuccinimide esters of NitroBIPS (34) or diarylethanes (39). Our group introduced a family of NitroBIPS and naphthoxazine probes for site-specific (cysteine) labeling of proteins (23, 40). These specific and uniquely labeled protein conjugates behave quite differently from the same probe in bulk water, as seen from the improved photochemical, spectroscopic, and functional properties, several of which are bulleted below:

- Reversible and controlled exclusively by light or opto-thermally
- Rapid photoisomerization: allows the study of fast physiological processes
- Clean and efficient: no photoproducts generated during switching
- Large change in absorption spectra between the two states
- Large change in dipole moment between the two states
- Large change in geometry/orientation between the two states
- Ability to trigger transitions by using two-photon excitation
- Tunable – for multiplexed control of proteins and other molecules/ions

- Easy to synthesize
- Able to incorporate diverse chemical substituent groups
- Incorporate an optical readout to determine the status of the switch in a sample

3.2. Nitrospirobenzopyrans and Related Optical Switches

Nearly six decades ago, Fischer and Hirshberg (41) showed that certain small molecules such as NitroBIPS (Fig. 3a) undergo reversible, optically driven, excited-state transitions between two distinct structural states. The intramolecular bond-breaking and bond-forming reactions that allow NitroBIPS to transition between these two states are shown in Fig. 3a. In almost all cases, the thermodynamically stable form of the optical switch is the spiro (SP) state, in which the two aromatic ring systems are orthogonal to each other, and so this state has a near-ultraviolet absorption spectrum with a small ground-state dipole moment of 2–5 Debye (D; (42)). NitroBIPS exhibits, to varying degrees, many of the ideal properties for an optical switch (listed above). As illustrated in Fig. 3a, b, a single cycle of the optical switching between the two states of NitroBIPS begins with excitation of SP using a short pulse of 365-nm light. The excited SP state undergoes a rapid (nanosecond), intramolecular spiro-bond cleavage reaction and a subsequent internal structural rearrangement, forming the metastable merocyanine (MC) state. MC is a planar molecule with an extensive π-electron conjugated ring system giving rise to a strong absorption band in the visible (around 550 nm) and a very large ground-state dipole of 20 D.

As we see in Fig. 3a–c, repeated irradiation on an optical switch with a defined sequence of UV or 720-nm (two-photon) and visible (543 nm) light provides a means to convert the SP to the MC state quantitatively and then return the MC back to the SP state over many cycles (35, 36, 38). The MC-fluorescence signal provides a convenient internal readout of the state of the switch at any moment in time. Since transitions between the SP and MC states are governed primarily by the quantum yields for the SP to MC and MC to SP transitions, the MC-fluorescence intensity exhibits a constant profile between each cycle of optical switching over many cycles – this condition holds so long as the energies of the UV and visible light remain constant. Modulation of the MC-fluorescence signal is depicted in Fig. 3c. The quantum yield for the SP to MC transition of NitroBIPS is as high as ~0.8 on labeled proteins. The SP to MC transition within NitroBIPS-labeled proteins is complete within 2 μs using a laser scanning microscope, or within 50 ms using the 365-nm line of an Hg-arc lamp (35). On the contrary, the MC to SP transition has a somewhat lower quantum yield, requiring multiple scans of an MC sample at 543 nm for quantitative conversion. The difference in the quantum yields for the SP to MC and MC to SP transitions, and the ability to control the amount of conversion by changing the energy of the laser provide a simple method to shape the waveform of optical switching as seen through the MC-fluorescence profile during a cycle of optical switching (Fig. 3c). For example, a quasi-square waveform profile is achieved by using a single high-intensity pulse of 365 nm (or 720 nm) and 543 nm light, whereas a saw-tooth waveform is realized by using irradiation energies that are typical for imaging studies on living cells (35).

The quantum yield for the SP to MC transition is the highest in apolar solvents such as dichloromethane and aprotic polar solvents such as DMF, and decreases with an increase in solvent polarity, with water being the worst solvent (43). While this property is not helpful for optical switching studies within biological samples, the same switches linked either covalently or noncovalently to proteins exhibit quantum yields for the SP to MC transition in aqueous buffer that are similar to that in aprotic solvents (23, 40). This property most likely arises because the protein-bound SP and MC states are largely shielded from bulk water via weak dipolar and H-bonding interactions with polar groups on the protein, creating an immediate solvent environment similar to formamide. Evidence for interactions between the protein and NitroBIPS was shown by analyzing MC-absorption spectra in different NitroBIPS protein conjugates (23). The quantum yield for the MC to SP transition is somewhat lower than that for the SP to MC transition, which most likely reflects the competition for decay of the excited MC state between photochemistry (MC to SP) and MC fluorescence. In other optical switches such as naphthoxazine (40), the quantum yield for the MC to SP state is very high and, correspondingly, the quantum yield for MC fluorescence is effectively zero.

3.3. Two-Photon-Mediated Cleavage of the Spiro Bond in Molecular Switches

In previous studies, we discovered that the SP state of NitroBIPS can efficiently be converted to MC upon two-photon irradiation at 720 nm (35, 36, 38), which provides an opportunity to control an optical switch within tissue samples. Although the actual extinction coefficient of the two-photon absorption cross-section for the SP to MC transitions is not known at this time, we surmise that it is high because the SP state of NitroBIPS can be quantitatively converted to the MC state within a single two-photon laser scan of a sample within the pixel dwell time (2 μs/pixel) using laser energy compatible with imaging studies in living cells (35). On the contrary, if the extinction coefficient for the SP to MC transition is low, then a far greater laser energy or multiple scans would be necessary to bring about quantitative conversion of the SP to the MC state during the pixel dwell time. This property of NitroBIPS provides a means to control the SP to MC transition within tissues and animals (38).

3.4. Spontaneous Reformation of the Spiro Bond

The spiro bond-forming reaction (MC to SP) in NitroBIPS occurs spontaneously in the ground state (Fig. 3b). The time constant for this reaction depends on chemical substitutions in the optical switch and can be measured by the time-dependent decrease in MC absorption. The decay of the MC state in NitroBIPS in ethanol is best fit by using a single exponential yielding a time constant of about 170 s (23). This time constant is very slow compared to that for the equivalent excited-state reaction, which is on the order of a microsecond or less (35). Interestingly, the time constant for the ground-state MC to SP transition is increased to >3,000 s when NitroBIPS is covalently linked to a protein such as G-actin (23). The stability of the MC-protein state on the protein arises from strong and specific dipolar interactions and/or steric effects within the conjugate. On the contrary, the MC state of spironaphthoxazine (NISO) is less polarized, and the time constant for its MC to SP transition on proteins is only about 6 s (40). We argue then that the remarkable stability of the MC ground state on proteins results from specific, ground-state dipolar interactions. These strong dipolar interactions and van der Waals bonds allow for an exclusive all-optical control of the two states within the conjugate. Interestingly, since the MC to SP reaction is spontaneous, we can control the two states of an optical switch using only a single perturbation to bring about the SP to MC transition. The time constant for the spontaneous MC to SP reaction can be fine-tuned by using slightly different substitutions on the NitroBIPS molecule (40).

3.5. Optical Spectroscopy of NitroBIPS in Solution

The ground- and excited-state dipole moments for polar aromatic probes such as MC are defined by the location of polar groups within the conjugated ring system. In the case of MC, the two monopoles of the dipole are most likely defined by the positive nitrogen atom and the negative nitro group (Fig. 3a (23)). The permanent charge on the nitrogen and the highly polarized nitro group probably account for the fact that MC has a very high ground-state dipole moment (20 D (42)). Absorption and fluorescence spectroscopic analysis of optical probes provides important information on the nature and strength of molecular interactions between the SP and MC states and solvent molecules (23, 42, 43). In particular, the average energy of the lowest energy absorption band (S_0–S_1) for the MC and SP states contains information on the dielectric constant of the solvent and the presence of specific solvent interactions, such as H bonds. For example, the similarity in the MC-absorption spectra for the five NitroBIPS probes (Fig. 3; **3**, **6**, **9**, **12**, **13**) dissolved in ethanol suggests that alkyl substitutions do not change specific interactions of the Mc state with solvent molecules. A comparative analysis of MC-absorption properties in polar and apolar solvents showed that blue-shifted spectra usually arise through specific MC-solvent hydrogen bonding (23). For example, the MC-absorption spectrum is blue-shifted in 1,2-propanediol compared to that in 2-propanol, even though their dielectric constants are almost identical. Conversely in apolar, low dielectric solvents, such as dichloromethane, the MC absorption is considerably shifted to the red compared to that in polar H-bonding solvents.

4. Design of Optical Switch Probes for Bioconjugation

High-fidelity optical switching of protein interactions and protein activity requires the introduction of a single optical switch probe at a unique site on the protein. This condition removes the probe from bulk water. Studies detailed herein show that the spectroscopic and photochemical transitions of the NitroBIPS protein conjugates are favorable for optical switching and similar to those reported for organic solvents. We have designed different types of optical switch probes for labeling to thiol and amino groups (40) in biomolecules, for click chemistry, and as suicide substrates for the Snap-tag (36, 44). Specific labeling of an optical switch to a protein can be achieved through covalent modification of single cysteine residues in a protein. The most commonly employed thiol-reactive groups are maleimides and haloalkanes.

4.1. Syntheses of Reactive Optical Switches

The combinatorial approach developed for the synthesis of thiol-reactive optical switches, including compounds **3**, **6**, **9**, **12**, and **13**, is shown in Fig. 4. This approach allows us to control the position of the thiol-reactive group on the optical switch and to control the flexibility of the linkage group. Thus we can position the thiol-reactive group full circle around the NitroBIPS probe and vary the flexibility of the probe on a protein by introducing short, stiff linkers in compounds **3**, **9**, **12**, and **13**, or floppy linkers such as the C5 group shown in compound **6**. This feature allows us to control both the orientation and dynamics of the highly polarized MC state of NitroBIPS on the protein.

4.2. Spectroscopic and Photochemical Properties of Optical Switch Protein Conjugates

As we indicated above, the usefulness of an optical switch is dependent on the molecular environment of the probe and the dielectric constant and H-bonding potential of the solvent. Unfortunately, water is one of the worst solvents for optical switching of NitroBIPS (43). Consequently, most studies on molecular switches are conducted within apolar organic solvents (see, for example, chapters in (37)). Only a few optical switching studies have been reported (34, 45). Willner et al. (45), for example, showed that sugar binding of concanavalin A that is randomly conjugated with multiple photochromes can be controlled by switching between the two states of the photochrome. Since

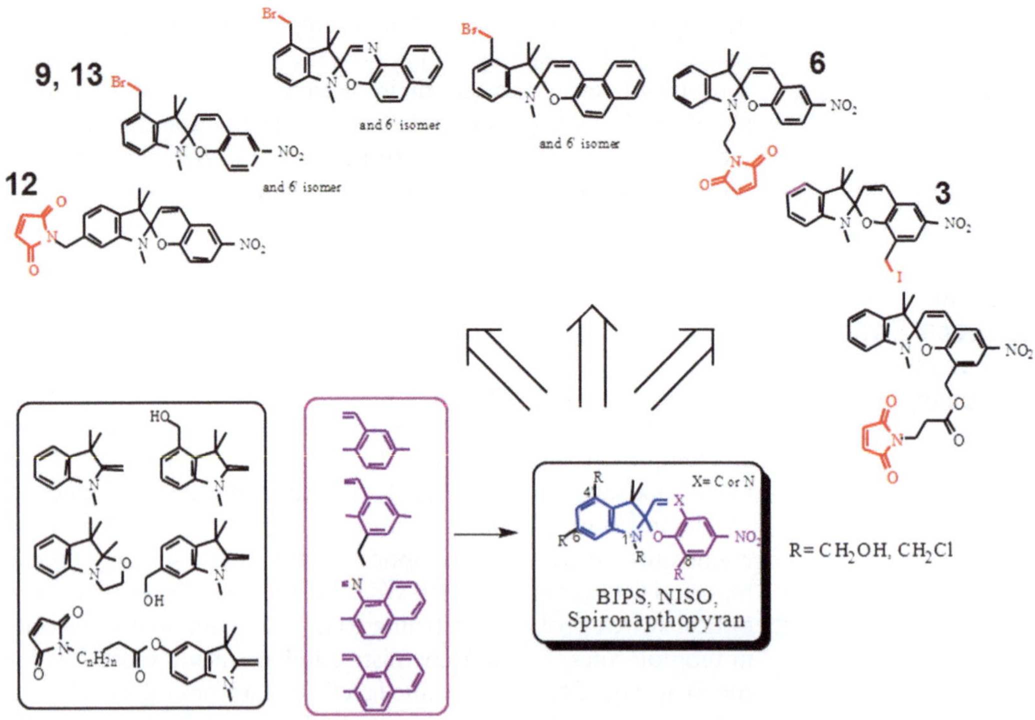

Fig. 4. Synthesis of optical switches. The methods for the synthesis of NitroBIPS and NISO using the Fischer–Hirscberg synthesis as well as our combinatorial approach for introducing pendant reactive groups were detailed in Sakata et al. (40). Here we show a family of thiol-reactive NitroBIPS probes in which the reactive group is marched around the molecule. In some probes, the linker between the aromatic scaffold and the protein is short (e.g., compounds **3**, **9**, **12**, and **13**) and for other probes, including compound **6**, the linker can be longer.

individual photochromes on the protein exhibit different MC-absorption spectra and quantum yields for transitions between their two states, it is quite difficult to arrive at the mechanism for this control. Moreover, the presence of multiple probes on a single protein often leads to the formation of intramolecular probe dimers having different spectroscopic and photochemical properties compared to the monomer. The ideal small molecule actuator of a protein interaction is best realized by linking a single photochrome to a defined location on the protein. Moreover, it would be preferable to use an optical switch exhibiting spectroscopic and photochemical properties listed earlier in this work. In the following section, we show that thiol-reactive forms of NitroBIPS go a long way in meeting these criteria.

4.3. Spectroscopic Analysis of Interactions Between MC and an Attached Protein

A common NitroBIPS aromatic scaffold was attached to a unique site (cysteine-374) on G-actin by using six different NitroBIPS–protein linkage groups that varied in terms of their location on the MC group and the length and flexibility of the linkage group between MC and the sulfur on cysteine-374 of actin. The ability to control the location of the NitroBIPS probe from a common attachment site allowed us to project the strong ground-state

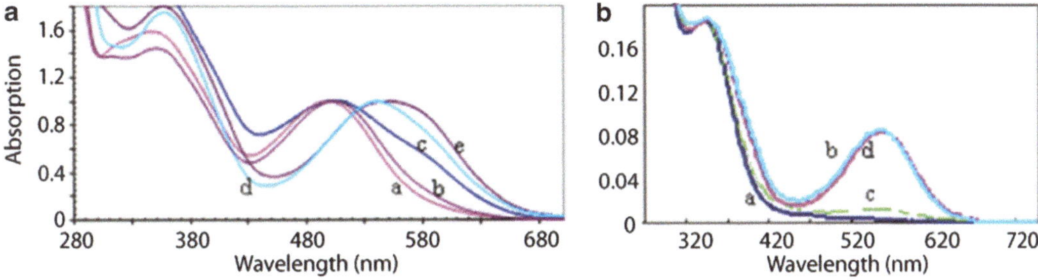

Fig. 5. (**a**) Normalized absorption spectra for the lowest energy transition of the MC state of five thiol-reactive probes differentially attached to cysteine-374 on actin. The *letters* (*a–e*) refer to different thiol-reactive NitroBIPS probes, **9**, **13**, **3**, **6**, and **12** (Fig. 4; (23)). (**b**) Optical switching between the SP and MC states on G-actin. Absorption spectra of compound **13** attached to cysteine-374 on G-actin in response to sequential irradiation with UV and visible light. The *letters* refer to (*a*) preirradiated SP state; (*b*) 30 s illumination of the conjugate with 365-nm light; (*c*) 30 s illumination of the MC conjugate (spectrum (*b*)) with 546-nm light; (d) 30 s illumination of the SP conjugate (curve *c*) with 365-nm.

dipole of the MC group to different sites around the cysteine-374 where it engaged in different interactions with dipolar elements on the protein, including the most common dipole on a protein, the peptide bond. Proof that the different labeling geometries resulted in slightly different interactions between the MC group and dipoles within the small volume element around Cys-374 was demonstrated through an absorption spectroscopic analysis of the MC differentially linked to the same cysteine-374 residue in G-actin (Fig. 5a). The five thiol-reactive probes used in this study are shown in Fig. 4 and described in more detail in the study by Sakata et al. (23). The average energy of the MC absorption in the G-actin conjugates varied from 20,000 cm^{-1} (500 nm) for compound **9**, to 18,132 cm^{-1} (551.5 nm) for compound **12**, corresponding to an energy difference of 1,868 $cm^{-1} \pm 27$ cm^{-1}. Interestingly, the difference in the average energy of the MC absorption between compounds **12** and **13** on G-actin is also large at 1,400 cm^{-1}, in spite of the fact that the thiol-reactive groups are linked to an identical atom on the probe. This result suggests that the further displacement of the MC dipole from Cys-374 allows the MC state to engage in far stronger dipolar interactions with the protein compared to compound **13**. The dipolar environment around Cys-374 of G-actin is, therefore, shown to be heterogeneous, changing dramatically from polar to apolar within just a few Angstroms.

As we argued earlier, if the protein interior were a homogeneous dielectric, then the average energy of the MC-absorption spectrum would be independent of the MC-protein linkage geometry. However, our studies using differently projected MC probes within G-actin show that the protein dipolar environment is in fact remarkably diverse, with the MC probe being exposed to molecular environments that are as polar as water, or as apolar as dichloromethane. On the contrary, the corresponding difference for the SP probe, which has a smaller dipole moment (5 D), is at least four times lower than that for the MC (23).

4.4. High-Fidelity Optical Switching on Proteins

Irradiation of all five NitroBIPS-labeled G-actin conjugates with 365-nm light for 30 s or less generates the highly colored MC conjugate as seen in the absorption spectra of the SP and MC states shown in curves a and b in Fig. 5b. Conversely, excitation of the MC conjugate with 546-nm light for 30 s or less leads to conversion of the MC state to the SP state (curve c in Fig. 5b). Optical control of the MC and SP states on G-actin is efficient and reversible and can be repeated over many cycles of 365-nm/546-nm irradiation. MC-absorption spectra between different optical switching cycles are almost super-imposable (curves b and d in Fig. 5b). Since the MC spectrum is highly sensitive to changes in specific dipolar interactions, these identical spectra strongly suggest that transitions between the SP and MC states on the protein occur with high fidelity, and this implies that the MC state at least engages in the same dipolar interactions with polar groups on the protein for each cycle of an optical switching study as schematized in Fig. 6a. Thus an optical transition from the MC to the SP state results in the dissolution of strong and specific dipolar interactions between MC and the protein

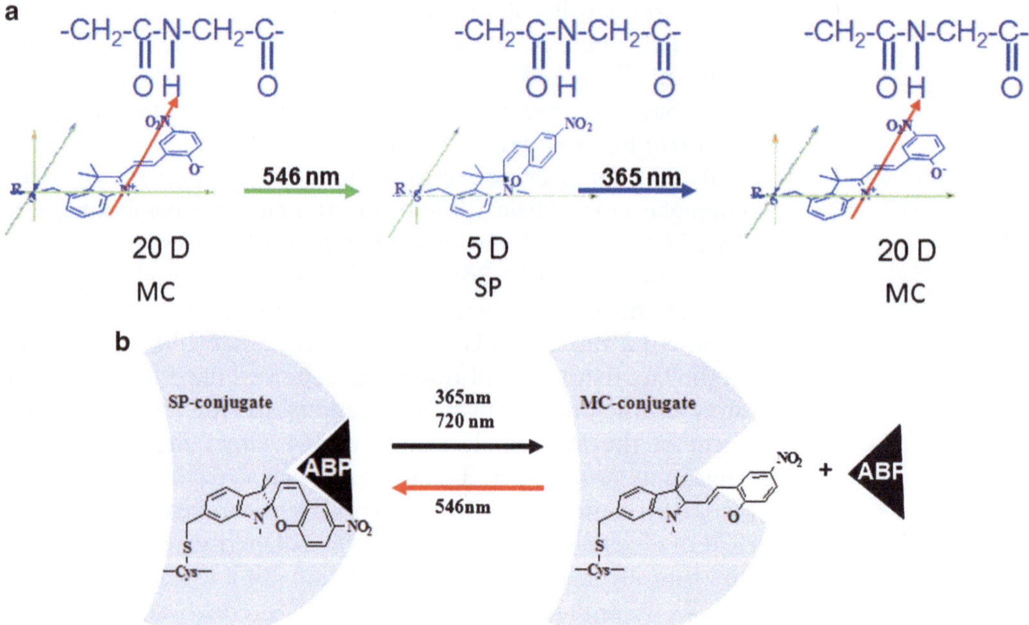

Fig. 6. (**a**) Schematic representation of the change in dipolar interactions between the SP and MC states of a NitroBIPS optical switch with a peptide dipole of a protein. The strong dipole moment of the MC state makes a strong ground-state interaction with specific peptide bonds in the protein. Illumination of the MC state with visible light converts the MC to the SP state, with an accompanying change in the dipole moment (~5 D) and weakening of the interaction with the peptide bond. Illumination of the SP state with 365-nm light generates the MC state again, which makes the same dipolar interaction with the peptide bond. (**b**) The highly polarized MC, but not the SP, state competes for specific polar groups on actin and effectively outcompetes the binding of an actin-binding protein (ABP) for those same G-actin interactions. Conversion of the MC to SP state exposes the protein dipoles, allowing the ABP to bind to G-actin.

that can release as much as ~6 kT of interaction energy, which is comparable to the binding energy of G-actin with regulatory proteins and ligands. Thus optical conversion of the MC state to the SP state rapidly exposes these dipolar groups on G-actin that may now engage in specific interactions with a ligand or an ABP. In this way, specific protein–ligand or protein–protein interactions can be made to compete with the MC state for key dipolar groups on the protein, and this competition can be controlled optically by manipulating the two states of NitroBIPS. In particular, protein conjugates can be engineered such that the MC, but not the SP, state effectively outcompetes other proteins for specific dipolar interactions in the conjugate, as schematized in Fig. 6a, b. Manipulation of specific dipolar interactions within a NitroBIPS–protein conjugate can, therefore, be achieved within a microsecond or less and, based on our knowledge of the field, this represents the fastest method for perturbing specific interactions on proteins.

Acknowledgments

We thank current and former members of the Marriott laboratory who contributed to the work described in this review, including Drs. Tomoyo Sakata, Shu Mao, and Chutima Petchprayoon, as well as those involved in collaborations with the Jacobson, Roy, and Loew laboratories. This work was supported in part by the NIH (5R01EB005217 and R01 GM086233-01).

References

1. Tsien R (1989) Fluorescent probes of cell signaling. Annu Rev Neurosci 12:227–253
2. Allen WE, Jones GE, Pollard JW, Ridley AJ (1997) Rho, Rac and Cdc42 regulate actin organization and cell adhesion in macrophages. J Cell Sci 110:707–720
3. Zhang J, Campbell RE, Ting A, Tsien RY (2002) Creating new fluorescent probes for cell biology. Nat Rev Mol Cell Biol 3:906–918
4. Betzig E, Patterson GH et al (2006) Imaging intracellular fluorescent proteins at nanometer resolution. Science 313(5793):1642–1645
5. Rust MJ, Bates M, Zhuang X (2006) Nanometer resolution imaging. Nat Methods 3:793–795
6. Bates M, Huang B, Dempsey GT, Zhuang X (2007) Multicolor super-resolution imaging with photo-switchable fluorescent probes. Science 317:1749–1753
7. Hell SW, Dyba M, Jakobs S (2004) Concepts for nanoscale resolution in fluorescence microscopy. Curr Opin Neurobiol 14:599–609
8. Eigen M, Hammes GG (1963) Elementary steps in enzyme reactions (as studied by relaxation spectrometry). Adv Enzymol Relat Areas Mol Biol 25:1–38
9. Yan Y, Marriott G (2003) Analysis of protein interactions using fluorescence technologies. Curr Opin Chem Biol 7:1–6
10. Marriott G (1994) Caged protein conjugates and light-directed generation of protein activity: photoactivation and spectroscopic characterization of caged G-actin. Biochemistry 33:9092
11. Gutfreund H (1997) Kinetics for the life sciences: receptors, transmitters and catalysts. Cambridge University Press, New York
12. Walker JW, Gilbert SH, Drummond RM, Yamada M, Sreekumar R, Carraway RE, Ikebe M,

Fay FS (1998) Signaling pathways underlying eosinophil cell motility revealed by using caged peptides. Proc Natl Acad Sci U S A 95:1568–1573

13. Roy P, Rajfur Z, Jones D, Marriott G, Loew L, Jacobson K (2001) Local photorelease of caged thymosin β4 in locomoting keratocytes causes cell turning. J Cell Biol 153:1035–1048
14. Denk W, Strickler J, Webb WW (1990) Two-photon laser scanning fluorescence microscopy. Science 248:73–76
15. Kaplan JH, Forbush B III, Hoffman JF (1978) Rapid photolytic release of adenosine 5′-triphosphate from a protected analogue: utilization by the Na:K pump of human red blood cell ghosts. Biochemistry 17:1929–1935
16. Marriott G, Heidecker M (1996) Light-directed generation of the actin-activated ATPase activity of caged heavy meromyosin. Biochemistry 35:3170–3174
17. Ghosh M, Song X, Mouneimne G, Sidani M, Lawrence DS, Condeelis JS (2004) Cofilin promotes actin polymerization and defines the direction of cell motility. Science 304:743–746
18. Zhang F, Wang LP, Brauner M, Liewald JF, Kay K, Watzke N, Wood PG, Bamberg E, Nagel G, Gottschalk A, Deisseroth K (2007) Multimodal fast optical interrogation of neural circuitry. Nature 446:617–619
19. Volgraf M, Gorostiza P, Numano R, Kramer RH, Isacoff EY, Trauner D (2006) Allosteric control of an ionotropic glutamate receptor with an optical switch. Nat Chem Biol 2:47–52
20. Marriott G, Miyata H, Kinosita K Jr (1992) Photomodulation of the nucleating activity of a photocleavable crosslinked actin dimer. Biochem Int 126:943–951
21. Marriott G, Jacobson K, Roy P (2003) Caging proteins through chemical modification. Methods Enzymol 360:274–288
22. Suh BC, Inoue T, Meyer T, Hille B (2006) Rapid chemically-induced changes of PtdIns(4,5)P2 gate KCNQ ion channels. Science 314:1454–1457
23. Sakata T, Yan Y, Marriott G (2005) Optical switching of dipolar interactions on proteins. Proc Natl Acad Sci U S A 102:4759–4764
24. McCray JA, Trentham DR (1989) Properties and uses of photoreactive caged compounds. Annu Rev Biophys Biophys Chem 18:239–270
25. Walker JW, Feeney J, Trentham DR (1989) Photolabile precursors of inositol phosphates. Preparation and properties of 1-(2-nitrophenyl)ethyl esters of myo-inositol 1,4,5-trisphosphate. Biochemistry 28:3272–3280
26. Ellis-Davies GCR (2007) Caged compounds: photorelease technology for control of cellular chemistry and physiology. Nat Methods 4:619–628
27. Marriott G (1998) Caged compounds. Methods enzymology, vol 291. Academic, New York
28. Sleep J, Herrmann C, Barman T, Travers F (1994) Inhibition of ATP binding to myofibrils and acto-myosin subfragment 1 by caged ATP. Biochemistry 33(20):6038–6042
29. Marriott G, Heidecker M, Diamandis E, Yan-Marriott Y (1994) Time-resolved delayed luminescence image microscopy using europium-ion chelates. Biophys J 67:957
30. Chang CY, Niblack B, Walker B, Bayley H (1995) A photogenerated pore-forming protein. Chem Biol 2:391–400
31. Cambridge SB, Davis RL, Minden JS (1997) Drosophila mitotic domain boundaries as cell fate boundaries. Science 277(5327):825–828
32. Kron SJ, Spudich JA (1986) Fluorescent actin filaments move on myosin fixed to a glass surface. Proc Natl Acad Sci USA 83: 6272–6276
33. Furuta T, Wang SS, Dantzker JL, Dore TM, Bybee WJ, Callaway EM, Denk W, Tsien RY (1999) Brominated 7-hydroxycoumarin-4-ylmethyls: photolabile protecting groups with biologically useful cross-sections for two photon photolysis. Proc Natl Acad Sci USA 96:1193–1200
34. Patolsky F, Filanovsky B, Katz E, Willner I (1998) Photoswitchable antigen–antibody interactions studied by impedance spectroscopy. J Phys Chem B 102:10359–10367
35. Marriott G, Mao S, Sakata T, Jackson D, Gomez T, Aaron H, Isacoff EY, Yan Y (2008) High contrast imaging based on optical lock-in detection imaging of synthetic and genetically encoded optical switches. Proc Natl Acad Sci U S A 46:17789–17794
36. Mao S, Benninger R, Yan Y, Petchprayoon C, Jackson D, Easley C, Piston D, Marriott G (2008) Optical lock-in detection of FRET using synthetic and genetically encoded optical switches. Biophys J 94:4515–4524
37. Dürr H, Bouas-Laurent H (1990) Photochromism: molecules and systems. Elsevier, Amsterdam
38. Sakata T, Jackson DK, Mao S, Marriott G (2008) Optically switchable chelates: optical control and sensing of metal ions. J Org Chem 73:227–233

39. Giordano L, Jovin TM, Irie M, Jares-Erijman EA (2002) Diheteroarylethenes as thermally stable photoswitchable acceptors in photochromic fluorescence resonance energy transfer (pcFRET). J Am Chem Soc 124:7481–7489
40. Sakata T, Yan Y, Marriott G (2005) A family of site selective optical switches. J Org Chem 70:2009–2013
41. Fischer E, Hirshberg Y (1952) Formation of coloured forms of spiropyrans by low temperature irradiation. J Chem Soc 4522–4524
42. Bletz M, Pfeifer-Fukumura U, Kolb U, Baumann W (2002) Ground- and first-excited-singlet-state electric dipole moments of some photochromic BIPSs in their spiropyran and merocyanine form. J Phys Chem A 106:2232–2236
43. Görner H (1997) Photoprocesses in spiropyrans and their merocyanine isomers: effects of temperature and viscosity. Chem Phys 222:315–329
44. Kindermann M, Sielaff I, Johnsson K (2004) Synthesis and characterization of bifunctional probes for the specific labeling of fusion proteins. Bioorg Med Chem Lett 7:2725–2728
45. Willner I, Rubin S, Wonner J, Effenberger F, Baeuerle P (1992) Photoswitchable binding of substrates to proteins: photoregulated binding of α-d-mannopyranose to concanavalin A modified by a thiophenefulgide dye. J Am Chem Soc 114:3150–3151

Chapter 13

Structure-Based Design of Light-Controlled Proteins

Harald Janovjak and Ehud Y. Isacoff

Abstract

Small photochromic molecules are widespread in nature and serve as switches for a plethora of light-controlled processes. In a typical photoreceptor, the different geometries and polarities of the photochrome isomers are tightly coupled to functionally relevant conformational changes in the proteins. The past decade has seen extensive efforts to mimic nature and create proteins controlled by synthetic photochromes in the laboratory. Here, we discuss the role of molecular modeling to gain a structural understanding of photochromes and to design light-controlled peptides and proteins. We address several fundamental questions: What are the molecular structures of photochromes, particularly for metastable isomers that cannot be addressed experimentally? How are the structures of bistable photoisomers coupled to the conformational states of peptides and proteins? Can we design light-controlled proteins rapidly and reliably? After an introduction to the principles of molecular modeling, we answer these questions by examining systems that range from the size of isolated photochromes, to that of peptides and large cell surface receptors, each from its unique computational perspective.

Key words: Quantum mechanics, Molecular mechanics, Optical switch, Azobenzene, SP, Hemi-thioindigo-Helix, Hairpin, Rational design, Glutamate receptor

1. Introduction

Light regulates many vital processes in all kingdoms of life such as energy generation in bacteria (1), spore production in fungi (2), phototropism in plants (3), and vision in animals (4). The photoreceptors that govern these diverse physiological responses share many functional principles. They all contain low-molecular weight photochromes (e.g., retinal or bilins) that undergo reversible, light-triggered conversion between distinct isomers (5, 6). These primary photochemical reactions induce conformational changes in receptor proteins to which they are bound, and directly or indirectly evoke the organism's response. For instance, in the prototypical rhodopsin photoreceptor and its prokaryotic homologs,

James J. Chambers and Richard H. Kramer (eds.), *Photosensitive Molecules for Controlling Biological Function*, Neuromethods, vol. 55, DOI 10.1007/978-1-61779-031-7_13, © Springer Science+Business Media, LLC 2011

isomerization of 11-*cis* retinal into all-*trans* retinal (or all-*trans* retinal to 13-*cis* retinal) changes the orientation of the receptor's transmembrane helices and thereby either activates a signaling cascade or allows passage of ions through the lipid bilayer membrane (5, 7).

From many perspectives, light may also be a researcher's premier choice to control biological structure and function artificially in vitro and in vivo. Light offers outstanding spatial and temporal resolution, precise control of intensity, and there is no need to connect stimulus and responding element physically given that the matrix is transparent. Inspired by nature, researchers have been coupling synthetic photochromes, e.g., azobenzenes or spiropyrans (SPs), to biological macromolecules for decades (8, 9). Pioneering work showed that synthetic photoreactive peptides and proteins can indeed respond to light by reversible conformational changes such as α-helix or β-hairpin (un-)folding, protein activation, and biomolecular assembly (10–15).

There is a rich literature on many aspects of photochromes and optically controlled biomolecules. Here, we focus on the role of computational techniques in understanding and designing light-controlled peptides and proteins. Molecular modeling is now routinely applied in this growing field by many groups. However, to the best of our knowledge, no collection and discussion of this work is available. We address isolated photochromes and optically controlled peptides and receptors. As we will show, every component in a protein–photochrome system requires its own computational strategy. Therefore, we begin in Sect. 2 with a concise introduction to molecular modeling techniques, many of which will be applied in the following sections. By definition, the isomers of photochromes alternate between two distinct geometries that couple differentially to proteins and peptides. In Sect. 3, we collected structural data for three classes of synthetic photochromes: azobenzenes, SPs, and hemithioindigos (HTI). For each photochrome, structural data from X-ray crystallography are combined with QM computing. We highlight that QM calculations can be essential to validate the experiments (e.g., to test for crystal artifacts) and to obtain models for metastable photochrome isomers. This section also includes a brief historical introduction to each photochrome. In Sect. 4, we focus on peptides with incorporated light switches. These peptides form α-helices and β-hairpins, and serve as model systems for protein folding. The quantitative understanding of light-induced conformational changes in these systems relies on molecular modeling. We review recent computational work that aims to decipher the interplay of photochromes and peptide structures and the design of these systems. Section 5 is a case study where techniques introduced in Sect. 2 and structural information from

Sect. 3 are applied toward the rational design of a light-gated receptor. The methods presented in detail in this section should be directly applicable to different classes of proteins. Finally, we close with an outlook in Sect. 6.

2. Principles of Molecular Modeling

The term molecular modeling refers to a wide range of computational techniques that theoretically describe biological systems on molecular scales (16). In the past decades, virtually all aspects of molecular biology have been investigated using modeling: Prominent examples are protein folding, enzyme catalysis, and molecular recognition (17–21). The most important concept in molecular modeling is the interplay between structure and energy that underlies all dynamic and energetic aspects of biological systems (16, 22). During spontaneous folding of a protein, e.g., the free energy of the protein–solvent system is minimized until the global energy minimum is reached in the native three-dimensional structure (23). It is clear that an understanding of the protein's energy function is required for correct modeling of this process. Similarly, the assembly of two proteins or the binding of a ligand to a receptor changes the free energy of these systems, and their conformations can be described only with a reliable, physical model for the relationship of structure and energy (24).

Two principle types of structure–energy models exist to describe chemical or biological processes. These methods are combined with sampling procedures, also called geometry algorithms, which will be discussed below. The term molecular mechanics (MM) refers to models based on classical, Newtonian mechanics, in which every atom is represented as a discrete particle with characteristic size, charge, and mass, while electrons are treated implicitly (16). In MM, the energy of a molecular conformation is defined in a potential energy function as the sum of individual energy terms for covalent interactions (bonds lengths, bond and dihedral angles) and non-covalent interactions (steric repulsion, electrostatics, van der Waals forces, and hydrogen bonds) (16, 22). The exact form of the potential energy function and its parameters are collectively referred to as a "force field". Many modern force fields exist that offer high-quality descriptions specific to different classes of biological molecules (e.g., proteins, nucleic acids, and lipids) (25). Each force field also contains one or several representations of a solvent, which can either be implicit (in the form of a continuous medium) or explicit (with discrete molecules) (22). Implicit solvation is computationally less intensive than explicit models, but only reproduces general solvent properties and not specific interactions between water and other molecules. The second class of structure–energy models is based on quantum mechanics (QM) and describes electronic properties in detail (26). QM models are based on either the wave function or the electron density. Wave function-based QM start with approximations to the Schrödinger equation, such as the commonly used Hartree–Fock (HF) method (27). In QM models, atomic orbitals are described using a combination of Gaussian functions. This set of Gaussian functions is called the "basis set" and its size determines one component of the accuracy of the calculation. HF is a mean-field theory in which electrons feel only the average repulsion of the other electrons. HF methods are thus typically extended beyond mean-field pictures to include "correlation" between the positions of the electrons (26, 27). The second type of QM descriptions is based on density functional theory (DFT) (28). In DFT methods, energy and all other molecular properties are a functional of the electron density with an intrinsic treatment of correlation. DFT calculations are computationally much more efficient than wave function-based methods but their accuracy can often not be controlled systematically (29). Implicit and explicit treatment of solvent also exists in the case of QM models.

What are the strengths and limitations of MM and QM models? MM simulations allow large molecules (such as proteins and their complexes) to be modeled, particularly if combined with implicit solvent. QM techniques, in contrast, are appropriate for small systems (typically less than 100 atoms in DFT models). MM force fields are well suited to reproduce equilibrium conformations and energies. However, they are based on predefined bonding geometries and thus provide little reliable information about transition states, and bonds can only be broken and formed in QM representations. Finally, many MM force fields are not parameterized for photochromes and their delocalized π-electron systems (Sect. 4 contains notable differences). Due to the small number of atoms in photochromes and constant improvements in computing power, QM calculations are now widely used to describe these molecules (Sect. 3) and reliably report dipole moments and spectral properties (30). In summary, the choice of molecular representation thus depends on the size of the system and the desired accuracy; combined MM/QM approaches can offer the best of the two worlds (31, 32).

In practice, MM and QM models are applied through a wide range of sampling procedures also called geometry algorithms (33, 34). Geometry algorithms are best introduced with the help of a schematic energy surface (Fig. 1). Biological and chemical molecules have large numbers of degrees of freedom. Therefore, these molecules can populate many conformations that are structurally similar but unique and consequently also have unique energies. The conceptual energy surface that

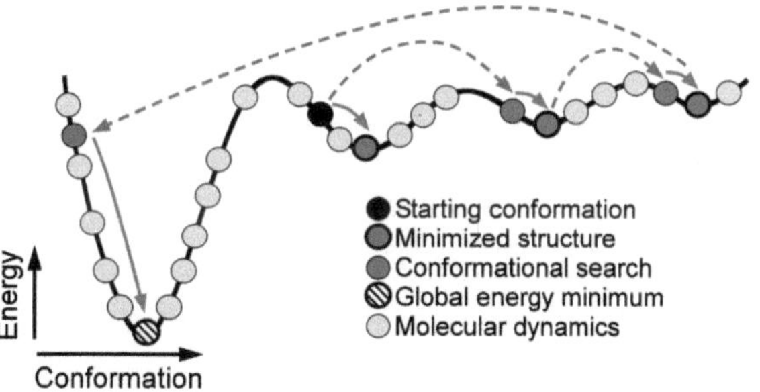

Fig. 1. Biological and chemical molecules have many degrees of freedom and consequently their rough energy landscapes contain many local maxima and minima. Geometry algorithms have been developed to probe energy surfaces. An initial starting conformation (*black sphere*) can be energy minimized (*solid arrows*) by optimizing the geometry of the molecule until a local energy minimum is reached. The energy surface can be explored globally using a conformational search (*dashed arrows*). In the search, structures in other potential wells are generated and subsequently minimized. If the energy surface is probed completely, then the global energy minimum can be localized (*dashed sphere*). Finally, molecular dynamics simulations sample the energy surface dynamically. In principle, the entire surface is visited but low-energy conformations are populated more than high-energy conformations.

describes such a system best is "rough," i.e., contains many local energy maxima and minima (Fig. 1). Minima represent stable conformations, while maxima are transition states. The goal of geometry algorithms is to explore energy surfaces either locally or globally. In a simple, local energy calculation, the energy of a starting geometry (Fig. 1, black sphere) is calculated based on the relationship of structure and energy. This algorithm does not alter the conformation but reports the energy in combination with solvation terms. To find local minima on the energy surface, geometry optimizations (energy minimizations) iteratively change the starting structure, while the molecular energy is calculated in each cycle (Fig. 1, solid arrows). Selected degrees of freedom can be excluded from the optimization. This is, for example, the case in MM simulations of protein–photochrome systems, where the coordinates of the photochrome are constrained to QM models or crystallographic structures (Sects. 3 and 4). The goal of conformational searching (Fig. 1, dashed arrows) is to sample the entire potential energy surface either systematically or randomly (e.g., in a Monte Carlo search, Sect. 5). Each "jump" on the energy surface is achieved by changing a degree of freedom and the energy is minimized for each new structure generated. This specific type of simulation within the large family of Monte Carlo methods that are all based on random-number decisions (35) is quite effective in searching the conformational space and finding the global minimum. In MD simulations (Fig. 1, open spheres), molecules are followed over time while Newton's laws of motion act on the MM system. In this way, MD simulations yield accurate and easily interpreted pictures of the dynamics and conformational changes during the function and interactions of biological molecules. However, typical all-atom MD simulations of even small proteins are limited to timescales shorter than microseconds.

3. Structural Analysis of Synthetic Photochromes: Experiments and Models

3.1. Introduction

Photochromes are small molecules that can be switched reversibly between two (meta)stable isomers with a light stimulus, and by definition these isomers exhibit unique spectral properties (36, 37). Photoconversion can be the result of *cis–trans* isomerization (e.g., in azobenzenes and HTIs, Sects. 3.2 and 3.4) or pericyclic reactions (cyclization/bond opening, e.g., in SPs, Sect. 3.3, or dithienylethenes (38)). In ideal photochromes, the two isomers have very different geometries and polarities that are exploited to control conformations of larger molecules. It is worth noting that structural changes in photochromes are small, particularly compared to those of proteins. As a consequence, photochromes need to be carefully coupled to macromolecules to achieve proper transduction and amplification of the optical signal. For instance, it was shown that only *meta*-substituted azobenzenes allowed light-controlled folding of β-hairpin peptides, while functionalization at *para*- or *ortho*-positions prohibited structure formation (Sect. 4.2). This example highlights that successful optical modulation of peptides and proteins is very sensitive to photochrome geometry (Sects. 4 and 5), and thus structural data are required for the rational design of these systems. In principle, structural information of small organic molecules can be obtained either experimentally (e.g., by X-ray crystallography) or through QM computing. The availability of experimental structures of synthetic photochromes ranges from many decades ago for azobenzene (Sect. 3.2) to still unknown for the metastable isomer of HTI (Sect. 3.4). Modern QM approaches, such as those discussed below, accurately reproduce experimental structures and their spectral properties.

In this section, we provide a comprehensive collection of the atomic structures of azobenzene, SP, and HTI photochromes. These structures were compiled with the Cambridge Structural Database (CSD) (39) and Science Citation Index (Thomson Reuters, Inc, Philadelphia, PA). For each class, we give a brief historical introduction and describe their structures in detail. A list of all available structures including substituents can be found in Tables 1–3, and their atomic coordinates can be obtained from the CSD (see the tables for accession identifiers). Tables 1–3 may be a particularly useful resource for photochromes with several

Table 1
List of experimentally determined structures for azobenzene. Here, only structures of unmodified azobenzene are listed as >1,400 structures of azobenzene derivatives can be found in the CSD

Isomer	CSD identifier	References
trans	N/A	(60)
cis	AZBENC	(61)
trans	AZOBEN01	(54)
cis	AZBENC01	(62)
trans	AZOBEN03	(63)
trans	AZOBEN04	(64)
trans	AZOBEN07	(65)
trans	N/A[a]	(56)
trans	N/A[a]	(55)

[a]Determined using gas-phase electron diffraction

isomers, whose structures depend on the position and nature of substituents. The second focus of this section is to highlight how QM calculations complement experimental results. For each photochrome, we discuss recent QM models, several of which helped resolve a controversy associated with experimental structures. For example, X-ray structures can be subject to crystallization artifacts because molecules are confined to conformations that are most stable in the lattice. As discussed in detail below, crystal packaging effects are significant and QM geometry optimization produces ground state models for solution structures. Finally, we use QM calculations to model the structure of the metastable *E*-isomer of the HTI photochrome for which experimental data are currently unavailable.

3.2. Azobenzene

Azobenzenes are undoubedtly the most studied photochromes. The first report of azobenzene synthesis dates back to more than 150 years (40). The strong light absorption was already noted in this initial work and Hartley was likely the first to observe azobenzene photoconversion (41). In the meantime, azobenzenes were applied in optical data storage (42), molecular motors (43), and photofunctional polymers and surfaces (44, 45). Most aspects of azobenzene photochemistry have received experimental and theoretical attention. In particular, the *cis–trans* isomerization around the N=N double bond has been studied extensively, along with the analogous reaction around the C=C bond of stilbenes

Table 2
SP derivatives with known molecular structure. This list includes all substituents of the SP ring system shown in Fig. 3 with structures that are published and/or deposited in the CSD (also see Sect. 3.1)

Substituents							
C_{18}	N_{11}	C_7	C_9	Other	CSD identifier	References	
di-Me	Me	NO_2	Br		BETGAI	(83, 86)	
di-Me	Me	NO_2			BEXLUL	(82)	
di-Me	Me	NO_2			BEXLUL01	(81)	
di-Me	Bn	NO_2			CEBKID10	(87, 88)	
di-Me	Me			C_{15} NO_2	DARYUQ	(89, 90)	
di-Me	Et	Br	[a]		DAZBOW	(91)	
di-Me	Me	OMe			ETURAM	(92)	
di-Me	Me			C_8 OMe	FUWHIO	(93, 94)	
di-Me	$(CH_2)_2OH$	NO_2			IHOFOA	(95)	
di-Me	Ph	Cl	NO_2		INSPCR	(79, 96)	[b]
di-Me	Me	COOH		C_4[c]	JASKUK	(97)	
di-Me	Me	NO_2		C_3 Me	JIKGAL	(81)	
di-Me	Ph	OEt	NO_2		MAXPEH	(98)	[b]
di-Me	Ph	OMe	NO_2	C_3 Me	MAXPIL	(98)	[b]
di-Me	Ph	OMe	NO_2		MAXPOR	(98)	[b]
di-Me	$(CH_2)COOH$	NO_2			NACBOJ	(99)	
di-Me	Ph		NO_2		NSPIBP	(80)	[b]
di-Me	Et	NO_2			OCEVUN	(100)	
di-Me	Me	Cl	Br		QETPOV	(101)	
di-Me	Me	Cl		C_{15} Me	SANVEJ	(102)	
di-Me	Me	SO_2CF_3			YIHRAI	(103, 104)	

[a] 1′-Ethyl-3′,3′-dimethyl-1′H-indolin-2′-ylidenemethyl
[b] N_{11} and O_1 are located on opposite side of the molecule
[c] 1′,3′,3′-Trimethyl-1′H-indolin-2′-ylidenemethyl

(45, 46). These transitions occur with a high quantum yield and on a picosecond time scale. Because the absorption spectra of *cis*- and *trans*-azobenzene overlap, irradiation typically produces photostationary states that are, at best, composed of ~80% *cis*-azobenzene or ~95% *trans*-isomers. In contrast, thermal isomerization yields >>99% *trans*-azobenzene. The azobenzene core, shown in Fig. 2, serves as the prototypical case for many

Table 3
MC derivatives with known molecular structure. This list includes all substituents of the MC ring system shown in Fig. 3 with structures that are published and/or deposited in the CSD (also see Sect. 3.1)

Substituents						
C_{18}	N_{11}	C_7	C_9	Isomer	CSD identifier	References
di-Me	Me	NO_2	NO_2	TTT	BAPNAH	(84)
di-Me	Me	NO_2	Br	TTC	BETGEM	(83, 86)
di-Me	$(CH_2)_2(CH)_2CH_3$	NO_2	Br	TTC	FAFPOR	(112, 113)
di-Me	$(CH_2)_2COOH$	NO_2	NO_2	TTT	GUWFEJ	(85)
di-Me	$(CH_2)OH$	NO_2		TTT	IHOFUG	(95)

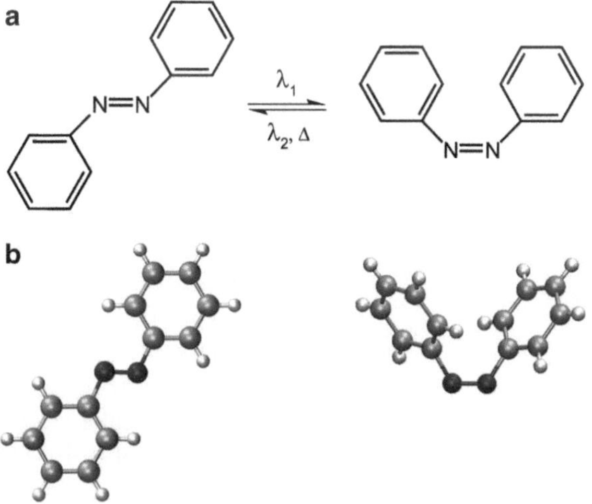

Fig. 2. Photoconversion (**a**) and three-dimensional structures (**b**) of azobenzene. Azobenzene is converted from the *trans*-isomer to the *cis*-isomer using UV light (λ_1) and vice versa with visible light (λ_2) or thermally. *trans*-Azobenzene is almost perfectly planar, while *cis*-azobenzene has a characteristic "kinked" structure. CSD identifiers are AZBENC01 (*cis*) and AZOBEN03 (*trans*; also refer to Table 1 for a list of azobenzene structures).

substituted compounds, which exhibit unique spectral properties, photostationary states, and thermal relaxation rates, ranging from seconds to days (46–48). Given their fast, reversible, and clean photoconversion and their straightforward synthesis, it may not be surprising that azobenzenes were the first photochromes applied to control biomolecules (8, 9, 49).

Experimental structures for azobenzene and its derivatives were determined using X-ray crystallography and gas-phase electron diffraction as early as 70 years ago (Table 1). More recently, extensive QM computations reported models that are in excellent agreement with these structural determinations and reveal the same general arrangement of the benzene rings (Fig. 2) (50–53). In the *cis* state, the benzene rings face each other in a "kinked" structure to minimize steric and electronic energy. A long-standing debate about whether *trans*-azobenzene is planar (54, 55) or if the phenyl rings are twisted relative to each other's N=N–C planes (53, 56) was resolved by QM computing, which indicated a planar conformation without a twist, with C–C–N=N dihedral angles of approximately ten degrees (Fig. 2b) (51, 52). Experimental and theoretical techniques both have shown that substituents have a small influence on azobenzene structure (57). Furthermore, azobenzene and most derivatives exhibit small dipole moments (<5 Debye), and changes in dipole moment upon photoisomerization are small (45, 58, 59). The conformational change in azobenzene is most commonly exploited through *para*- or *meta*-substitutions (Sect. 4). During *cis–trans* isomerization, the distance between the *para*-positions (*meta*-positions) changes considerably from ~6 to 9 Å (~4.8–7.8 Å). These changes in distance are comparable to the unit length of an amino acid backbone and can be amplified by substituting the photochrome with rigid linkers (Sect. 4.2).

3.3. Spiropyran and Merocyanine

Photochromism in SPs was discovered by Hirshberg in 1952 (66) and Hirshberg was the first to propose the application of this phenomenon for optical data storage in his "Photochemical Memory Model" (67). SPs are composed of two ring systems joined at a quaternary spiro-carbon atom. Photoisomerization from the weakly polarized, weakly colored SP state to the violet-red MC involves breakage of the spiro-carbon–oxygen bond (Fig. 3a) (68, 69). Not only is the back reaction from the MC to the SP state accelerated by visible light, but it also occurs spontaneously with a half-life ranging from several minutes (at room temperature in polar solutions) to hours (e.g., in polymer matrices) (70, 71). MCs have long been used as dyes in photography and photovoltaics (72, 73) and SPs became preferred photoswitches for optical memory devices (69). One should note that the MC state can populate several isomers (Fig. 3c) (74), which can also exist in zwitterionic resonance structures with large dipole moments (~20 Debye) (75). Because of large changes in geometry and dipole moment and their robust photoswitching (also in two-photon excitation; (76)), SPs have become popular compounds to control the structure and function of biological molecules (9, 77, 78). Furthermore, MCs are fluorescent (69) and in this way, their localization and photostate can be monitored in large biological systems. This remarkable feature also opened avenues to new imaging techniques (see contribution by G. Marriott).

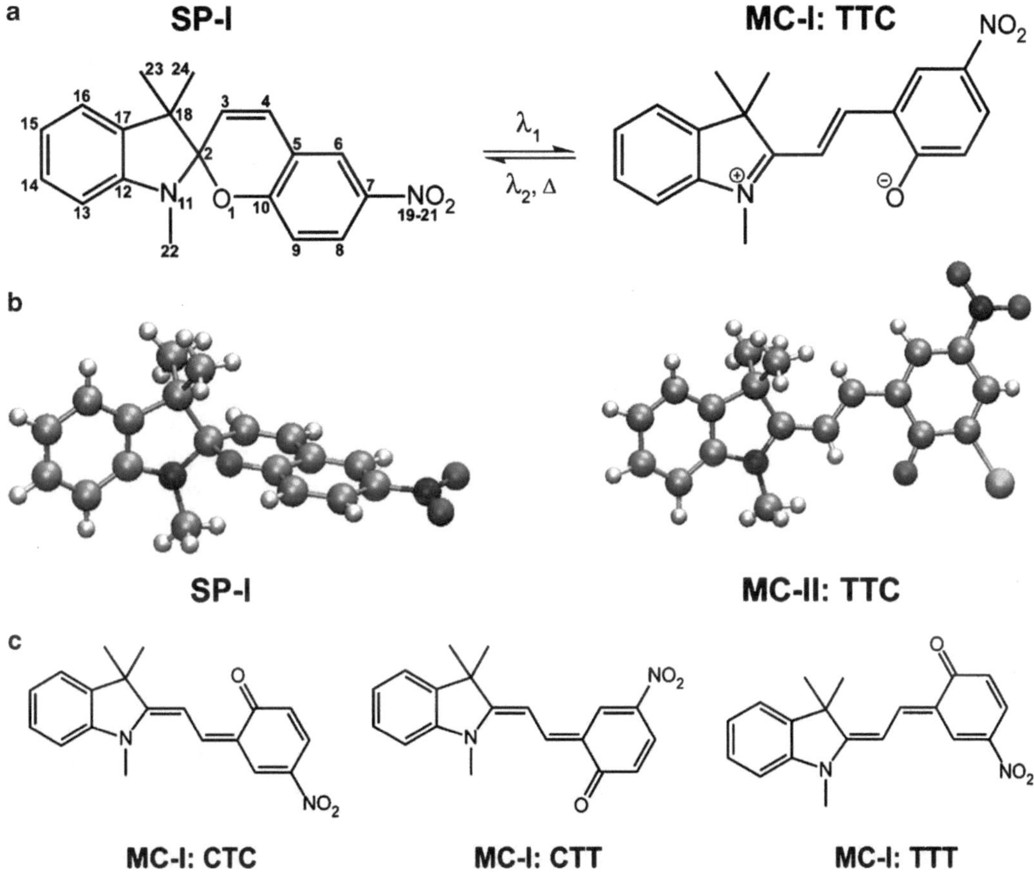

Fig. 3. (**a**) SP and merocyanine and their photoconversion with UV (λ_1) and visible light (λ_2). (**b**) Three-dimensional structures of **SP-I** and **MC-II**. Note that the structure of **MC-II** is shown here since no coordinates are available for **MC-I** (see text for details and QM models of **MC-I** structures). CSD identifiers are BEXLUL01 (**SP-I**) and BETGEM (**MC-II**; also refer to Tables 2 and 3). (**c**) MC exists in a total of four isomers (also see (**a**)) that are defined by *cis–trans* isomerization around the three central bonds. Population of the MC isomers depends on substituents and solvent, and the TTC isomer represents the ground state isomer of **MC-I** as described in the text.

Structures of many SP derivatives have been solved by X-ray crystallography since early reports in the 1970s (79, 80) (Table 2). Here we focus on **SP-I** (11,18,18-trimethyl-7-nitro-spirobenzopyran-2-indole, Fig. 3) since this particular compound has been widely used in biological applications (9, 77, 78). In all SPs, an orthogonal orientation between the ring systems is enforced by the spiro-carbon (Fig. 3b) (68). Interestingly, the crystal structures of **SP-I** highlighted that the conformation of the six-membered heterocycle, and hence the relative orientation of the two ring systems, is determined not only by intramolecular factors but also by crystal packing (81, 82). Two structures were obtained in the unit cell of the crystals. In the first structure, the benzopyran ring was planar (Fig. 3), while in the second structure, this and the adjacent ring were bent toward the 18-position. Structural information on MC are collected by irradiation of a SP solution and crystallization of the photoproduct (68). Structures of **MC-I** are not available but exist for a related compound with similar substituents (**MC-II**, Fig. 3, (83)). **MC-II** has an almost perfectly planar structure (Fig. 3). Interestingly, both the TTC and TTT isomers of merocyanines can be observed in crystals depending on the substituent at the 9-position (Table 3, (74, 83–85)).

QM calculations on SP/MC were initiated by earlier work on model compounds such as pyrans or cyanines (105–107). These calculations focused on solving ground state geometries and intermediates during SP/MC photoconversion (108–110). QM models proved powerful to understand crystallography experiments and predict solution structures. For **SP-I**, theoretical models were in good general agreement with the X-ray data and specifically showed a planar conformation of the pyran ring (108, 110). In the case of **MC-I**, for which no experimental data are available, computational techniques determined that the TTC isomer likely represents the solution ground state (108–110). Spectroscopic data supported the above finding (110), but notably TTC and TTT isomers can coexist in solution for **MC-II** (111). These examples highlight that the structures of SP/MC compounds significantly depend on solvent and the nature and position of substituents (68, 81, 84, 85, 111). In a similar way, the structures of these photochromes could be influenced by the local environment they experience when coupled to proteins.

3.4. Hemithioindigo

HTI compounds are appealing new candidates as components of photoswitchable biomolecules. In HTIs, the combination of hemistilbene and HTI parts forms robust photochromes that isomerize at wavelengths significantly longer than in azobenzenes and SPs (114, 115). For instance, illumination of the ω-amino acid **HTI-I** (Fig. 4a) with blue light leads to a photostationary state containing ~85% of the molecules in the *E*-form (116). The *E*-form then converts thermally or accelerated by light of slightly longer wavelengths to the stable *Z*-isomer with a small change in dipole moment (see below; (117)). Despite the fact that HTIs were first synthesized many decades ago (114, 115), they are less well studied and have seen fewer modifications than azobenzenes or SPs. Thus, it may not be surprising that only a few applications of HTI photochromes exist, which include lipids with HTI chains (118) and HTI amino acids (116, 119).

Structural information of HTIs is sparse, with only a single available crystal structure. In this structure, the *Z*-isomer of **HTI-II** assumes an extended, planar conformation (Fig. 4b) (119). We obtained coordinates for the *E*-isomer of the ω-amino acid **HTI-I** with DFT calculations (see Table 4 for coordinates and experimental details). The theoretical structure of **E-HTI-I** is also extended and corresponds to the *Z*-form after rotation around the central double bond (Fig. 4b). In addition, we noticed

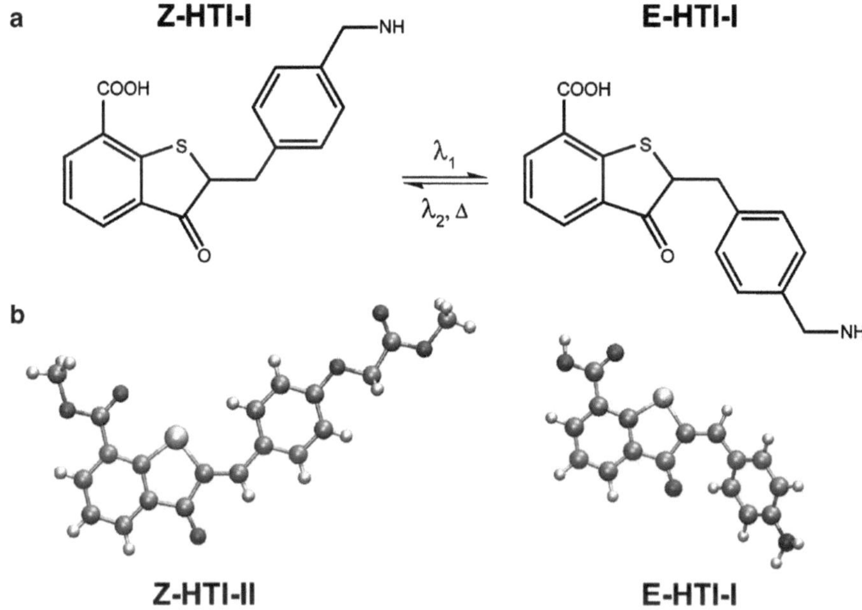

Fig. 4. Photoconversion (**a**) and structure (**b**) of HTI. Blue light (λ_1, λ_2) converts planar Z-HTI to a twisted E-HTI form. Note that **Z-HTI-II** is shown here since this is the only available HTI structure (CSD identifier WAFQOK; (119)). In the QM model, **E-HTI-I** deviates from planarity and the hemistilbene group is rotated out of the plane of the HTI moiety (see Table 4 for atomic coordinates and experimental details).

a tilt of the hemistilbene group relative to the plane of the HTI moiety. Lougheed et al. used DFT methods to calculate the dipole moment of **HTI-II** and found it to be 1.2D for **E-HTI-II** and 2.8D for **Z-HTI-II** (119). Comparing the HTI-isomers reveals that the photochrome undergoes light-induced distance changes of the same magnitude as azobenzenes or SPs. For instance, the first atoms of the two substituents on **HTI-I** are separated by 9.5 Å in the *Z*-isomer and 11.2 Å in the *E*-isomer.

4. Understanding Light-Controlled α-Helices and β-Hairpins: Model Systems for Protein Folding

4.1. Introduction

Folding is the key biological process during which a linear amino acid chain assembles into a three-dimensional, functional protein structure (23). While a universal understanding of folding pathways and kinetics may still be missing, significant progress has been achieved for selected model proteins and secondary structure building blocks (120). In particular, small designed peptides have made it possible to measure the folding dynamics and intermediates of α-helices and β-strands (121–123). An interesting class of these model peptides contains light switches (8, 9, 124, 125). The unique advantage of these systems is that optical folding stimuli

Table 4
Cartesian atomic coordinates of the ω-amino acid HTI-I. The atoms are listed in the order of their linkage. The QM model was generated using the Jaguar (Version 7.5, Schrödinger, LLC, New York, NY). The geometry was optimized with the DFT method using the B3LYP++ basis set and Poisson–Boltzmann water**

	X	*Y*	*Z*
S	-0.23	-7.77	-3.26
C	0.29	-9.13	-5.48
N	-2.74	-3.34	-9.78
C	-1.87	-5.15	-6.09
C	-0.81	-7.07	-4.83
C	0.48	-9.17	-4.09
C	-0.45	-7.95	-5.97
C	-1.59	-3.78	-6.30
C	-2.51	-3.91	-8.55
C	-1.32	-5.81	-4.91
O	-0.66	-7.75	-7.15
O	1.99	-11.37	-1.58
C	-2.61	-5.83	-7.10
C	-2.95	-5.22	-8.30
C	-1.88	-3.17	-7.50
C	1.15	-10.25	-3.50
C	0.77	-10.16	-6.29
O	0.99	-9.38	-1.27
C	1.36	-10.28	-2.03
C	1.63	-11.27	-4.34
C	1.44	-11.23	-5.72

can be applied with greater speed than can temperature jumps or mixing by stopped flow (126–128).

This section summarizes recent work on α-helical and β-hairpin peptides with incorporated photochromes. Many excellent review articles are available (8, 9, 49, 125) and our special emphasis is the design and understanding of these systems with the help of modeling. As we will show, peptide systems can be linked to photochromes either in their backbone (Sect. 4.2) or in their side chains (Sect. 4.3). Azobenzenes are the most popular photoswitches for applications in peptides due to their fast kinetics and relatively large conformational changes (Sect. 3.2). However, several notable applications of other photochromes exist (9, 129, 130). In general, however, the structural effects of photoisomerization are difficult to understand and in many cases, these are either not large or irreversible (125, 131).

Work over the past few years has shown that MM simulations can address these problems and resolve structural intermediates and their kinetics during the light-controlled (un-)folding of peptides. In their seminal work, Zinth, Moroder, and Tavan investigated picosecond dynamics in a cyclic peptide system by combining absorption spectroscopy and all-atom MD simulations (128, 132). In their elegant simulation, a realistic implementation of the azobenzene *cis–trans* photoisomerization was achieved by injecting energy into the protein–solvent system and applying torque to the photochrome. Careful treatment of solvent and detailed analysis of the trajectories allowed tracing the time course of conformational relaxation and dissipation of vibrational energy. The excellent agreement between simulations and experiments highlighted two valuable contributions of combined theoretical–experimental approaches. On the one hand, computer simulations permit detailed interpretation of experimental observations, and on the other hand, experimental data can be used to assess the theoretical descriptions.

4.2. α-Helices

A versatile strategy to light-control peptides and proteins with novel photochromic crosslinkers was pioneered by Woolley and coworkers (125). Non-photochromic crosslinking reagents have been used in molecular and structural biology for many years (133, 134). These small linear molecules bind amino acids through two (or more) functional groups located at their two ends (Fig. 5a). In principle, any two solvent-accessible residues can be connected if they are located at a distance that matches the length of the crosslinker and there is an unhindered through space pathway between them. The reactive groups of crosslinkers are most commonly specific for cysteines or lysines. Cysteines are typically rare on the surface of proteins but can be readily introduced at desired locations through mutagenesis. In the most typical experiments, crosslinkers function as "molecular rulers" and measure the distance between the residues that they are attached to. In this way, conformational changes during protein function and the arrangement of quaternary structure can be mapped. Photochromic crosslinkers such as those introduced by Woolley additionally contain a central azobenzene (12) or SP (129) moiety (Fig. 5a). Photoisomerization of this group changes the

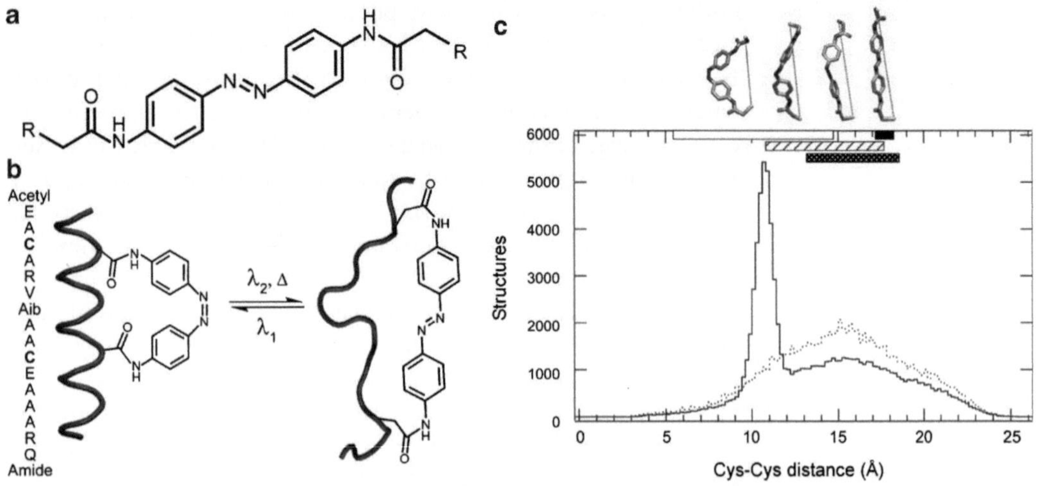

Fig. 5. (**a**) Photoisomerizable crosslinkers contain a central photochromic group and cysteine reactive groups such as are halogen acetates (R=–I (12) or –Cl (138)) and methanthiosulfonates (R=–S–SO_2CH_3 (139)). (**b**) Cartoon of the light-controlled, reversible helix-coil transition in a peptide with seven-residue cysteine–cysteine distance (*Aib* aminoisobutyric acid). (**c**) Histograms showing of the distance distribution between cysteine residues for an uncrosslinked helix with the sequence shown in (**b**). The *solid line* shows the expected distribution for a helical peptide (~25% helical content), while the *dashed line* shows the distribution of a coiled peptide with smaller helical content (~5%). The distance range allowed by *cis-* or *trans*-crosslinker is indicated by the *white* or *black bar*, respectively. Azobenzenes are photoconverted either through inversion or rotation of transition states. The distance range allowed by crosslinkers undergoing inversion or rotation is shown with a *gray* and *dashed bar*. *Panel* (**c**) is adapted with permission from G.A. Woolley (125), Copyright 2005 American Chemical Society.

length of the molecule, which then operates as light-sensitive "molecular tweezers."

The structural effect of crosslinker photoisomerization on a peptide or protein depends on the choice of the attachment site and the structural propensity of the target molecule. Light-controlled (un-)folding of α-helical peptides functionalized with photochromic crosslinkers is the prototypical example (Fig. 5b) (12). Photoisomerization to *cis* increases helical content of the peptide if the two cysteines are spaced four or seven residues apart (Fig. 5b) (12, 135). In contrast, helical content is reduced for a peptide with larger spacing between the cysteines (135). In both cases, photocontrolled (un-)folding is reversible. This elegant and rapid approach to initiate the formation of an α-helix has allowed the folding pathway to be dissected with spectacular time resolution. Time-resolved optical rotatory dispersion and infrared (IR) spectroscopy showed that the helix folds through several intermediates on time scales of 100 ns to a few microseconds, while the crosslinker isomerizes within a few picoseconds (126, 136, 137). Success with model helices also initiated exciting work to photocontrolled secondary structure elements in proteins (Sect. 4.4).

The design of photocontrolled α-helices appears straightforward at first glance: The length of the *cis*-crosslinker must match the distance between cysteines in the folded α-helix, and the same should be true for the *trans*-isomer and an unfolded helix. In reality, however, the situation is complex. The crosslinker contains many rotatable bonds and thus can exist in many conformations with different lengths. Similarly, an α-helical peptide populates a heterogeneous ensemble of folded and unfolded structures that are in rapid equilibrium with each other. For these reasons, it is not trivial to understand the structures of the two molecules and their interplay.

To systematically design the light-controlled α-helices, Woolley and coworkers combined several classes of molecular modeling (125, 140). A conformational search was applied to determine the dimensions of *cis*- and *trans*-crosslinkers as a function of rotations around single bonds. A model of the crosslinker was constructed based on crystal structures of azobenzene (Sect. 3), and all single bonds were systematically rotated in 30° increments (140). For every structure generated in the conformational search, the length of the molecule was measured. End-to-end distances ranged from ~6 to 14 Å for the *cis*-isomer and ~17 to 19 Å for the *trans*-isomer (Fig. 5c, white and black bars) (140). MD simulations revealed a similar range of distances for a slightly longer photochromic crosslinker with different functional groups and thermal isomerization kinetics (139). These results indicate that crosslinkers can span a broad range of lengths, particularly in the *cis*-isomer. Subsequent work showed that the range of distances can be further increased by attaching longer groups to the photochrome. Particularly, the addition of rigid elements allows the *cis*- and *trans*-isomers to populate very different lengths (141, 142).

In a next step, the dimensions of the crosslinker must be matched to those of a folded and unfolded helix. We are particularly interested here in the distances between two sites that are spaced seven residues apart, since this substitution pattern allowed efficient photocontrol of helix content (see above). Using the tool FOLDTRAJ (143), a large ensemble of all-atom peptide models can be generated based on helix content. Once these structures are obtained, it is straightforward to measure the distances between selected residues of the helix (Fig. 5c). In the case of a folded peptide (solid line, 25% helical content), the distribution of distance for the structural ensemble is broad, with a sharp peak at ~11 Å. A peptide with low helix content (~5%, dashed line) shows a broad distribution with a maximum at longer distances (Fig. 5c). These distributions combined with the lengths of the crosslinkers explain the observed decrease and increase in helix content upon crosslinker photoconversion. For each photoisomer, the crosslinker linker limits the peptide to a specific range of residue–residue distances. In case of the peptide shown in Fig. 5, the *cis* range encompasses a large number of structures in the distribution of folded helices, whereas the *trans*-isomer does not. Therefore, *cis–trans* isomerization initiates helix folding. This pioneering experimental and computational work also helped to understand why many photocontrolled systems are limited in their reversibility (131). Photochromes can be stabilized by interactions with the peptides or proteins that they are attached to. Any interaction that is formed to the photochrome in only one photostate changes the relative energy of the isomers and thereby effectively increases the energy barrier separating them. This increase in energy barrier reduces the transition rates, possibly to an extent where reversibility is lost. In the case of α-helices, the rate of thermal isomerization offers a direct reporter of reversibility (131). The extent to which this rate is reduced depends on the overlap of folded and unfolded helix conformations with the azobenzene moiety in the inversion or rotation transition state (Fig. 5c, dashed bars), since in this region an unrestricted *cis–trans* isomerization of the crosslinker is possible. Generally speaking, it seems desirable to maximize the conformational compatibility of photochrome transition states and conformations of the target protein to avoid loss in reversibility. A second factor influencing reversibility, reduced solubility of a peptide–photochrome system in only one photostate, will be discussed below.

4.3. β-Hairpins

β-Sheets are the second structural motif common to many proteins. In contrast to α-helices, which are stabilized by interactions between residues located close in sequence, the folding of β-strands depends on the formation of tertiary contacts. For example, β-hairpins consist of two antiparallel strands linked by a short loop, and even this simple β-motif is stabilized by interstrand hydrogen bonds and hydrophobic interactions (123, 144). β-type structures often aggregate or have marginal stability when removed from their native structural contexts. Nevertheless, β-hairpin model sequences that fold autonomously and robustly in aqueous solution were successfully designed recently (123, 144) and served as starting points for peptides with incorporated photochromes (10, 11, 145).

In their seminal work, Hilvert, van Gunsteren, and coworkers presented a light-controlled β-hairpin peptide (10, 146) that was derived from the protein GB1 (147). This peptide was designed computationally (see below) and contained a central, *meta*-substituted azobenzene moiety in the position of the turn residues (Fig. 6a). At about the same time, Renner, Moroder, Zinth, and coworkers characterized a β-hairpin that resembles a tryptophan zipper (148) and also contained a *meta*-substituted azobenzene group (Fig. 6b) (11, 127). These two β-hairpins shared the same overall geometry and performed similarly in experiments. *cis*-Peptides showed a compact, well-defined ensemble of folded hairpins as

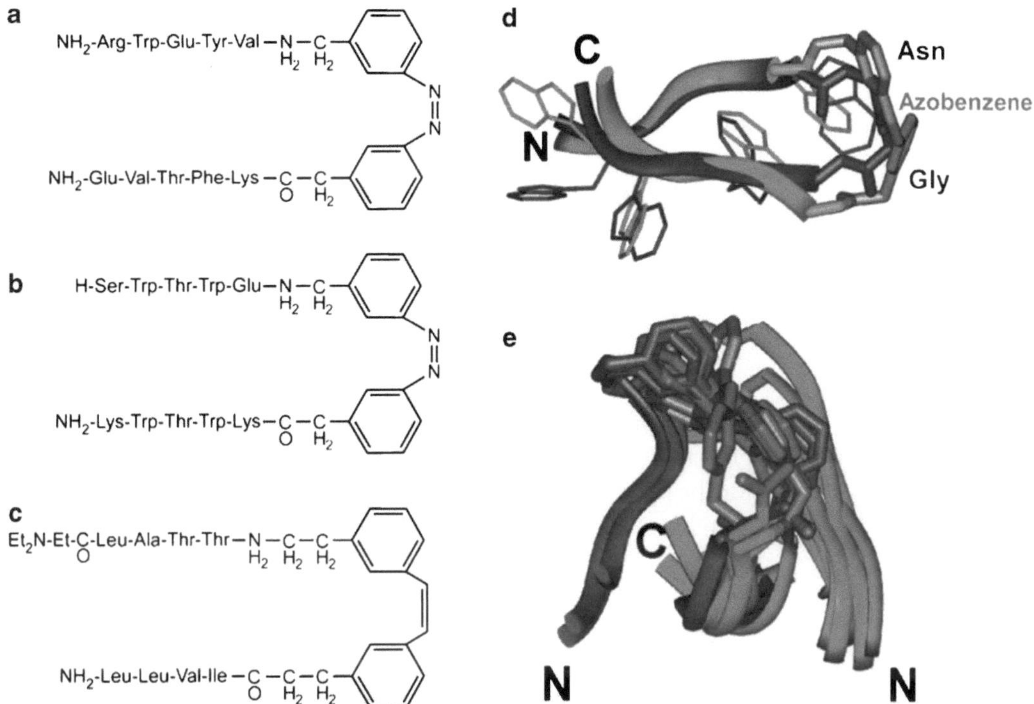

Fig. 6. (**a–c**) Photochromic β-hairpin model peptides derived from protein GB1 (**a**) (10), tryptophan zipper (**b**) (11), and the TATA box-binding protein (**c**) (145). (**d**) Superposition of the NMR structure of the *cis*-isomer of the peptide shown in (**b**, *light gray structure*) and the parent tryptophan zipper (*dark structure*, PDB-ID 1LE1, (148)). The azobenzene photochrome replaced the Asn-Gly turn residues. Only side chains of the tryptophans are shown. The structures of the photocontrolled peptide and the parent peptide agreed almost perfectly, even in their side chain orientations. (**e**) Ensemble of the ten NMR structures with the lowest energy for the *trans*-peptide shown in (**b**). Individual molecules were overlayed on the carboxy-linked peptide strand (Lys-Trp-Thr-Trp-Lys). Many disordered structures are observed compared to the *trans* state. Two classes of conformations exist shown in *light* or *dark gray*. Figures (**d**) and (**e**) are modified from Dong et al. (11) with permission from Wiley-VCH, Weinheim, Germany.

determined by circular dichroism and nuclear magnetic resonance (NMR) spectroscopy. Most if not all interstrand hydrogen bonds present in *cis*-peptides are lost in the *trans* state. The tryptophan zipper provided outstanding solubility in both isomers, allowing complete characterization of light-controlled (un)folding cycles. Solubility is one of two important factor influencing reversibility in light-controlled peptide systems, in addition to isomer-specific photochrome stabilization (13, 125).

Both peptides were subjected to extensive molecular modeling, which included simulations for their design, structural refinement, probing of their structural stability and interpretation of spectroscopic experiments. In their initial design simulations, Kräutler et al. probed the ability of different azobenzene substitution patterns to replace turn residues for the GB1-derived β-hairpin (146). Specifically, azobenzenes were functionalized with peptide segments at *ortho-*, *meta-*, or *para*-positions and either through one or two methylene groups. Spontaneous folding of these peptides from an extended starting conformation was then monitored in MD simulations. It was found that single *meta*-linked methylene spacers offer superior geometry and number of degrees of freedom for hairpin formation. In agreement with the simulations, both groups observed robust folding when using this substitution pattern. Furthermore, spacers with two methylene groups did not yield stable hairpins in the experiments of Erdélyi et al. (Fig. 6c) (145).

In a next phase, both groups applied MD simulations to refine NMR structures and explore their stability and folding. During structural refinement of the peptide derived from protein GB1, two roughly isoenergetic groups of *cis* conformations were noticed (10, 146). The backbone of the hairpin resembled those reported for the parent peptide but the side-chain packaging and the hydrogen bonding pattern was substantially different. Unrestrained MD simulations starting either from the NMR model or from fully extended structures converged to a single class of models. The conformation of this class agreed well with the NMR data and impressively showed that hairpin structures can be predicted with MM computer models. In case of the tryptophan zipper, Dong et al. observed one and two classes of structures for *cis*-peptides and *trans*-peptides, respectively (11). The *cis* conformations compare well to the structure of the parent peptides (Fig. 6d). In MD simulations without experimental constraints, the lowest energy structure derived from the NMR measurements was extremely stable even at elevated temperatures. In contrast, their *trans* forms exhibited disordered and irregular conformations that clustered in two classes (Fig. 6e). Likely additional disordered conformations exist that are not resolved by NMR (11).

A final, elegant example highlights the role of computational techniques in the quantitative interpretation of spectroscopic experiments on light-controlled peptides (127). Following their previous work on cyclic peptides (128, 149), Zinth, Moroder, Tavan, and coworkers monitored backbone conformations and hydrogen bonding of the tryptophan zipper hairpin with IR spectroscopy. They used MD simulations followed by QM/MM normal mode analysis to decode time-resolved IR spectra recorded during (un-)folding (127). Unfolding was first simulated by MD in deuterated methanol, and selected *cis* and *trans* structures were stored during the simulations. In a next step, the IR absorbance intensity of the amide bands was computed for these structures in a hybrid DFT/MM approach. Remarkably, a simple subset of structures with either maximal or minimal numbers of hydrogen bonds was sufficient to decipher the experimental spectra. During β-hairpin folding, several sub-nanosecond intermediates were observed, while microsecond rearrangement of a hydrophobic cluster was the rate-limiting step to the correct hairpin structure.

This step is most likely also rate limiting during folding of the parent peptide and β-type structures in proteins (127).

4.4. Conclusions

The past decades showed that researchers can design α- and β-type peptides that fold robustly and reversibly under the control of synthetic photochromes. These peptides enabled folding studies in the absence of denaturants and on picosecond time scales. The seminal work reviewed above highlighted that due to their small size, the folding of the same molecules can be investigated both experimentally and theoretically to understand their structures and dynamics. Furthermore, it became clear that complete QM or photochemical description was not necessary to design these systems computationally. The prerequisites are, however, that realistic force fields and atomic structures of the photochrome are used.

Motivated by this success, it has been suggested that photochromic peptides may be incorporated into proteins with the goal to control their structures and function (49, 125). Indeed, first reports along these lines exist (125, 150–153). These results also raise the intriguing possibility that optical switches might transform proteins deliberately from one defined state into another state and thereby specifically "gate" the interaction of the target protein with a ligand and or another protein. The ongoing challenge is to extend these experiments to a broader range of proteins and apply them within living cells and organisms. However, certain limitations to in vivo applications of azobenzene photochromes may exist, given their stability in reducing intracellular environments (154).

5. Modeling Protein–Photochrome Complexes: Design of a Light-Gated Receptor

5.1. Introduction

A major goal in biology is to understand the functional role of proteins in both cellular networks and in the behavior of organisms. Recent work has shown that optical techniques can be used to activate proteins and peptides in vitro with outstanding spatial and temporal resolution (Sect. 4). At least three complementary methods exist to light-control proteins in cells or in vivo: "Caged" ligands (also see contribution of G. Ellis-Davies), photochromic ligands, and photochromic tethered ligands (see below). Furthermore, natural photoreceptors can be applied to control biological functions in heterologous systems (also see contributions of D. Gutierrez and X. Han).

Photochromic tethered ligands have been developed to light-control several classes of ion channels, including nicotinic acetylcholine receptors (155), glutamate receptors (also see contribution of P. Gorostiza) (156), and the Shaker K^+ channels (also see to contribution of D. Fortin) (157). Tethered ligands are linear

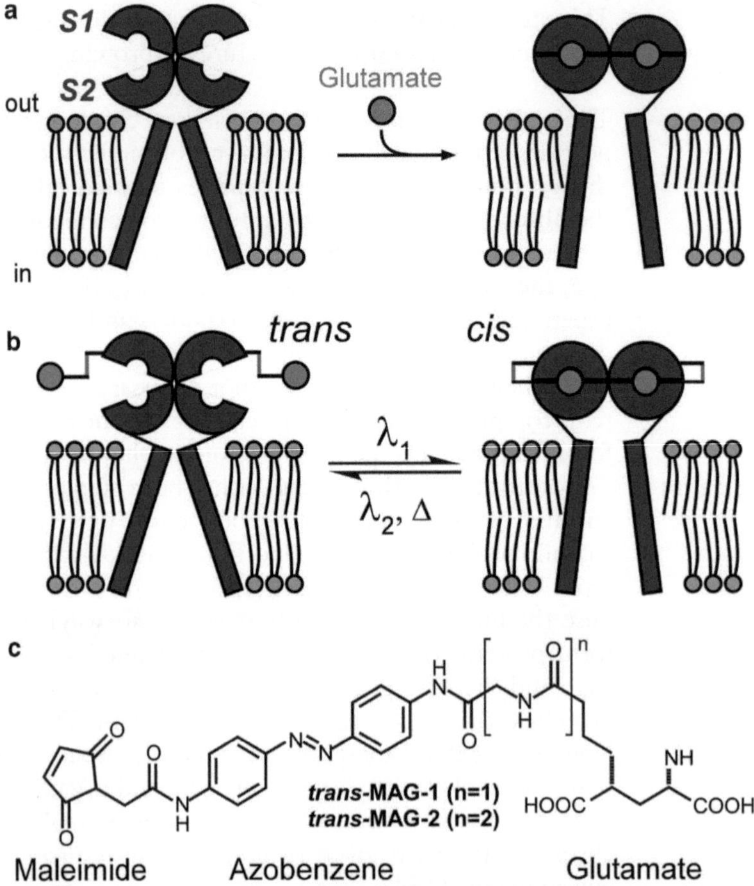

Fig. 7. (**a**) Ionotropic glutamate receptors are activated by glutamate binding to the extracellular LBD formed by S1 and S2 lobes. Ligand binding induces a conformational change in the LBD that is transmitted to the transmembrane segments opening the ion channel pore. (**b**) Azobenzene tethered ligands are anchored in a site-directed manner to the glutamate receptor. They confer light sensitivity to the protein by presenting the agonist to the binding pocket. (**c**) MAG tethered ligands consist of a glutamate moiety, a photoisomerizable core in the center of the molecule, and a maleimide group that attaches to a cysteine residue introduced into the protein.

molecules that consist of a ligand moiety, a photoisomerizable core in the center, and a reactive group that attaches to the protein (Fig. 7c). Through this reactive group, tethered ligands functionalize the target protein in a site-directed manner. This is typically achieved by genetically introducing a cysteine residue near the ligand-binding site to which the tethered ligand binds covalently (158). After attachment, optical control is realized by photoisomerizing the center of the tethered ligand. The agonist or antagonist located at the end of the tether is presented to the binding site of the receptor in a light-dependent manner (Fig. 7b). A precise knowledge of the geometry of the photochrome ensures that the ligand can only bind in one of the photostates, and photocontrol is bidirectional through reversible conversion of the photochrome (Sect. 3). Tethered ligands have unique strengths to control a selected protein specifically since the cysteine substitution is required for light sensitivity. Receptors with substitutions can then be genetically targeted to specific cell types or organs in living organisms. Finally, due to covalent attachment, there is little spill over of the optical signal from its initial location.

The design of tethered ligands can be challenging. Since the natural ligand is chemically coupled to the photochrome moiety, it must be tested whether this modification interferes with the capability of the molecule to activate the target protein (158). First, this potential complication does not exist in caged ligands as a covalent ligand–photochrome bond is broken by light. Second, the function of a tethered ligand strongly depends on the choice of attachment site on the protein. This is especially true when controlling regulatory domains, such as in the ligand-binding domain (LBD) of ionotropic glutamate receptors (Fig. 7a) (159). In this particular case, the ligand must stabilize a closed conformation of the LBD to activate the protein, and the attachment site should be placed on the surface of the LBD to avoid altering residues that contribute to receptor folding and function. Furthermore, only a narrow exit tunnel connects the ligand-binding site and the surface of the receptor. Thus only attachment to selected residues allows the tethered ligand to reach the ligand-binding site and at the same time preserve protein function.

In this section, we present our recent computational method that allows the rational design of light-gated proteins with tethered ligands (manuscript in preparation). The light-gated glutamate receptor LiGluR serves as a model system to introduce and validate the design. The ionotropic glutamate receptor iGluR6 can be specifically photocontrolled with tethered ligands of the maleimide azobenzene glutamate family (MAG, Fig. 7c, also see contribution of P. Gorostiza). In the first version of LiGluR, a cysteine residue introduced at position 439 was functionalized with the ligands **MAG1** and **MAG2** (156, 160). The *cis* state of MAGs activate the receptor (Fig. 7b), which is consistent with the fact that the side chain of residue 439 faces away from the glutamate-binding site. The LiGluR photoresponse modeled as described in the following sections is in good agreement with experimental results, and this simple and rapid protocol is directly transferable to other classes of proteins.

5.2. Modeling a Light-Gated Receptor

We developed a computational approach to understand and design proteins gated by tethered ligands, particularly the proteins with geometrically challenging geometries such as the LBD of LiGluR (manuscript in preparation). In our three-step procedure (Sect. 5.5), we take advantage of the known position of the glutamate ligand in the binding pocket of the active iGluR6 receptor, as determined by X-ray crystallography (159). Docking tools can be applied if the position of the ligand is not known (161). The first step consists of aligning the glutamate end of the *cis*- and *trans*-MAGs with the ligand that occupies the binding pocket. In the second step, the tether of each MAG was threaded through the narrow exit tunnel to reach the surface of the protein. To avoid steric clashes between MAG and the protein, we based our simulations on an open conformation of the LBD. However, since the structure of iGluR6 LBD has only been solved in closed states (162, 163), we used a model based on an open LBD structure of the homologous iGluR2 receptor (164). An open conformation may also be required since the mouth of the fully closed LBD cannot accommodate MAG molecules (165). In a final step, the conformational space accessible to the maleimide group was explored in a Monte Carlo (MC) simulation. In this rapid conformational search, all single bonds of the MAG molecule were defined as degrees of freedom and rotated by a random amount to generate an ensemble of thousands of unique MAG structures (Fig. 8a). After removing nonphysical structures, such as those with high energy or steric clashes between MAG and the protein, the position of the maleimide group is measured for all structures. The distance from the maleimide group to residue 439 for each structure (Fig. 8b, c) reveals the probability that MAG simultaneously binds to the residue and fits into the binding site. Specifically, we counted the number of structures where the distance between the

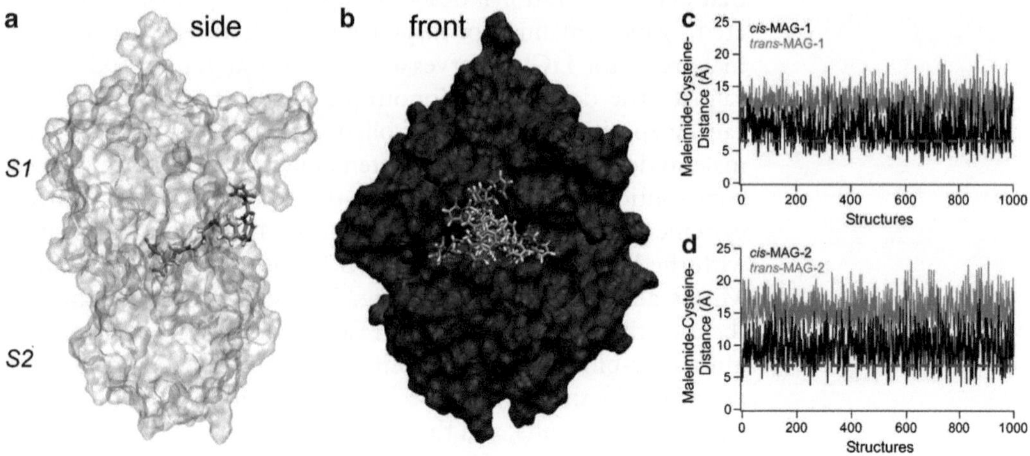

Fig. 8. Results of a conformational search. (**a**) Transparent side view of the LiGluR LBD with a bound MAG molecule. It becomes clear that the binding site of glutamate is located deep in the protein. (**b**) Ten representative structures generated by the Monte Carlo search on the LBD (*front view*). During the conformational search, the glutamate end of the photochrome was fixed in the protein, while the maleimide moiety explores the surface of the protein. (**c**) Distances between the maleimide and residue 439 for *cis*- (*black trace*) and *trans*-**MAG1** (*gray trace*). (**d**) Distances between the maleimide and residues 439 for *cis*- (*black trace*) and *trans*-**MAG2** (*gray trace*). *Lines* indicate the distance threshold (6.5 Å), which corresponds to the approximate length of a maleimide–cysteine conjugate.

nitrogen atom of the maleimide group and residue 439 is smaller than a 6.5 Å threshold, the approximate length of a cysteine–maleimide conjugate.

5.3. Understanding LiGluR Photoresponses

The results of this simple MC simulation agree well with experiments on LiGluR. For *cis*-**MAG1**, the conformational search reported a large number of structures with maleimide–C439 distances below the 6.5 Å threshold (28.4% of the structure satisfied this criterion, Fig. 8c, black trace). In contrast, only 0.1% of the *trans*-**MAG1** structures had comparable maleimide–residue distances (Fig. 8c, gray trace). This finding agrees with the observations made in two experiments on **MAG1**–LiGluR. In the first experiment, a large activation of the receptor by *cis*-**MAG1** but not *trans*-**MAG1** was measured (156). The experimental structure indicated that as suggested by the simulations, few *trans*-**MAG1** conformations must exist in which the glutamate moiety reaches the binding site and turns the receptor on. In the second experiment, the receptor was functionalized with either *cis*-**MAG1** or *trans*-**MAG1**. The simulation accurately mimics this type of experiment, since in both cases glutamate binding precedes the maleimide–cysteine reaction. A higher-labeling efficiency was detected for *cis*-**MAG1** in accordance with the prediction of the simulation. For **MAG2**, it was found experimentally that this molecule elicits ≈50% smaller photoresponses than the shorter **MAG1** ligand (160). This experimental observation was also traced by the MC simulations: 12.3% of *cis*-**MAG2** structures fall below the distance threshold (Fig. 8d, black trace) which is approximately two-fold smaller than the probability of *cis*-**MAG1** (28.4%). No such structures were detected for *trans*-**MAG2** (Fig. 8d, gray trace).

5.4. Conclusions

We developed a simple and rapid computational technique to design light-gated receptors based on tethered photochromic ligands. Our approach is efficient as it takes advantage of the known position of the natural ligand in the structure of the receptor. One short simulation is sufficient to the test the reach of a tethered ligand to all of the residues in the protein that could represent cysteine attachment sites. In contrast, the more labor-intensive alternative would be to attach the tethered ligand to the protein and in this case, one simulation would be required for each candidate site of attachment. Our method captures essential details of the MAG photoresponse, such as directionality and how receptor activation depends on the length of the MAG molecule. We have recently shown that this technique can also be used to screen anchoring sites on glutamate receptors and predict their photoresponses (manuscript in preparation). This approach is an attractive initial screen that should be directly transferable to different classes of proteins. It allows a large experimental parameter space to be explored rapidly, and in particular, different ligand molecules can be tested using computer simulations before their laborious synthesis. An important extension will be to include functionally relevant conformational changes of the receptor in the simulation. These refined models may in turn allow us to gain deeper understanding of receptor structure and function, since tethered and natural ligands may activate receptors differently.

5.5. Simulation Details

Models of MAG molecules were built in Maestro 6.5 (Schrödinger Inc., New York, NY) starting with the crystal structures of *cis*- and *trans*-azobenzene (Sect. 3.2). MAGs were placed into an open structure of the iGluR6 LBD build using the unliganded LBD domain of iGluR2 (162, 164). This structure was generated by superimposing the lobes of the closed iGluR6 LBD (PDB-ID 1S50) onto the corresponding domains of the iGluR2 structure (PDB-ID 1FTO). The glutamate analog in MAG was superimposed onto the co-crystallized glutamate after adjusting its dihedral angles to match the bound ligand. Then the remaining portion of MAG was manually threaded to reach the surface of the protein without visible steric clashes. A Monte Carlo multiple minimum conformational search ((166); Macromodel 9.1, Schrödinger) generated 20,000 new structures by randomly turning dihedral angles of all single bonds in MAG (except those in the glutamate analog and azobenzene). Solvent was treated implicitly using a generalized Born/surface area (GB/SA) water model in the context of the OPLS force field (167, 168). Side-chain atoms were allowed to fluctuate, while backbone atoms were frozen. Bonds lengths and (dihedral) angles of the azobenzene moiety were constrained to the coordinates of the crystal structure. In the course of this conformational search, any bias from the manual starting conformation was removed. Structures were filtered by energy and for each structure, the distance between the maleimide group and the backbone of the protein was measured for residue 439 (this residue is numbered 12 in the crystal structure). To exclude the influence of side-chain rotamers, distances were measured between the nitrogen atom of the maleimide moiety and the C-β atom (Fig. 8b, c).

6. Outlook

The optical control of protein structure and function has become a key objective in molecular biosciences. The performance of an optical switch mechanism depends on the structures of its components and their dynamic interplay, just as it is the case for other conformational changes in protein such as those triggered by ligand binding. It is evident that the better we are able to tailor molecular properties, the more successful we will be in controlling the system as a whole. In this outlook, we envision challenges for the molecular design of photocontrolled proteins. We propose how computational tools may help to overcome these challenges, also with the objective to summarize the previous sections.

Current progress in combinatorial chemistry and photochemistry will challenge us with a large repertoire of novel photochromes with unique structures, colors, and reaction rates. The application of these new compounds will benefit from computational methods in several ways. First, QM calculations will reliably and rapidly characterize structural and spectral properties of new compounds (Sect. 3 and especially (30)). Second, macromolecular modeling (Sects. 4 and 5) will efficiently test new photochromes in the structural context of target proteins, potentially even before their synthesis in the laboratory. Many tools have been developed for virtual drug screening over the past decades, which could be adapted for this purpose (161). Finally, it is important to understand the influence of local environments, e.g., those near or in proteins, on the structure and chemistry of photochromes. Combined QM/MM models are best suited to address these questions. Our design of light-controlled proteins is inspired by nature and, in the same way, we can learn much from QM/MM simulations performed on natural photoreceptors (31).

The second challenge addresses remarkable systems that exhibit "gated" photoresponses (95). Gated photosystems either only respond to light if an additional, external stimulus is present, or modulate an external stimulus by the conformational effect of light. In the simplest case, we envision a receptor in which a light-controlled molecule serves as a reversible competitive antagonist and blocks the binding site for the natural ligand. Due to large requirements in the structural design of such systems, we expect computational techniques to play a key role in their development.

It is commonly noted that the fascinating new field of synthetic optical nanomaterials and biomolecules requires the collaboration of specialists from many different disciplines. Chemistry, physics, and biology are commonly emphasized. This chapter has highlighted the role that computational sciences may play. Theoretical work has already proven essential in the design and application of photosystems with high efficiency and reversibility. It was our goal to provide a complete picture of combined theoretical and experimental contributions, and the many valuable lessons learned highlight their large future potential.

Acknowledgments

We are thankful to K. Durkin, B. Kane, A. Woolley, L. Moroder, C. Renner, J. Patti, A. Sodt, K. Dubay, L. Kirkby, D. Fortin and C. Duggan for many valuable discussions and for contributing figures. We thank the College of Chemistry for computing resources funded by the NSF (CHE-0233882 and CHE-0840505). The work was supported by NIH grants R01 NS35549-12 and PN2EY018241-03, FIBR, by Human Frontier Science Program Grant RPG23-2005, and a postdoctoral fellowship of the European Molecular Biology Organization.

References

1. Bryant DA, Frigaard NU (2006) Prokaryotic photosynthesis and phototrophy illuminated. Trends Microbiol 14:488–496
2. Purschwitz J, Muller S, Kastner C, Fischer R (2006) Seeing the rainbow: light sensing in fungi. Curr Opin Microbiol 9:566–571
3. Chen M, Chory J, Fankhauser C (2004) Light signal transduction in higher plants. Annu Rev Genet 38:87–117
4. Boycott BB, Gegenfurtner KR, Sharpe LT (eds) (2001) Color vision: from genes to perception. Cambridge University Press, Cambridge
5. Ridge KD, Palczewski K (2007) Visual rhodopsin sees the light: structure and mechanism of G protein signaling. J Biol Chem 282:9297–9301
6. Rockwell NC, Lagarias JC (2006) The structure of phytochrome: a picture is worth a thousand spectra. Plant Cell 18:4–14
7. Fuhrman JA, Schwalbach MS, Stingl U (2008) Proteorhodopsins: an array of physiological roles? Nat Rev Microbiol 6:488–494
8. Ciardelli F, Pieroni O, Fissi A, Houben JL (1984) Azobenzene-containing polypeptides – photoregulation of conformation in solution. Biopolymers 23:1423–1437
9. Willner I, Rubin S (1996) Control of the structure and functions of biomaterials by light. Angew Chem Int Ed Engl 35:367–385
10. Aemissegger A, Krautler V, van Gunsteren WF, Hilvert D (2005) A photoinducible beta-hairpin. J Am Chem Soc 127: 2929–2936
11. Dong SL, Loweneck M, Schrader TE et al (2006) A photocontrolled beta-hairpin peptide. Chemistry 12:1114–1120
12. Kumita JR, Smart OS, Woolley GA (2000) Photo-control of helix content in a short peptide. Proc Natl Acad Sci U S A 97:3803–3808
13. Pieroni O, Fissi A, Houben JL, Ciardelli F (1985) Photoinduced aggregation changes in photochromic polypeptides. J Am Chem Soc 107:2990–2991
14. Ueno A, Takahashi K, Anzai JI, Osa T (1981) Photocontrol of polypeptide helix sense by cis–trans isomerism of side-chain azobenzene moieties. J Am Chem Soc 103:6410–6415
15. Willner I, Rubin S, Riklin A (1991) Photoregulation of papain activity through anchoring photochromic azo groups to the enzyme backbone. J Am Chem Soc 113:3321–3325
16. Hincliffe A (2000) Modelling molecular structures, 2nd edn. Wiley, New York
17. Daggett V, Fersht A (2003) The present view of the mechanism of protein folding. Nat Rev Mol Cell Biol 4:497–502
18. Garcia-Viloca M, Gao J, Karplus M, Truhlar DG (2004) How enzymes work: analysis by modern rate theory and computer simulations. Science 303:186–195
19. Karplus M, Kuriyan J (2005) Molecular dynamics and protein function. Proc Natl Acad Sci U S A 102:6679–6685
20. Lu D, Aksimentiev A, Shih AY et al (2006) The role of molecular modeling in bionanotechnology. Phys Biol 3:S40–S53
21. Simonson T, Archontis G, Karplus M (2002) Free energy simulations come of age: protein–ligand recognition. Acc Chem Res 35:430–437
22. Halgren TA (1995) Potential energy functions. Curr Opin Struct Biol 5:205–210
23. Dill KA (1999) Polymer principles and protein folding. Protein Sci 8:1166–1180
24. Gilson MK, Zhou HX (2007) Calculation of protein–ligand binding affinities. Annu Rev Biophys Biomol Struct 36:21–42
25. Mackerell AD Jr (2004) Empirical force fields for biological macromolecules: overview and issues. J Comput Chem 25:1584–1604
26. Cramer CJ (2004) Essentials of computational chemistry: theories and models, 2nd edn. Wiley, New York

27. Neese F, Petrenko T, Ganyushin D, Obrich G (2007) Advanced aspects of ab initio theoretical optical spectroscopy of transition metal complexes: multiplets, spin–orbit coupling and resonance Raman intensities. Coord Chem Rev 251:288–327
28. Koch W, Holthausen MC (2001) A chemist's guide to density functional theory, 2nd edn. Wiley VCH Verlag GmbH, Weinheim, Germany
29. Cohen AJ, Mori-Sanchez P, Yang W (2008) Insights into current limitations of density functional theory. Science 321:792–794
30. Jacquemin D, Preat J, Wathelet V, Fontaine M, Perpete EA (2006) Thioindigo dyes: highly accurate visible spectra with TD-DFT. J Am Chem Soc 128:2072–2083
31. Altun A, Yokoyama S, Morokuma K (2008) Quantum mechanical/molecular mechanical studies on spectral tuning mechanisms of visual pigments and other photoactive proteins. Photochem Photobiol 84:845–854
32. Friesner RA, Guallar V (2005) Ab initio quantum chemical and mixed quantum mechanics/molecular mechanics (QM/MM) methods for studying enzymatic catalysis. Annu Rev Phys Chem 56:389–427
33. Leach A (2001) Molecular modelling: principles and applications, 2nd edn. Prentice Hall, Inc., Upper Saddle River, NJ
34. Liwo A, Czaplewski C, Oldziej S, Scheraga HA (2008) Computational techniques for efficient conformational sampling of proteins. Curr Opin Struct Biol 18:134–139
35. Frenkel D, Smit B (2001) Understanding molecular simulation, 2nd edn. Academic, San Diego, CA
36. Hirshberg Y, Fischer E (1953) Multiple reversible color changes initiated by irradation at low temperature. J Chem Phys 21:1619–1620
37. Irie M (2000) Photochromism: memories and switches – introduction. Chem Rev 100:1683
38. Irie M (2000) Diarylethenes for memories and switches. Chem Rev 100:1685–1716
39. Allen FH (2002) The Cambridge Structural Database: a quarter of a million crystal structures and rising. Acta Crystallogr C B58:380–388
40. Noble A (1856) Zur geschichte des azobenzols und benzidins. Ann Chem Pharm 98:253–256
41. Hartley GS (1937) The cis-form of azobenzene. Nature 140:281
42. Liu ZF, Hashimoto K, Fujishima A (1990) Photoelectrochemical information storage using an azobenzene derivative. Nature 347:658–660
43. Credi A (2006) Artificial molecular motors powered by light. Aust J Chem 59:157–169
44. Katsonis N, Lubomska M, Pollard MM, Feringa BL, Rudolf P (2007) Synthetic light-activated molecular switches and motors on surfaces. Prog Surf Sci 82:407–434
45. Kumar GS, Neckers DC (1989) Photochemistry of azobenzene-containing polymers. Chem Rev 89:1915–1925
46. Rau H (1989) Photoisomerization of azobenzene. In: Rabek JF, Scott GW (eds) Photochemistry and photophysics. CRC Press, Inc., Boca Raton, FL, pp 119–142
47. Griffiths J II (1972) Photochemistry of azobenzene and its derivatives. Chem Soc Rev 1:481–493
48. Pozhidaeva N, Cormier ME, Chaudhari A, Woolley GA (2004) Reversible photocontrol of peptide helix content: adjusting thermal stability of the cis state. Bioconjug Chem 15:1297–1303
49. Renner C, Moroder L (2006) Azobenzene as conformational switch in model peptides. Chembiochem 7:868–878
50. Biswas N, Umapathy S (2000) Structures, vibrational frequencies, and normal modes of substituted azo dyes: infrared, Raman, and density functional calculations. J Phys Chem A 104:2734–2745
51. Briquet L, Vercauteren DP, Perpete EA, Jacquemin D (2006) Is solvated trans-azobenzene twisted or planar? Chem Phys Lett 417:190–195
52. Hattig C, Hald K (2002) Implementation of RI-CC2 triplet excitation energies with an application to trans-azobenzene. Phys Chem Chem Phys 4:2111–2118
53. Kurita N, Tanaka S, Itoh S (2000) Ab initio molecular orbital and density functional studies on the stable structures and vibrational properties of trans- and cis-azobenzenes. J Phys Chem A 104:8114–8120
54. Brown CJ (1966) A refinement of the crystal structure of azobenzene. Acta Crystallogr 21:146–152
55. Tsuji T, Takashima H, Takeuchi H, Egawa T, Konaka S (2001) Molecular structure and torsional potential of trans-azobenzene. A gas electron diffraction study. J Phys Chem A 105:9347–9353
56. Traetteberg M, Hilmo I, Hagen K (1977) Gas electron diffraction study of the molecular structure of trans-azobenzene. J Mol Struct 39:231–239

57. Briquet L, Vercauteren DP, Andre J, Perpete EA, Jacquemin D (2007) On the geometries and UV/Vis spectra of substituted trans-azobenzenes. Chem Phys Lett 435: 257–262
58. Minisini B, Fayet G, Tsobnang F, Bardeau JF (2007) Density functional theory characterisation of 4-hydroxyazobenzene. J Mol Model 13:1227–1235
59. Raduge C, Papastavrou G, Kurth DG, Motschmann H (2003) Controlling wettability by light: illuminating the molecular mechanism. Eur Phys J E Soft Matter 10:103–114
60. de Lange JJ, Robertson JM, Woodward I (1939) X-ray crystal analysis of trans-azobenzene. Proc R Soc Lond A Math Phys Sci 171:398–410
61. Hampson GC, Robertson JM (1941) Bond lengths and resonance in the cis-azobenzene molecule. J Chem Soc 2:409–413
62. Mostad A, Romming C (1971) Refinement of crystal structure of cis-azobenzene. Acta Chem Scand 25:3561–3568
63. Bouwstra JA, Schouten A, Kroon J (1983) Structural Studies of the system trans-azobenzene/trans-stilbene. I. A reinvestigation of the disorder in the crystal structure of trans-azobenzene, C12H10N2. Acta Crystallogr C C39:1121–1123
64. Harada J, Ogawa K, Tomoda S (1997) Molecular motion and conformational interconversion of azobenzenes in crystals as studied by X-ray diffraction. Acta Crystallogr B 53:662–672
65. Harada J, Ogawa K (2004) X-ray diffraction analysis of nonequilibrium states in crystals: observation of an unstable conformer in flash-cooled crystals. J Am Chem Soc 126:3539–3544
66. Fischer E, Hirshberg Y (1952) Formation of coloured forms of spirans by low-temperature irradation. J Chem Soc 4518–4548
67. Hirshberg Y (1956) Reversible formation and eradication of colors by irradiation at low temperatures. A photochemical memory model. J Am Chem Soc 78:2304–2312
68. Aldoshin SM (1990) Spiropyrans: structural features and photochemical properties. Russ Chem Rev 59:663–684
69. Berkovic G, Krongauz V, Weiss V (2000) Spiropyrans and spirooxazines for memories and switches. Chem Rev 100:1741–1754
70. Malkin J, Dvornikov AS, Straub KD, Rentzepis PM (1993) Photochemistry of molecular systems for optical 3D storage memory. Res Chem Intermediates 19:159–189
71. Shumburo A, Biewer MC (2002) Stabilization of an organic photochromic material by incorporation in an organogel. Chem Mater 14:3745–3750
72. Morel D, Ghosh AK, Feng T et al (1978) High-efficiency organic solar cells. Appl Phys Lett 32:495–497
73. West W, Carroll BH, Whitcomb DH (1952) The adsorption of sensitizing dyes in photographic emulsions. J Phys Chem 56: 1054–1067
74. Minkin VI (2004) Photo-, thermo-, solvato-, and electrochromic spiroheterocyclic compounds. Chem Rev 104:2751–2776
75. Bletz M, Pfeifer-Fukumura U, Kolb U, Baumann W (2002) Ground- and first-excited-singlet-state electric dipole moments of some photochromic spirobenzopyrans in their spiropyran and merocyanine form. J Phys Chem A 106:2232–2236
76. Parthenopoulos DA, Rentzepis PM (1989) Three-dimensional optical storage memory. Science 245:843–845
77. Aizawa M, Namba K, Suzuki S (1977) Light-induced enzyme activity changes associated with the photoisomerization of bound spiropyran. Biochim Biophys Acta 182: 305–310
78. Sakata T, Yan YL, Marriott G (2005) Family of site-selective molecular optical switches. J Org Chem 70:2009–2013
79. Shibaeva RP, Lobkovskaya RM (1979) Crystal and molecular-structure of the photochromic spinopyran, C24N2O3H19Cl. J Struct Chem 20:316–319
80. Shibaeva RP, Rozenberg LP, Kholmanskii AS, Zubkov AV (1976) Crystalline and molecular-structure of photochromic spiropyrane, C-24N-2O-3H-2O. Dokl Akad Nauk SSSR 226:1374–1377
81. Clegg W, Norman NC, Flood T et al (1991) Structures of 3 photochromic compounds and 3 non-photochromic derivatives, the effect of methyl substituents. Acta Crystallogr C 47:817–824
82. Karaev KS, Furmanova NG, Belov NV (1982) Crystal-structure and molecular-structure of photochromic spiropyrane, C19H18N2O3. Dokl Akad Nauk SSSR 262: 877–880
83. Aldoshin SM, Atovmyan LO, Dyachenko OA, Galbershtam MA (1981) Molecular and crystal-structure of 1,3,3-trimethyl-6′-nitro-8′-bromospiro(indolin-2,2′-[2H-1]-benzopyran) and products of its photochemical conversion. Russ Chem Bull 30:2262–2270

84. Hobley J, Malatesta V, Millini R, Montanari L, Parker WON (1999) Proton exchange and isomerisation reactions of photochromic and reverse photochromic spiro-pyrans and their merocyanine forms. Phys Chem Chem Phys 1:3259–3297
85. Zou WX, Chen PL, Gao Y, Meng JB (2003) 3-{2-[2-(3,5-Dinitro-2-oxidophenyl)ethyl]-3,3-dimethyl-3H-indol-1-ium-1-yl}propanoic acid bis(dimethylformamide) solvate. Acta Crystallogr Sect E Struct Rep Online 59:o337–o339
86. Aldoshin SM, Atovmyan LO, Dyachenko OA, Galbershtam MA (1981) Molecular and crystal-structure of 1,3,3-trimethyl-6′-nitro-8′-bromospiro(indolin-2,2′-[2H-1]-benzopyran) and products of its photochemical conversion. Izv Akad Nauk SSSR 31: 2720–2729
87. Karaev KS, Furmanova NG (1984) Crystal and molecular-structure of the photochromic spiropyran 1-benzyl-3,3-dimethyl-6′-nitrospiro([2H-1]-benzopyran-2′,2-indoline). Zh Strukt Khim 25:182–184
88. Karaev KS, Furmanova NG (1984) Crystal and molecular-structure of the photochromic spiropyran 1-benzyl-3,3-dimethyl-6′-nitrospiro([2H-1]-benzopyran-2′,2-indoline). J Struct Chem 25:168–170
89. Aldoshin SM, Atovmyan LO (1985) Structure of a new photochromic indoline spiropyran. Russ Chem Bull 34:180–182
90. Aldoshin SM, Atovmyan LO (1985) Structure of a new photochromic indoline spiropyran. Izv Akad Nauk Ser Khim 34:191–194
91. Zhang F, Jin D, Zhang Y, Zhang DC (2005) 6-Bromo-1′-ethyl-4-(1-ethyl-3,3-dimethyl-1H-indolin-2-ylidenemethyl)-3′,3′-dimethylspiro[3,4-dihydro-2H-1-benzopyran-2,2′(3′H)-1′H-indoline], a dicondensed spiropyran. Acta Crystallogr Sect E Struct Rep Online 61:o4985–o4987
92. Raic-Malic S, Tomaskovic L, Mrvos-Sermek D et al (2004) Spirobipyridopyrans, spirobinaphthopyrans, indolinospiropyridopyrans, indolinospironaphthopyrans and indolinospironaphtho-1,4-oxazines: synthesis, study of X-ray crystal structure, antitumoral and antiviral evaluation. Bioorg Med Chem 12: 1037–1045
93. Aldoshin SM, Atovmyan LO, Kozina OA (1987) Structure of photochromic 1′,3′,3′-trimethyl-7-methoxyspiro(indoline-2,2′-[2H-1]benzopyran). Russ Chem Bull 36:169–171
94. Aldoshin SM, Atovmyan LO, Kozina OA (1987) Structure of photochromic 1′,3′,3′-trimethyl-7-methoxyspiro(indoline-2,2′-[2H-1]benzopyran). Izv Akad Nauk Ser Khim 36:190–192
95. Raymo FM, Giordani S, White AJ, Williams DJ (2003) Digital processing with a three-state molecular switch. J Org Chem 68:4158–4169
96. Shibaeva RP, Lobkovskaya RM (1979) Crystal and molecular-structure of the photochromic spinopyran, C24N2O3H19Cl. Zh Strukt Khim 20:369–372
97. Keum SR, Ku BS, Shin JT, Ko JJ, Buncel E (2005) Stereoselective formation of dicondensed spiropyran product obtained from the reaction of excess Fischer base with salicylaldehydes: first full characterization by X-ray crystal structure analysis of a DC center dot acetone crystal. Tetrahedron 61:6720–6725
98. Vary MW, McBride JM (2005) Using spiropyran fragment superposition to assign the EPR spectrum of localized triplet excitation on a distorted aromatic nitro group. Cryst Growth Des 5:2036–2042
99. Zou WX, Huang HM, Gao Y, Matsuura T, Meng JB (2004) Structures of two spiropyrans in the open and closed form. Struct Chem 15:317–321
100. Godsi O, Peskin U, Kapon M, Natan E, Eichen Y (2001) Site effects in controlling the chemical reactivity in crystals: solid-state photochromism of *N*-(*n*-propyl)nitrospiropyrane. Chem Commun (Camb) 2132–2133
101. Wang AQ, Wu Y, Guo H (2006) Crystal structure of 8-bromo-6-chloro-l′,3′,3′-trimethylspiro[2H-1-benzopyran-2,2′-indoline], C19H17BrClNO. Z Kristallogr New Cryst Struct 221:321–322
102. Guo H, Gao YB, Han J, Meng JB (2005) 8-Bromo-6-chloro-1′,3′,3′,5′-tetramethylspiro[2H-1-benzopyran-2,2′-indoline]. Acta Crystallogr Sect E Struct Rep Online 61:o1461–o1462
103. Chuev II, Filipenko OS, Aldoshin SM, Kozina OA (1994) Molecular-structure of 1′,3′,3′-trimethyl-6-trifluoromethylsulfonyl-spiro-(indoline-2,2′-benzo[B]pyran). Izv Akad Nauk Ser Khim 43:1969–1971
104. Chuev II, Filipenko OS, Aldoshin SM, Kozina OA (1994) Molecular-structure of 1′,3′,3′-trimethyl-6-trifluoromethylsulfonyl-spiro-(indoline-2,2′-benzo[B]pyran). Russ Chem Bull 43:1857–1859
105. Day PN, Wang ZQ, Pachter R (1995) Ab-initio study of the ring-opening reactions of pyran, nitrochromene, and spiropyran. J Phys Chem 99:9730–9738
106. Kuthan J, Bohm S (1981) MO study of molecular and electronic-structure of 2H-pyran and 4H-pyran. Collect Czechoslov Chem Commun 46:759–771

107. Zerbetto F, Monti S, Orlandi G (1984) Thermal and photochemical interconversion of spiropyrans and merocyanines – a semi-empirical all-valence-electron study. J Chem Soc, Faraday Trans 2 80:1513–1527
108. Broo A (2000) Theoretical characterization of a multifunctional electrooptical molecular device: photochemical ring-opening mechanism of indolinospirobenzopyran. Int J Quantum Chem 77:454–467
109. Cottone G, Noto R, La Manna G (2004) Theoretical study of spiropyran–merocyanine thermal isomerization. Chem Phys Lett 388:218–222
110. Futami Y, Chin M, Kudoh S, Takayanagi M, Nakata M (2003) Conformations of nitro-substituted spiropyran and merocyanine studied by low-temperature matrix-isolation infrared spectroscopy and density-functional-theory calculation. Chem Phys Lett 370: 460–468
111. Hobley J, Malatesta V (2000) Energy barrier to TTC-TTT isomerisation for the merocyanine of a photochromic spiropyran. Phys Chem Chem Phys 2:57–59
112. Aldoshin SM, Atovmyan LO (1985) Crystal and molecular structure of the product of the photochemical transformation of 1′-pentyl-3′,3′-dimethyl-6-nitro-8-bromospiro(indoline-2,2′-[2H-1]benzopyran). Izv Akad Nauk SSSR 34:2016–2023
113. Aldoshin SM, Atovmyan LO (1985) Crystal and molecular structure of the product of the photochemical transformation of 1′-pentyl-3′,3′-dimethyl-6-nitro-8-bromospiro(indoline-2,2′-[2H-1]benzopyran). Russ Chem Bull 34:1859–1865
114. Mostoslavskii MA, Kravchenko MD (1970) Absorption spectra of 3-oxo-2,3-dihydrothionaphthene and its derivatives. Chem Heterocycl Comp 4:45–47
115. Yamaguchi T, Seki T, Tamaki T, Ichimura K (1992) Photochromism of hemithioindigo derivatives I. Preparation and photochromic properties in organic solvents. Bull Chem Soc Jpn 65:649–656
116. Cordes T, Heinz B, Regner N et al (2007) Photochemical Z→E isomerization of a hemithioindigo/hemistilbene omega-amino acid. Chemphyschem 8:1713–1721
117. Cordes T, Weinrich D, Kempa S et al (2006) Hemithioindigo-based photoswitches as ultrafast light trigger in chromopeptides. Chem Phys Lett 428:167–173
118. Eggers K, Fyles TM, Montoya-Pelaez PJ (2001) Synthesis and characterization of photoswitchable lipids containing hemithioindigo chromophores. J Org Chem 66:2966–2977
119. Lougheed T, Borisenko V, Hennig T, Ruck-Braun K, Woolley GA (2004) Photomodulation of ionic current through hemithioindigo-modified gramicidin channels. Org Biomol Chem 2:2798–2801
120. Schaeffer RD, Fersht A, Daggett V (2008) Combining experiment and simulation in protein folding: closing the gap for small model systems. Curr Opin Struct Biol 18:4–9
121. Osterhout JJ (2005) Understanding protein folding through peptide models. Protein Pept Lett 12:159–164
122. Scholtz JM, Baldwin RL (1992) The mechanism of alpha-helix formation by peptides. Annu Rev Biophys Biomol Struct 21:95–118
123. Searle MS, Ciani B (2004) Design of beta-sheet systems for understanding the thermodynamics and kinetics of protein folding. Curr Opin Struct Biol 14:458–464
124. Pieroni O, Fissi A, Angelini N, Lenci F (2001) Photoresponsive polypeptides. Acc Chem Res 34:9–17
125. Woolley GA (2005) Photocontrolling peptide alpha helices. Acc Chem Res 38:486–493
126. Ihalainen JA, Paoli B, Muff S et al (2008) Alpha-helix folding in the presence of structural constraints. Proc Natl Acad Sci USA 105:9588–9593
127. Schrader TE, Schreier WJ, Cordes T et al (2007) Light-triggered beta-hairpin folding and unfolding. Proc Natl Acad Sci USA 104:15729–15734
128. Sporlein S, Carstens H, Satzger H et al (2002) Ultrafast spectroscopy reveals subnanosecond peptide conformational dynamics and validates molecular dynamics simulation. Proc Natl Acad Sci U S A 99:7998–8002
129. Fujimoto K, Amano M, Horibe Y, Inouye M (2006) Reversible photoregulation of helical structures in short peptides under indoor lighting/dark conditions. Org Lett 8:285–287
130. Helbing J, Bregy H, Bredenbeck J et al (2004) A fast photoswitch for minimally perturbed peptides: investigation of the trans→cis photoisomerization of *N*-methylthioacetamide. J Am Chem Soc 126:8823–8834
131. Borisenko V, Woolley GA (2005) Reversibility of conformational switching in light-sensitive peptides. J Photochem Photobiol A Chem 173:21–28
132. Nguyen PH, Gorbunov RD, Stock G (2006) Photoinduced conformational dynamics of a photoswitchable peptide: a nonequilibrium

molecular dynamics simulation study. Biophys J 91:1224–1234
133. Fasold H, Klappenberger J, Meyer C, Remold H (1971) Bifunctional reagents for the cross-linking of proteins. Angew Chem Int Ed Engl 10:795–801
134. Ji TH (1983) Bifunctional reagents. Methods Enzymol 91:580–609
135. Flint DG, Kumita JR, Smart OS, Woolley GA (2002) Using an azobenzene cross-linker to either increase or decrease peptide helix content upon trans-to-cis photoisomerization. Chem Biol 9:391–397
136. Bredenbeck J, Helbing J, Kumita JR, Woolley GA, Hamm P (2005) Alpha-helix formation in a photoswitchable peptide tracked from picoseconds to microseconds by time-resolved IR spectroscopy. Proc Natl Acad Sci USA 102:2379–2384
137. Chen E, Kumita JR, Woolley GA, Kliger DS (2003) The kinetics of helix unfolding of an azobenzene cross-linked peptide probed by nanosecond time-resolved optical rotatory dispersion. J Am Chem Soc 125:12443–12449
138. Zhang ZH, Burns DC, Kumita JR, Smart OS, Woolley GA (2003) A water-soluble azobenzene cross-linker for photocontrol of peptide conformation. Bioconjug Chem 14:824–829
139. Chi L, Sadovski O, Woolley GA (2006) A blue-green absorbing cross-linker for rapid photoswitching of peptide helix content. Bioconjug Chem 17:670–676
140. Burns DC, Flint DG, Kumita JR et al (2004) Origins of helix-coil switching in a light-sensitive peptide. Biochemistry 43:15329–15338
141. Standaert RF, Park SB (2006) Abc amino acids: design, synthesis, and properties of new photoelastic amino acids. J Org Chem 71:7952–7966
142. Zhang F, Sadovski O, Woolley GA (2008) Synthesis and characterization of a long, rigid photoswitchable cross-linker for promoting peptide and protein conformational change. Chembiochem 9:2147–2154
143. Feldman HJ, Hogue CW (2002) Probabilistic sampling of protein conformations: new hope for brute force? Proteins 46:8–23
144. Gellman SH (1998) Minimal model systems for beta sheet secondary structure in proteins. Curr Opin Chem Biol 2:717–725
145. Erdelyi M, Karlen A, Gogoll A (2005) A new tool in peptide engineering: a photoswitchable stilbene-type beta-hairpin mimetic. Chemistry 12:403–412
146. Krautler V, Aemissegger A, Hunenberger PH, Hilvert D, Hansson T, van Gunsteren WF (2005) Use of molecular dynamics in the design and structure determination of a photoinducible beta-hairpin. J Am Chem Soc 127:4935–4942
147. Espinosa JF, Gellman SH (2000) A designed beta-hairpin containing a natural hydrophobic cluster. Angew Chem Int Ed Engl 39:2330–2333
148. Cochran AG, Skelton NJ, Starovasnik MA (2001) Tryptophan zippers: stable, monomeric beta-hairpins. Proc Natl Acad Sci USA 98:5578–5583
149. Renner C, Kusebauch U, Loweneck M, Milbradt AG, Moroder L (2005) Azobenzene as photoresponsive conformational switch in cyclic peptides. J Pept Res 65:4–14
150. Guerrero L, Smart OS, Weston CJ, Burns DC, Woolley GA, Allemann RK (2005) Photochemical regulation of DNA-binding specificity of MyoD. Angew Chem Int Ed Engl 44:7778–7782
151. Guerrero L, Smart OS, Woolley GA, Allemann RK (2005) Photocontrol of DNA binding specificity of a miniature engrailed homeodomain. J Am Chem Soc 127:15624–15629
152. Jurt S, Aemissegger A, Guntert P, Zerbe O, Hilvert D (2006) A photoswitchable mini-protein based on the sequence of avian pancreatic polypeptide. Angew Chem Int Ed Engl 45:6297–6300
153. Kusebauch U, Cadamuro SA, Musiol HJ, Moroder L, Renner C (2007) Photocontrol of the collagen triple helix: synthesis and conformational characterization of bis-cysteinyl collagenous peptides with an azobenzene clamp. Chemistry 13:2966–2973
154. Boulegue C, Loweneck M, Renner C, Moroder L (2007) Redox potential of azobenzene as an amino acid residue in peptides. Chembiochem 8:591–594
155. Chabala LD, Lester HA (1986) Activation of acetylcholine receptor channels by covalently bound agonists in cultured rat myoballs. J Physiol 379:83–108
156. Volgraf M, Gorostiza P, Numano R, Kramer RH, Isacoff EY, Trauner D (2006) Allosteric control of an ionotropic glutamate receptor with an optical switch. Nat Chem Biol 2:47–52
157. Banghart M, Borges K, Isacoff E, Trauner D, Kramer RH (2004) Light-activated ion channels for remote control of neuronal firing. Nat Neurosci 7:1381–1386
158. Gorostiza P, Isacoff E (2007) Optical switches and triggers for the manipulation of ion channels and pores. Mol Biosyst 3:686–704
159. Mayer ML (2005) Glutamate receptor ion channels. Curr Opin Neurobiol 15:282–288

160. Gorostiza P, Volgraf M, Numano R, Szobota S, Trauner D, Isacoff EY (2007) Mechanisms of photoswitch conjugation and light activation of an ionotropic glutamate receptor. Proc Natl Acad Sci USA 104: 10865–10870
161. Kitchen DB, Decornez H, Furr JR, Bajorath J (2004) Docking and scoring in virtual screening for drug discovery: methods and applications. Nat Rev Drug Discov 3:935–949
162. Mayer ML (2005) Crystal structures of the GluR5 and GluR6 ligand binding cores: molecular mechanisms underlying kainate receptor selectivity. Neuron 45:539–552
163. Nanao MH, Green T, Stern-Bach Y, Heinemann SF, Choe S (2005) Structure of the kainate receptor subunit GluR6 agonist-binding domain complexed with domoic acid. Proc Natl Acad Sci USA 102:1708–1713
164. Armstrong N, Gouaux E (2000) Mechanisms for activation and antagonism of an AMPA-sensitive glutamate receptor: crystal structures of the GluR2 ligand binding core. Neuron 28:165–181
165. Numano R, Szobota S, Lau AY, Gorostiza P, Volgraf M, Roux B, Trauner D, Isacoff EY (2009) Nanosculpting reversed wavelength sensitivity into a photoswitchable iGluR. Proc Natl Acad Sci USA 106(16):6814–6819
166. Chang G, Guida WC, Still WC (1989) An internal coordinate Monte-Carlo method for searching conformational space. J Am Chem Soc 111:4379–4386
167. Jorgensen WL, Maxwell DS, Tirado-Rives J (1996) Development and testing of the OPLS all-atom force field on conformational energetics and properties of organic liquids. J Am Chem Soc 118:11225–11236
168. Still WC, Tempczyk A, Hawley RC, Hendrickson T (1900) Semianalytical treatment of solvation for molecular mechanics and dynamics. J Am Chem Soc 112:6127–6129

Chapter 14

Photoswitchable Ligand-Gated Ion Channels

Pau Gorostiza and Ehud Y. Isacoff

Abstract

Ligand-activated proteins can be controlled with light by means of synthetic photoisomerizable tethered ligands (PTLs). The application of PTLs to ligand-gated ion channels, including the nicotinic acetylcholine receptor and ionotropic glutamate receptors, is reviewed with emphasis on rational photoswitch design and the mechanisms of optical switching. Recently reported molecular dynamic methods allow simulation with high reliability of novel PTLs for any ligand-activated protein whose structure is known.

Key words: Nicotinic acetylcholine receptor, Kainate receptor, Glutamate receptor, Photoisomerizable tether ligand (PTL), Optical switch, Nanotoggle, Azobenzene, Neurobiology, Nanoengineering, Nanomedicine

1. Introduction

The development of methods for remote control of protein function should make it possible to model and understand how specific macromolecular signaling systems operate in their native environment, in protein complexes, organelles, cells, and tissues, up to entire organisms. These methods also open the door for building complex, programmable biochemical circuits in vitro and in synthetic cells that could be exploited technologically and medically (1). The unique properties of light (noninvasiveness, ease of achieving precise spatiotemporal control, and orthogonality to biological stimuli in most instances) have made it a successful means of transmitting control signals to proteins in biological systems. This chapter describes the chemical genetic engineering of optical control into the function of ligand-gated receptors using synthetic photoisomerizable tethered ligands (PTLs).

PTL compounds or "optical nanotoggles" contain at one end a ligand (agonist, antagonist, or blocker), at the other end a

James J. Chambers and Richard H. Kramer (eds.), *Photosensitive Molecules for Controlling Biological Function*, Neuromethods, vol. 55, DOI 10.1007/978-1-61779-031-7_14, © Springer Science+Business Media, LLC 2011

reactive group that mediates covalent attachment to the protein, and in the middle a linker containing a photoisomerizable moiety, which changes its geometry in response to light. The entire compound is attached site specifically to the target protein at a precise position near the ligand-binding site. Receptor switching is based on the differential ability of *trans* and *cis* isomers of the linker to permit ligand binding.

Many photoisomerizable compounds are known, but so far only azobenzene (2, 3) has been used in PTLs due to its chemical robustness (low photobleaching rate) and large end-to-end length change upon isomerization – an important requirement of PTLs (4). The dark, thermally relaxed *trans* isomer of azobenzene adopts an extended configuration that is ~0.7 nm longer than the higher energy *cis* or "bent" isomer. Illumination with near-ultraviolet (UV) irradiation (~380 nm) leads to accumulation of the *cis* isomer, and visible light irradiation (500–600 nm) switches it back to the *trans* form. Photoisomerization cycles can be repeated many times, with the population of azobenzene molecules typically displaying a maximum of ~85% *cis* at the optimal UV wavelength, and a minimum of ~15% *cis* at the optimal visible wavelength, with full conversion to the *trans* form only after thermal relaxation. Wavelengths between the optima lead an ensemble of molecules to distribute themselves in proportional fractions between the two states, resulting in a wavelength-dependent photostationary state (2).

Although most PTL developments have been carried out on membrane-bound receptors including ionotropic acetylcholine receptors (nAChRs) (5) and ionotropic glutamate receptors (iGluRs) (6) (see Fig. 1), the only requisite is ligand gating and thus the strategy has been successfully extended to soluble proteins (7). PTLs of nAChRs and iGluRs will be presented in the next sections, first introducing the current understanding of their structure and gating mechanisms, and then describing the reported PTL compounds, their studies, and applications. The mechanisms of receptor photoswitching and conjugation of the photoswitch to the receptor will be reviewed in the following section, with special emphasis on the development of new computational methods for PTL design. The final section will discuss the modular structure of the compounds (reactant-switch-ligand) and the opportunities to engineer each of the modules independently in order to customize the performance of the PTL.

2. Structure and Gating of nAChRs

Nicotinic acetylcholine receptors are integral membrane, ligand-gated ion channels in the central and peripheral nervous systems and in muscle, which mediate fast responses to synaptically released

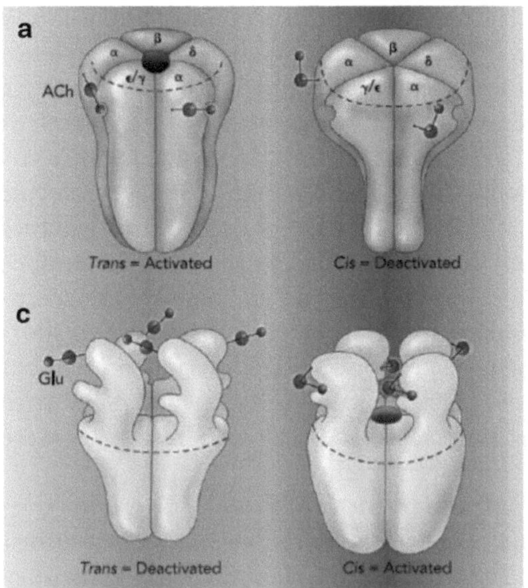

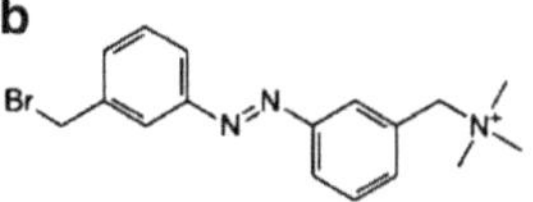

Fig. 1. Photoswitching of ligand-gated ion channels using photoisomerizable tethered ligands. (**a**) Optical control of nAChR is achieved by attaching a choline agonist group via a photoisomerizable tether near the ligand-binding sites of receptor (5). Compound QBr (**b**) was the first reported example of the PTL approach. The receptor is activated with azobenzene in *trans* (under visible light, *left side* of the figure) and deactivated in *cis* (under UV light, *right side* of the figure). (**c**) In LiGluR, the photoisomerizable tethered glutamate agonist MAG (**d**) is attached near the ligand-binding sites in receptor (6). The receptor is deactivated with azobenzene in *trans* and activated in *cis*. Reproduced with permission from (62).

acetylcholine (for reviews, see (8, 9)). They are pentameric proteins belonging to the family of Cys-loop receptors together with serotonin, glycine, $GABA_A$, and $GABA_C$ receptors, which possess a characteristic loop formed by a disulfide bond between two cysteine residues. The neurotransmitters bind to these receptors at the interface between subunits in the extracellular domain, specifically at two sites located in the clefts between subunits α–δ, and α–ε/γ in each nAChR, as indicated in Fig. 1a. Ligand binding induces structural changes in the membrane-spanning domain and opens the transmembrane pore. Nicotinic AChRs are cation-selective ion channels, which permeate Na^+, K^+, and Ca^{2+} and evoke membrane depolarization. Prolonged exposure to agonist leads the receptors to enter a closed desensitized state. Insights into the structural basis of nAChRs' function have come from electron diffraction (10, 11) and from the X-ray structure of AChBP (acetylcholine binding protein), a soluble protein homolog of nAChR extracellular domain (12), as well as from recent X-ray structures of prokaryotic homologs ELIC and GLIC, which have provided a high resolution view of the receptor and the conformational changes of gating (13, 14).

2.1. PTLs of nAChR

More than 35 years before the structures of nAChR were obtained, a quest was launched to make synthetic ligands and modulators of the nAChR that could help in their functional characterization.

Among the compounds developed, tethered agonists were found to conjugate to native cysteines in the vicinity of the acetylcholine-binding site of the receptor, yielding permanently active receptors with the degree of activation being proportional to tether length (15, 16). Introducing an azobenzene into the tether, while maintaining the extended length of the molecule (QBr, Fig. 1b), made it possible to use light to control whether the agonist could "reach" its binding site. In the *trans* (extended) configuration, the ligand was able to bind and activate the receptor, while in the *cis* (bent) configuration, the tether was too short for the agonist to bind (5). Photoisomerization and subsequent induction of nAChR currents could be achieved reversibly with millisecond long light pulses (17–19). Optical responses were studied in detail and compared to those of a free photoisomerizable ligand (bis-Q) (18) and to those of a non-PTL (bromoacetylcholine) (17). Macroscopic and single-channel currents and their pharmacology were characterized together with the functional stoichiometry of tethered QBr, which corresponded to one QBr molecule per receptor. In these studies, the photoisomerizable tether length was not varied and conjugation sites other than the native cysteines were not accessible. The conjugating residues were later identified to be C192 or C193 (20, 21). With the ensuing discovery of both homomeric nAChRs and slowly desensitizing mutants (22), the way is now open to reengineer the photoswitchable nAChR more systematically.

The PTL of nAChR was studied in primary cell cultures (rat myoballs) (17) and in tissue preparations (electroplaques) (18). However, because of the need to reduce the disulfide bond to open up the native cysteines to QBr conjugation, this system was not suited for use in vivo.

3. Structure and Gating of iGluRs

Glutamate receptor ion channels mediate responses at the majority of excitatory synapses in the central nervous system (for review, see (23)). They form four major families: the AMPA, kainate, NMDA, and delta receptors, all of which have a tetrameric architecture with one extracellular ligand-binding domain (LBD) in each subunit (Fig. 1c). The transmembrane region has not been crystallized but it is known to be a homolog of the KcsA potassium channel, although it has an inverted topology the structure of an intact ionotropic glutamate receptor was reported by (63). The LBD structures have been solved for AMPA (24), kainate (25), and NMDA (26, 27) receptors, revealing nearly identical folds. The ligand-binding core of each subunit resembles an open clamshell in the absence of an agonist (28).

The LBDs assemble as dimers, and thus the tetrameric channel is thought to function as dimer of dimers (29). Agonist-stabilized domain closure of the individual subunits in a dimer leads to channel opening, with the degree of closure in the agonist determining the efficacy of the ligand in opening the channel, and partial LBD closure producing partial agonism (30). Desensitization results from a weakening of dimer contacts that enables the channel to close even though the individual LBDs remain in the active, agonist-bound conformation (29). Recently, membrane proteins have been found to interact with AMPA (TARPs) (31) and kainate (NETO2) (32) receptors and to modify both their trafficking and gating properties. The interaction of AMPARs with cornichon CNIH proteins was also discovered early this year (33), providing an even more elaborate perspective on the protein complexes built around glutamate receptors, and opening the possibility that optical control could be engineered not only into the gating control mechanism of the LBD of the channel forming subunit, but also, perhaps, into regulatory proteins.

3.1. PTLs of GluRs

In contrast to the photoswitchable nAChR described above, photoswitchable glutamate receptors were designed directly on the basis of the structure of the LBD (25, 28) without previous knowledge of tethered ligands (Fig. 1d). A systematic method (outlined in Fig. 2 in the form of a flow diagram) was developed for the empirical structure-based design of PTLs (4). The approach was used with the homotetrameric iGluR6 kainate receptor, but, in principle, the approach is expected to generalize to other ligand-activated proteins of known structure.

Examination of the structure of the glutamate-binding site of iGluR6 with an agonist bound (25) reveals a narrow, solvent-exposed "exit tunnel" that could potentially host a molecular tether to the ligand. In order to test whether a tether could indeed occupy this tunnel without interfering with the conformational changes of the LBD, a tethered agonist model compound can be designed. In some cases, such compounds already exist or else structure activity relation studies can provide clues about their viability (e.g., affinity labels (34), biotinylated or fluorescent ligands that are known to function as agonists). In iGluR6, a free (not attaching) "tether model" compound was based on the known agonists (2S,4R)-4-allyl glutamate 35 and (2S,4R)-4-methyl glutamate with which the LBD was co-crystallized (25). The structure and position of the agonist in the LBD suggested a site for extending a tether off the glutamate without altering affinity or potency (6).

Once a model compound has been identified and shown to act as a potent agonist (i.e., to drive the conformational change that activates the LBD), then a complete PTL can be synthesized.

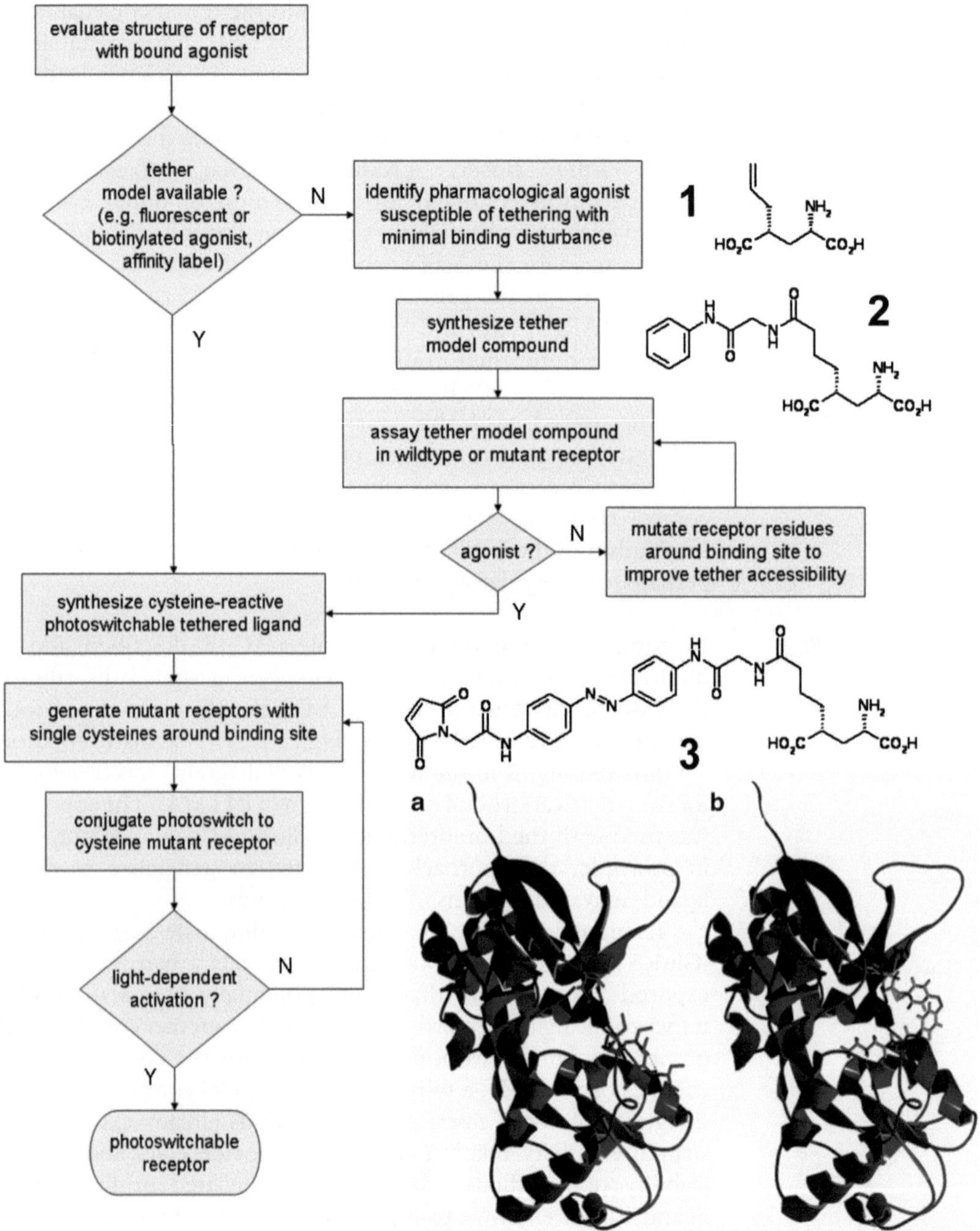

Fig. 2. Flow chart describing a rationally designed approach to receptor photoswitching using PTLs. The method was first applied to a glutamate receptor (6) based on the known agonist (2S,4R)-4-allyl glutamate (**1**) (35). A tether model compound (**2**) was designed, leading to the full photoswitchable tethered agonist MAG (**3**). In order to find the optimal attachment site for MAG, receptor mutants were generated bearing a single cysteine at different positions near the glutamate-binding site (eight cysteine side chains shown simultaneously in *inset* ***a***). When conjugated to the optimal site (residue 439, *above*), the glutamate end of *cis*-MAG docks perfectly in the binding of the receptor (*inset* ***b***). Reproduced with permission from (4).

For iGluR6, the first PTL was a maleimide–azobenzene–glutamate (MAG) of intermediate length (MAG1) (6). This was followed by two homologous compounds with longer (MAG2) (36) and shorter linkers (MAG0) (37). These compounds can be used to screen a series of single cysteine mutants of the receptor at positions around the ligand-binding site, in order to determine the optimum location to attach the PTL. In iGluR6, 16 exposed residues were individually substituted with cysteine, of which 11 yielded functional receptors, enabling the evaluation of photoresponses due to each of the three MAGs (6, 36, 37). This screen identified six mutants that showed optical responses to one or more MAGs. Most of them were activated more strongly by the *cis*-MAGs than by the *trans*-MAGs, and, of these, L439C displayed the largest responses and was extensively characterized (6, 36). Interestingly, site 486 and two nearby sites in the upper lobe, 482 and 484, displayed stronger activation by *trans*-MAG0 than by *cis*-MAG0. This compound also shows the largest reported optical change in local effective concentration of glutamate (see next section), which is 123-fold higher in *trans* than in *cis* (37). Overall, these results provided a unique workbench to understand the mechanisms of receptor photoswitching by PTLs, as well as to illustrate the remarkably high success rate of structure-based PTL nanoengineering: optical responses were achieved for roughly half of the cysteine attachment sites attempted, with 60% of them displaying strong differences between *cis* and *trans* for at least one of the MAGs, and 40% of them showing maximal responses (to a theoretical maximum of 75% of the saturating glutamate response for a tetramer with a maximum of 93% *cis*-azobenzene at the optimum wavelength of 380 nm) (36). Since the geometry of the iGluR LBD is challenging, with the deep narrow access to the ligand-binding site in the bound state, one would expect that the success rate in other proteins may be even higher.

The exit tunnel found in GluR6 LBD is also present in the structures of GluR5 (25), GluR0 (38), and the NR2a subunit of NMDA receptors (27), making them good candidates for photoswitching using MAG compounds. It was recently reported that the tether model compound (2) of Fig. 2 does not activate GluR0 (64). However, the authors were able to replace the GluR0 LBD by that of LiGluR and obtained a potassium-selective, light-gated channel (HyLighter). Indeed, the tether model compound used in GluR6 acts as a partial agonist in GluR5, but initial experiments using MAG1 and the 455C site homolog of 439C in GluR6 only produced weak photoswitching responses with MAG3 (39). Better success may be obtained at neighboring attachment sites, or with other MAG compounds.

In contrast, the exit tunnel is absent in the structure of glutamate-bound LBD of the iGluR2 subunit of the AMPA receptor (24), and, indeed, the tether model compound has no agonist effect on homotetramers of this receptor. As indicated in the flow diagram of Fig. 2, it may be possible to engineer an exit tunnel from the ligand-binding site to bulk water by mutating side chains that are in the way. In iGluR2, these are residues 471 and 671 on the lips of the LBD clamshell. In fact, the triple iGluR2 mutant 422C 471F 671V partially responds to the tether model, although photoswitching has not yet been obtained for MAG (39). Here too, further work should make it possible to identify a suitable anchoring residue that supports in photoswitching.

The optical activation of iGluR6 induces large depolarizations of HEK293 cells and of cultured primary neurons, and can drive neurons to fire action potentials at high frequencies (40). Action potentials can be triggered by millisecond long light pulses with a high reliability and in reproducible spatial and temporal patterns (40, 41). This can be used to drive activity in specific neurons in free-swimming zebrafish larvae, and to, thereby, gate the behaviors that those cells underpin (40). Given the distinct geometric constraints from MAGs tethered at different positions, a single MAG (MAG0) is able to activate the receptor at 380 nm when attached at one site, while at a second attachment site it deactivates the receptor at that wavelength (37). As a result, neurons in a culture that express either iGluR6 439C (*cis*-on with MAG0) or iGluR6 486C (*trans*-on with MAG0) have opposite responses to light, with the first turning on and firing neurons when the second turns off, and vice versa. In this way, opponency could be engineered in two cell populations which selectively express one or another of the variant iGluR6s but are exposed to a single PTL.

4. Multiple Factors Drive Switching of PTLs

The mechanism of photoswitching appeared to follow straightforward rules for the first PTL designs reported for the agonist of the nAChR (5) and the blocker of the K^+ channels (42), with the ligand binding only in the *trans* or extended configuration of azobenzene in agreement with the "reach" model demonstrated for non-PTLs (i.e., with an effect only in the extended form of the linker because only in this configuration can the ligand reach the binding site, otherwise it is too short) (15, 16) (Fig. 3a). The properties of PTLs and the mechanism of receptor photoswitching were also investigated using free agonists, and competitive and noncompetitive antagonists. In nAChR, *trans*-QBr inhibits the binding of the agonist carbamylcholine since the two

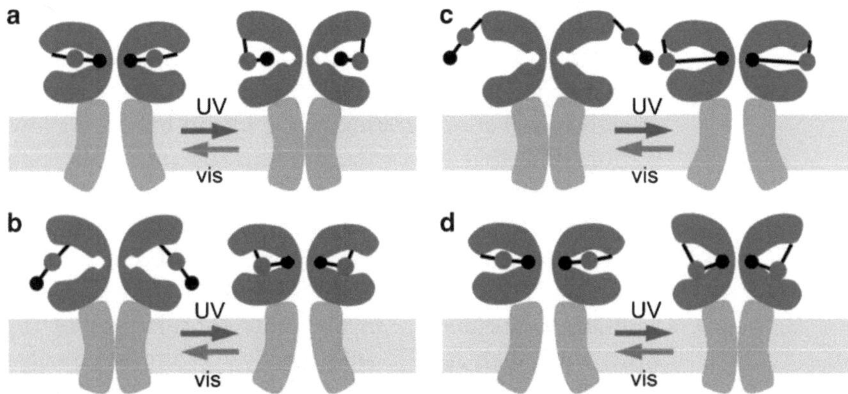

Fig. 3. Models of state-dependent liganding by PTLs, illustrated for MAG and iGluR6. (**a**, **b**) Line-of-sight and matched length. Photoisomerization adjusts the length of MAG, enabling the glutamate end to reach the binding site. (**c**) Turning the corner. The bent state is required to point the glutamate end into the deep binding pocket from the superficial attachment position. (**d**) Even when glutamate points into the binding pocket, certain linker geometry may interfere with clamshell closure. Reproduced with permission from (37).

compete for the same binding site (5). Similarly, the competitive antagonist d-tubocurarine (dTC) inhibits the QBr photoresponse (18), but its sensitivity is lower than that of the free photoswitchable agonist bis-Q (5). This difference was attributed to the higher local effective concentration of the conjugated QBr (43), whose ligand end is confined to a volume of a few nm^3. On the contrary, nAChR photoresponses generated by QBr and bis-Q were blocked by the same concentrations of a pore blocker, QX-222, as expected for the pore located far from the ligand-binding site (18).

The first reports on the photoswitching behavior of iGluR6-L439C-MAG1 were opposite to what had been observed with the nAChR, with the ligand binding in *cis* and unbinding in *trans* (6). This behavior was attributed to the complex geometry of the iGluR LBD. A model of MAG1 docked into the iGluR6 LBD showed that the switch is anchored at the lip of the LBD clamshell and the ligand must "turn a corner" in order to bind (6). The mechanism of photoswitching was investigated by inhibiting the responses to UV and visible light using the competitive antagonist DNQX (36). Inhibition of photoresponses was found to be proportional to DNQX concentration and could be compared to the DNQX inhibition of responses to a range of concentrations of free agonists, notably the tether model compound whose affinity is expected to resemble that of MAG1. DNQX competition of MAG1 photoresponses, calibrated to the tether model, yielded estimates of effective local concentrations of 12 mM for *cis*-MAG1 and 0.3 mM for *trans*-MAG1, generally agreeing with the smaller spatial volume sampled by the shorter *cis* configuration (36). The longer compound MAG2, which is expected to sweep through a larger volume, had a threefold lower effective local concentration in the *cis* state and a lower basal activation in the *trans* state (36). This led to the conclusion that in iGluR6-439C, the change in azobenzene length of MAG1 and MAG2 allows toggling of the effective local concentration of the ligand between a low concentration state that leaves the binding site practically unoccupied, and a higher concentration state that saturates the binding site, thus enabling robust activation of the channel.

However, further studies of more MAGs at a larger number of positions (37) revealed that while some sites, such as 439, activate almost exclusively in one state for all three of the MAGs, other sites, such as 482, are strongly and almost equally activated by both isomers for all three of the MAGs. Site 439 emerged as the best of the *cis* activators and site 486 as the best of the *trans* activators, while the shortest MAG, MAG0, showed the biggest isomer preference, with a 123-fold higher DNQX IC50 in *cis* than in *trans* at 439C.

In order to rationalize all these results and to build a photoswitching model that can become a predictive tool in future PTL designs, simple geometric assumptions must be replaced by structure-based simulations of ligand docking and LBD clamshell closure (37). Recently, molecular dynamics calculations were used to compare the relative free energy of undocked MAG0 and MAG0 docked in the iGluR6-binding site, in order to obtain a measure of the probability of finding MAG0 in those states. The iGluR6 LBD was constrained to an open conformation and the MAG0 azobenzene was constrained to either a *cis* or a *trans* configuration. These simulations showed that *cis*-MAG0 is more likely to bind than *trans*-MAG0 when attached not only to 439C, as expected, but also to 486C, despite its better experimental activation in *trans*. Further modeling revealed that the degree of LBD closure, which is proportional to the degree of receptor activation (30) and which increases the apparent agonist-binding affinity (44), could be influenced by the configuration of the linker. In 439C, the extent of LBD closure calculated for *cis*-MAG0 was higher than that for *trans*-MAG0, reinforcing the extra agonist potency of the *cis* state. However, in iGluR6 486C, the opposite result was obtained and although the *cis* configuration permitted better docking, the *cis* linker interfered with LBD closure. Thus, one can account for the polarity of photoswitching of iGluR6 by MAG by taking into account two factors: the isomer dependence of ligand docking into the open LBD (Fig. 3a) and the ability of the LBD clamshell to close on the bound ligand, given the disposition of the linker (Fig. 3d), with greater closure increasing both channel activation and ligand-binding free energy.

It would be interesting to assess these design principles by testing several cysteine mutants and PTLs in nAChR, whose ligand-binding site likely has a simpler geometry than iGluRs, as suggested by the structure of AChBP (45). In addition, the

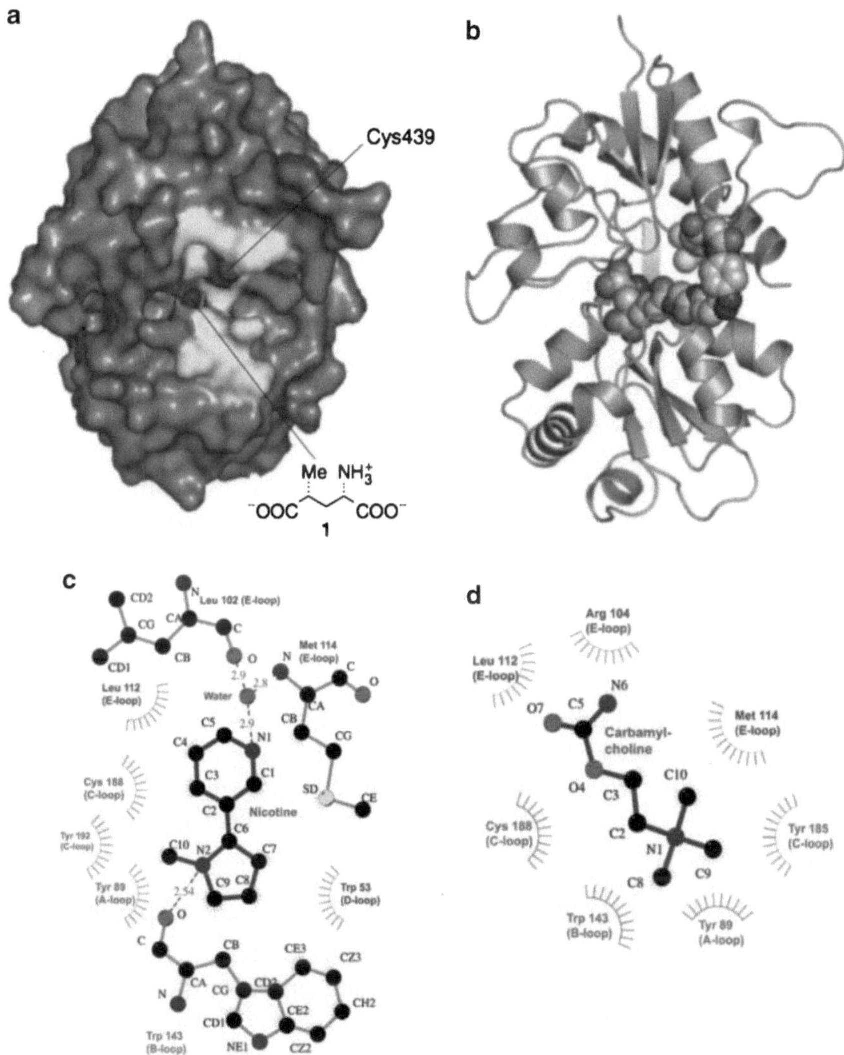

Fig. 4. Structures and fit of photoswitched agonist and iGluR6 LBD. (**a**) View looking into the "mouth" of iGluR6 LBD in complex with (2S,4R)-4-allyl glutamate (25). Residues of clamshell "lips" that were individually mutated to cysteine are *highlighted*, and the position 439 is indicated. The methyl group of (2S,4R)-4-allyl glutamate can be seen at the bottom of the "exit channel." Reproduced with permission from (6). (**b**) Docking model of MAG in the *cis* state attached at L439C and bound to the activated (closed) conformation of the LBD. Reproduced with permission from (6). The crystal structure of AChBP in complex with nicotine (**c**) and carbamylcholine (**d**) shows the interactions of these agonists with the surrounding residues in the ligand-binding site (45), and provides insights into how QBr (Fig. 1b) may bind to its site in nAChR. Reproduced with permission from (45).

location of nicotine and carbamylcholine binding in AChBP provides insights into how QBr may bind to its site in nAChR (Fig. 4). The tetramethyl ammonium end group of QBr is most likely bound as in carbamylcholine, while the proximal ring of azobenzene probably interacts in a similar way to the pyridine in nicotine (45). The participation of azobenzene in ligand binding would be in contrast to that of MAG1 in iGluR6-L439C, whose longer tether projects azobenzene out of the LBD according to a docking model (6). However, it cannot be ruled out that azobenzene interacts with the receptor for shorter tethers like that of MAG0 or attachment sites other than 439 (37).

Another aspect of the light-gating mechanism that has been studied is the stoichiometry of the PTL and receptor, i.e., how many switches control the activation of a single channel. In nAChR, only one QBr molecule is conjugated per receptor, as was reported for non-photoswitchable tethered ligands (16, 46, 47), in contrast with the two free agonist molecules that are required to activate the receptor fully (17, 18). In iGluR6, comparison of the agonist- and light-induced currents after saturating the conjugation of MAG yielded a degree of agonism that approximated the photostationary state at the optimal wavelength for isomerizing to the *cis* conformation (36), suggesting that one MAG molecule is attached to each subunit of the tetrameric receptor.

Just as macroscopic devices can be improved by adjusting their parts, it is possible to tune PTL nanodevices rationally by adapting the molecular modules (ligand, switch, and reactive group) to suit different requirements. This is an intrinsic advantage of PTLs over naturally photosensitive proteins such as channelrhodopsin-2, where the optical switch (retinal) is snugly fit in a deep protein pocket and performance improvements have been obtained by finding suitable homologous proteins (48) or by modifying the protein itself (49).

In PTLs, the identity of the ligand can be changed based on known pharmacological compounds, which allow varying affinity and efficacy. The length and probably the molecular rigidity of the tether provide additional design parameters. It has been shown that increasing MAG tether length leads to a reduced local effective concentration of the glutamate ligand when conjugated to iGluR6-L439C (36), and an interplay of MAG tether length and attachment site can result in some cases in partial antagonism, where the ligand is bound but does not allow closure of the LBD (37). Furthermore, certain combinations of tether length and attachment site favor ligand binding with azobenzene in *cis* and unbinding in *trans*, while others result in the opposite behavior (binding in *trans* and unbinding in *cis*) (37). In principle, it should also be possible to adjust the character of the ligand by tethering antagonists, but so far progress in this direction in iGluR6 has been hindered by low affinity and solubility of DNQX derivatives (50).

Since the chromophore in PTLs is usually exposed to the solution and is not sterically restricted, there is much room for adjusting its chemical and optical properties. In azobenzene, charged substituents can be added to improve solubility (51) and to tune the absorption spectrum of the *trans* form finely (52), usually to shift the absorption maximum to longer wavelengths that penetrate tissue better and damage it less. These manipulations often reduce the stability of the *cis* state (53), but recently blue-absorbing azobenzenes with relaxation rates of a few seconds have been reported (54). As a comparison, UV-absorbing azobenzenes can reach relaxation rates of about 2 days (55). The modular nature of PTLs should actually make it possible to replace azobenzene with another chromophore. Several candidates exist and have been used in other contexts, including spiropyran (56), hemithioindigo (57), and stilbene (58), each having distinct properties that might make it possible to achieve bistable, red-shifted switching.

The reactive moieties in the PTLs of nAChR and iGluR were designed to be thiol reactive in order to target them to extracellular cysteine residues in the receptor, either native (as in nAChR) or introduced by site-directed mutagenesis (iGluR). Since in both cases the ligand-binding site and the attachment site are located in the extracellular domain, membrane-impermeant PTL compounds were simply perfused over cells expressing the receptors. This likely leads to conjugation to all of the exposed cysteine residues both in the target channel and in other externally exposed proteins. However, the requirement of a precise geometric relation between the attachment site and the ligand-binding site provides for an extremely high selectivity for optical gating, and light responses are obtained only at locations in close proximity and proper orientation to the ligand-binding site.

One expectation for the conjugation process is that the ligand end of the free (as yet unbound) PTL will bind to the ligand-binding site and place the reactive end near the introduced cysteine anchoring site. The proximity of the reactive maleimide to the cysteine will be greater in the same configuration that favors ligand binding after cysteine conjugation. Thus, two regimes arise, depending on the affinity of the ligand relative to the concentration of the free PTL. At high PTL concentrations, conjugation kinetics is usually dominated by the direct reaction, but at lower concentrations, conjugation can be enhanced by docking of the ligand at the binding site, which places the reactive end near the cysteine. The latter process is known as affinity labeling, and is widely employed in pharmacological research to identify drug targets and to localize binding sites (34). All PTLs reported to date can be conjugated by affinity labeling, including the ligand-gated receptors reviewed here, as well as voltage-gated ion channels (see Chap. 11). In nAChR, it was found that 100 nM *trans*-QBr (active isomer) conjugates much faster than *cis*-QBr (inactive isomer) (18). A similar effect was observed in LiGluR, where 100 nM *cis*-MAG (active isomer) conjugated more efficiently than *trans*-MAG (Fig. 5a, c) (36). In addition, as noted in

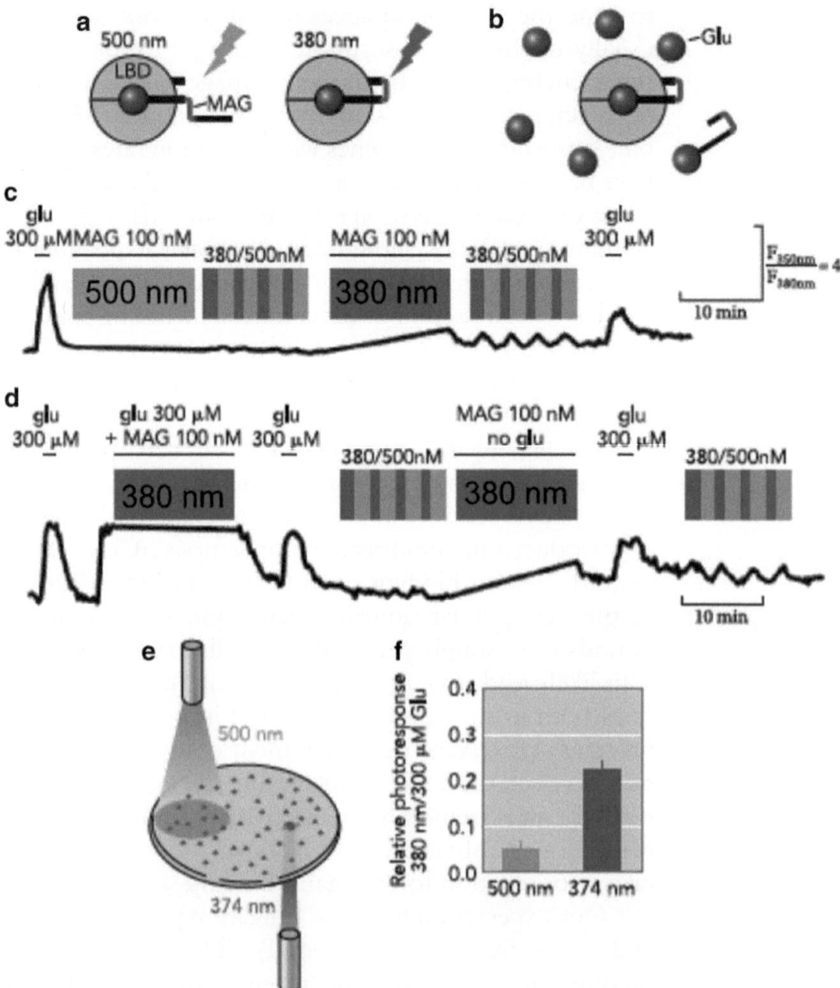

Fig. 5. PTL conjugation occurs by affinity labeling. MAG-1 conjugation to iGluR6-L439C can be interfered by two means (36): (**a**) favoring the *trans* configuration with 500 nm illumination, which puts the maleimide group away from cysteine 439 when the glutamate is bound to the LBD and (**b**) occupying the binding site with a competing concentration of glutamate, thus preventing docking of MAG1 in a configuration that favors conjugation of the maleimide to the introduced cysteine. (**c**) Photoresponses obtained by calcium imaging in cells transfected with iGluR6-L439C and conjugated to MAG1 under the conditions shown in (**a**). Weak responses are obtained after MAG-1 conjugation at 100 nM under visible illumination (*trans*-MAG; maleimide group facing away from cysteine 439), but a substantial increase in photoresponses is observed after conjugation under UV (*cis*-MAG; maleimide group facing cysteine 439). (**d**) Weak responses are obtained after MAG-1 conjugation at 100 nM in the presence of 300 μM glutamate (ligand-binding pocket occupied), but they are significantly increased after MAG-1 conjugation at 100 nM in the absence of glutamate (binding pocket free to dock glutamate end of MAG1 and place maleimide near cysteine 439). (**e**) Spatial patterning of MAG conjugation with patterned illumination ("optical lithography") (36). Cells are adhered on a glass coverslip and exposed to MAG under an illumination pattern consisting of a 0.5-mm spot of 374 nm light and 500 nm light on the remaining area of the coverslip. After a 10-min exposure to MAG and the designated light pattern, MAG was washed away before subsequent patch clamp recording. (**f**) *Bar graph* showing amplitude of inward currents evoked by illumination at 380 nm normalized to amplitude of current evoked by 300 μM glutamate. Photocurrents are ~4.5-fold larger in cells illuminated at the shorter wavelength during MAG exposure, indicating that affinity labeling could be biased spatially with patterns of light. Reproduced with permission from (36).

the previous section, *trans*-QBr competes with carbamylcholine for binding to nAChR (5), and conjugation of *trans*-QBr is inhibited by the competitive antagonist dTC (18). Correspondingly, in iGluR6-439C, free glutamate inhibits the conjugation of *cis*-MAG (active isomer) (36) by occupying the binding pocket and preventing docking of the glutamate end of MAG (Fig. 5b, d). A similar competition for the binding site was previously found in the non-photoswitchable tethered compounds upon which photoswitchable nAChRs were based (46, 47). It is surprising that QBr and MAG display affinity labeling within the same concentration range (100 nM), although their properties suggest a much stronger effect in QBr: The affinity of QBr is higher than that of MAG, as can be, respectively, estimated from the related compound bisQ (~0.1 μM) (5) and the "tether model" compound (~200 μM) (6). In addition, MAG is about twice as long as QBr, which should produce a much lower local effective concentration of the reactive end when the ligand is bound. One reason for this apparent compensation may be the lower cysteine reactivity of the bromomethyl group in QBr compared to that of the maleimide in MAG. It is also interesting to compare with affinity labeling conditions of MAQ/AAQ in voltage-gated K^+ channels, which span up to the mM range (59) despite the similar length of MAQ and MAG. The reason in this case may be the very low affinity (>150 mM) (60, 61) of the tetraethyl ammonium end in the Shaker T449V mutant used in these experiments (42). Interestingly, optical responses (and presumably affinity labeling characteristics) are different in channel mutants with altered affinity to TEA binding (42). It has been suggested that AAQ photomodulates Shaker potassium channels by reversibly binding to the internal TEA site rather than by covalent conjugation to the extracellular side, i.e. it acts as a photochromic ligand (PCL) instead of a PTL (65). No modeling or simulation of affinity labeling has been yet reported, but when it is available, it should be useful for rationally designing this useful feature of PTLs.

The state dependence of affinity labeling (preferred conjugation for the activating isomer) was used to pattern MAG conjugation over cells expressing iGluR6-L439C optically, such that in regions illuminated with 380 nm light, the activating *cis*-MAG isomer was favored, thus favoring conjugation, whereas in regions exposed to 500 nm light, *trans*-MAG was favored, thereby hindering conjugation (Fig. 5e, f) (36). Affinity labeling can be further exploited when the requirement for selective chemistry is reduced and the PTL conjugates to the native amino acids in the protein around the ligand-binding site. Thus reactive groups that are more promiscuous than maleimide (such as acrylamide or epoxide) can be used, rendering the introduction of cysteines unnecessary and opening the possibility of using wild-type proteins (see photoisomerizable affinity labels (PALs) in Chap. 11).

Indeed, the "optical lithography" method (selective light-patterned regional favoring of PTL conjugation) applied to the PTL blocker of K^+ channels produced an enhanced contrast between regions illuminated at 380 nm (to favor the non-liganding *cis* state) and ones that were allowed to remain fully in the *trans* state in the dark (59).

5. Conclusion

Synthetic photoswitches of protein function have come of age. While the first reports of light-activated proteins relied on geometrically simple and readily accessible binding sites and available native cysteines for PTL attachment, it is now possible to design and simulate in silico with high reliability novel PTLs for any ligand-gated protein whose structure is known. Using molecular dynamics tools, the reach, local effective concentration, and accommodation to geometric constraints of the PTL can be evaluated and scored under different optical configurations, as well as how they allow subsequent conformational rearrangements of the protein. In iGluRs, channel activation by an optical switch is found to depend on two main factors: (a) the fraction of rotamers of an anchored MAG that enables the glutamate to orient in the binding site of the open LBD and (b) the ability of the LBD clamshell to close, given the location of the linker.

The modular structure of PTL compounds (reactive group-switch-ligand) makes it possible to adjust each of the modules by design independently. The ligand affinity and character (agonist, antagonist, blocker, or modulator), tether length, conjugation selectivity, anchoring site, photoswitch absorption spectra, and thermal relaxation rates can all be tuned to meet specific requirements. In addition, PTLs conjugate to the receptor following an affinity labeling process, which can be exploited to achieve region-selective conjugation with light, and receptor-specific conjugation without the need for selective reactive groups.

Our fundamental understanding of the nanoscale mechanisms underlying light activation has greatly advanced in recent years following the detailed characterization of photoswitchable nAChRs, K^+ channels, and iGluRs, providing a solid basis for further nanoengineering. The development of the optical engineering methods and the wide variety of applications for the study of the photoswitched proteins and for applications in neurobiology suggest that these approaches for remote control of biological function will continue to expand to other areas of biomedicine.

Acknowledgments

P.G. is supported by the Human Frontier Science Program (HFSP) through a Career Development Award, by the European Research Council (ERC) through a Starting Grant, by the FET-ICT programme of the European Commission and by the Ministry of Science and Innovation (Spain). This work was supported by the NIH Nanomedicine Development Center for the Optical Control of Biological Function (5PN2EY018241) and by Human Frontier Science Program Grant RPG23-2005. The authors are grateful to H. Lester for providing useful references and comments.

References

1. Gorostiza P, Isacoff EY (2008) Optical switches for remote and noninvasive control of cell signaling. Science 322(5900):395–399
2. Rau H (1990) Azo compounds. In: Dürr H, Bouas-Laurent H (eds) Photochromism: molecules and systems. Elsevier, Amsterdam, pp 165–192
3. Rau H (1990) Photoisomerization of azobenzenes. In: Rabek JF (ed) Photochemistry and photophysics. CRC Press, Boca Raton, FL, pp 119–142
4. Gorostiza P, Isacoff E (2007) Optical switches and triggers for the manipulation of ion channels and pores. Mol Biosyst 3(10):686–704
5. Bartels E, Wassermann NH, Erlanger BF (1971) Photochromic activators of the acetylcholine receptor. Proc Natl Acad Sci U S A 68(8):1820–1823
6. Volgraf M, Gorostiza P, Numano R, Kramer RH, Isacoff EY, Trauner D (2006) Allosteric control of an ionotropic glutamate receptor with an optical switch. Nat Chem Biol 2(1):47–52
7. Harvey JH, Trauner D (2008) Regulating enzymatic activity with a photoswitchable affinity label. Chembiochem 9(2):191–193
8. Karlin A (2002) Emerging structure of the nicotinic acetylcholine receptors. Nat Rev Neurosci 3(2):102–114
9. Lester HA, Dibas MI, Dahan DS, Leite JF, Dougherty DA (2004) Cys-loop receptors: new twists and turns. Trends Neurosci 27(6): 329–336
10. Miyazawa A, Fujiyoshi Y, Unwin N (2003) Structure and gating mechanism of the acetylcholine receptor pore. Nature 423(6943): 949–955
11. Unwin N (2005) Refined structure of the nicotinic acetylcholine receptor at 4 Å resolution. J Mol Biol 346(4):967–989
12. Smit AB, Brejc K, Syed N, Sixma TK (2003) Structure and function of AChBP, homologue of the ligand-binding domain of the nicotinic acetylcholine receptor. Ann N Y Acad Sci 998:81–92
13. Hilf RJ, Dutzler R (2008) X-ray structure of a prokaryotic pentameric ligand-gated ion channel. Nature 452(7185):375–379
14. Hilf RJ, Dutzler R (2009) Structure of a potentially open state of a proton-activated pentameric ligand-gated ion channel. Nature 457(7225):115–118
15. Karlin A, Winnik M (1968) Reduction and specific alkylation of the receptor for acetylcholine. Proc Natl Acad Sci U S A 60(2):668–674
16. Silman I, Karlin A (1969) Acetylcholine receptor: covalent attachment of depolarizing groups at the active site. Science 164(886): 1420–1421
17. Chabala LD, Lester HA (1986) Activation of acetylcholine receptor channels by covalently bound agonists in cultured rat myoballs. J Physiol 379:83–108
18. Lester HA, Krouse ME, Nass MM, Wassermann NH, Erlanger BF (1980) A covalently bound photoisomerizable agonist: comparison with reversibly bound agonists at *Electrophorus electroplaques.* J Gen Physiol 75(2):207–232
19. Lester HA, Nass MM, Krouse ME, Nerbonne JM, Wassermann NH, Erlanger BF (1980) Electrophysiological experiments with

photoisomerizable cholinergic compounds: review and progress report. Ann N Y Acad Sci 346:475–490

20. Kao PN, Dwork AJ, Kaldany RR et al (1984) Identification of the alpha subunit half-cystine specifically labeled by an affinity reagent for the acetylcholine receptor binding site. J Biol Chem 259(19):11662–11665
21. Kao PN, Karlin A (1986) Acetylcholine receptor binding site contains a disulfide cross-link between adjacent half-cystinyl residues. J Biol Chem 261(18):8085–8088
22. Placzek AN, Grassi F, Meyer EM, Papke RL (2005) An alpha7 nicotinic acetylcholine receptor gain-of-function mutant that retains pharmacological fidelity. Mol Pharmacol 68(6):1863–1876
23. Mayer ML (2005) Glutamate receptor ion channels. Curr Opin Neurobiol 15(3): 282–288
24. Armstrong N, Sun Y, Chen GQ, Gouaux E (1998) Structure of a glutamate-receptor ligand-binding core in complex with kainate. Nature 395(6705):913–917
25. Mayer ML (2005) Crystal structures of the GluR5 and GluR6 ligand binding cores: molecular mechanisms underlying kainate receptor selectivity. Neuron 45(4):539–552
26. Furukawa H, Gouaux E (2003) Mechanisms of activation, inhibition and specificity: crystal structures of the NMDA receptor NR1 ligand-binding core. EMBO J 22(12): 2873–2885
27. Furukawa H, Singh SK, Mancusso R, Gouaux E (2005) Subunit arrangement and function in NMDA receptors. Nature 438(7065):185–192
28. Armstrong N, Gouaux E (2000) Mechanisms for activation and antagonism of an AMPA-sensitive glutamate receptor: crystal structures of the GluR2 ligand binding core. Neuron 28(1):165–181
29. Sun Y, Olson R, Horning M, Armstrong N, Mayer M, Gouaux E (2002) Mechanism of glutamate receptor desensitization. Nature 417(6886):245–253
30. Jin R, Banke TG, Mayer ML, Traynelis SF, Gouaux E (2003) Structural basis for partial agonist action at ionotropic glutamate receptors. Nat Neurosci 6(8):803–810
31. Tomita S, Adesnik H, Sekiguchi M et al (2005) Stargazin modulates AMPA receptor gating and trafficking by distinct domains. Nature 435(7045):1052–1058
32. Zhang W, St-Gelais F, Grabner CP et al (2009) A transmembrane accessory subunit that modulates kainate-type glutamate receptors. Neuron 61(3):385–396
33. Schwenk J, Harmel N, Zolles G et al (2009) Functional proteomics identify cornichon proteins as auxiliary subunits of AMPA receptors. Science 323(5919):1313–1319
34. Dorman G, Prestwich GD (2000) Using photolabile ligands in drug discovery and development. Trends Biotechnol 18(2):64–77
35. Pedregal C, Collado I, Escribano A et al (2000) 4-Alkyl- and 4-cinnamylglutamic acid analogues are potent GluR5 kainate receptor agonists. J Med Chem 43(10):1958–1968
36. Gorostiza P, Volgraf M, Numano R, Szobota S, Trauner D, Isacoff E (2007) Mechanisms of photoswitch conjugation and light activation of an ionotropic glutamate receptor. Proc Natl Acad Sci U S A 104(26):10865–10870
37. Numano R, Szobota S, Lau AY et al (2009) Nanosculpting a Yin/Yang photoswitch for an ionotropic glutamate receptor. Proc Natl Acad Sci U S A 106(16):6814–6819
38. Mayer ML, Olson R, Gouaux E (2001) Mechanisms for ligand binding to GluR0 ion channels: crystal structures of the glutamate and serine complexes and a closed apo state. J Mol Biol 311(4):815–836
39. Numano R, Szobota S, Lau AY, Gorostiza P, Volgraf M, Roux B, Trauner D, Isacoff EY (2009) Nanosculpting reversed wavelength sensitivity into a photoswitchable iGluR. Proc Natl Acad Sci U S A 106(16):6814–6819
40. Szobota S, Gorostiza P, Del Bene F et al (2007) Remote control of neuronal activity with a light-gated glutamate receptor. Neuron 54(4):535–545
41. Wang S, Szobota S, Wang Y et al (2007) All optical interface for parallel, remote, and spatiotemporal control of neuronal activity. Nano Lett 7(12):3859–3863
42. Banghart M, Borges K, Isacoff E, Trauner D, Kramer RH (2004) Light-activated ion channels for remote control of neuronal firing. Nat Neurosci 7(12):1381–1386
43. Colquhoun D, Dreyer F, Sheridan RE (1979) The actions of tubocurarine at the frog neuromuscular junction. J Physiol 293:247–284
44. Lau AY, Roux B (2007) The free energy landscapes governing conformational changes in a glutamate receptor ligand-binding domain. Structure 15(10):1203–1214
45. Celie PH, van Rossum-Fikkert SE, van Dijk WJ, Brejc K, Smit AB, Sixma TK (2004) Nicotine and carbamylcholine binding to nicotinic acetylcholine receptors as studied in AChBP crystal structures. Neuron 41(6):907–914

46. Damle VN, Karlin A (1978) Affinity labeling of one of two alpha-neurotoxin binding sites in acetylcholine receptor from *Torpedo californica*. Biochemistry 17(11):2039–2045
47. Damle VN, McLaughlin M, Karlin A (1978) Bromoacetylcholine as an affinity label of the acetylcholine receptor from *Torpedo californica*. Biochem Biophys Res Commun 84(4):845–851
48. Zhang F, Prigge M, Beyriere F et al (2008) Red-shifted optogenetic excitation: a tool for fast neural control derived from *Volvox carteri*. Nat Neurosci 11(6):631–633
49. Berndt A, Yizhar O, Gunaydin LA, Hegemann P, Deisseroth K (2009) Bi-stable neural state switches. Nat Neurosci 12(2):229–234
50. Volgraf et al Unpublished
51. Burns DC, Zhang F, Woolley GA (2007) Synthesis of 3,3′-bis(sulfonato)-4,4′-bis (chloroacetamido)azobenzene and cysteine cross-linking for photo-control of protein conformation and activity. Nat Protoc 2(2): 251–258
52. Sadovski O, Beharry AA, Zhang F, Woolley GA (2009) Spectral tuning of azobenzene photoswitches for biological applications. Angew Chem Int Ed Engl 48(8):1484–1486
53. Chi L, Sadovski O, Woolley GA (2006) A blue-green absorbing cross-linker for rapid photoswitching of peptide helix content. Bioconjug Chem 17(3):670–676
54. Beharry AA, Sadovski O, Woolley GA (2008) Photo-control of peptide conformation on a timescale of seconds with a conformationally constrained, blue-absorbing, photo-switchable linker. Org Biomol Chem 6(23):4323–4332
55. Pozhidaeva N, Cormier ME, Chaudhari A, Woolley GA (2004) Reversible photocontrol of peptide helix content: adjusting thermal stability of the cis state. Bioconjug Chem 15(6):1297–1303
56. Koçer A, Walko M, Meijberg W, Feringa BL (2005) A light-actuated nanovalve derived from a channel protein. Science 309(5735): 755–758
57. Lougheed T, Borisenko V, Hennig T, Ruck-Braun K, Woolley GA (2004) Photomodulation of ionic current through hemithioindigo-modified gramicidin channels. Org Biomol Chem 2(19):2798–2801
58. Erdelyi M, Karlen A, Gogoll A (2005) A new tool in peptide engineering: a photoswitchable stilbene-type beta-hairpin mimetic. Chemistry 12(2):403–412
59. Fortin DL, Banghart MR, Dunn TW et al (2008) Photochemical control of endogenous ion channels and cellular excitability. Nat Methods 5(4):331–338
60. Heginbotham L, MacKinnon R (1992) The aromatic binding site for tetraethylammonium ion on potassium channels. Neuron 8(3):483–491
61. MacKinnon R, Yellen G (1990) Mutations affecting TEA blockade and ion permeation in voltage-activated K^+ channels. Science 250(4978):276–279
62. Gorostiza P, Isacoff EY (2008) Nanoengineering ion channels for optical control. Physiology (Bethesda) 23:238–247
63. Sobolevsky AI, Rosconi MP, Gouaux E (2009). Nature 462(7274):745–756
64. Janovjak H, Szobota S, Wyart C, Trauner D, Isacoff EY (2010). Nature Neuroscience 13:1027–1032
65. Banghart MR, Mourot A, Fortin DL, Yao JZ, Kramer RH, Trauner D (2009). Angew. Chem. Int. Ed. 48:9097–9101

Index

A

James J. Chambers and Richard H. Kramer (eds.), *Photosensitive Molecules for Controlling Biological Function*, Neuromethods, vol. 55, DOI 10.1007/978-1-61779-031-7, © Springer Science+Business Media, LLC 2011

D

E

F

G

H

O

P

Q

V

W

X

Y

Z

MIX
Papier aus verantwortungsvollen Quellen
Paper from responsible sources
FSC® C105338

If you have any concerns about our products,
you can contact us on
ProductSafety@springernature.com

In case Publisher is established outside the EU,
the EU authorized representative is:
Springer Nature Customer Service Center GmbH
Europaplatz 3, 69115 Heidelberg, Germany

Printed by Libri Plureos GmbH
in Hamburg, Germany